Hot Springs of Bath

Investigations of the thermal waters of the Avon Valley

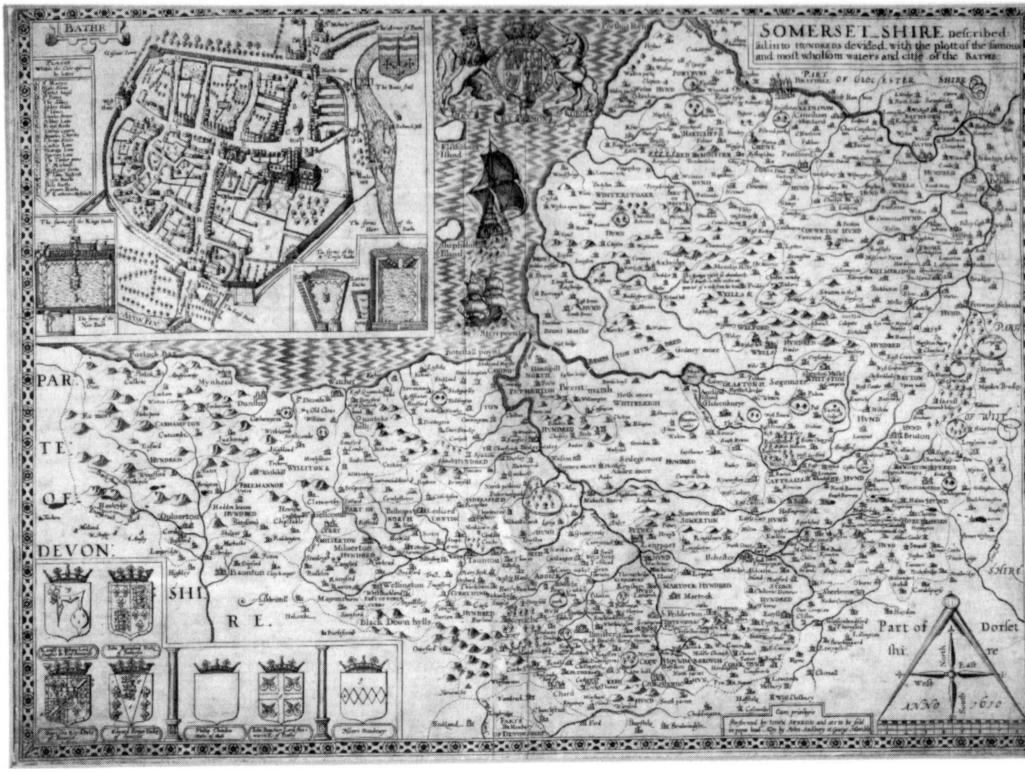

John Speed's Map of Somersetshire 1610. © 1985 Bath Archaeological Trust. Reproduced by permission.

Hot Springs of Bath

Investigations of the thermal waters of the Avon Valley

EDITED BY

G. A. Kellaway

1991

Published by

Bath City Council

First published 1991 by Bath City Council, Guildhall, High Street, Bath BA1 5AW

A catalogue record for this book is available from the British Library

ISBN 0-901303-25-9

Typeset by Bath Typesetting Limited, Bath
Printed by the Alden Press, Oxford

Contents

Contributors

J. N. Andrews, School of Chemistry, University of Bath

J. H. Black, Golder Associates (UK) Ltd, Landmere Lane, Edwalton, Nottingham

W. G. Burgess, Department of Geological Sciences, University College London, Gower Street, London

A. J. Cook, 8 Press Lane, Norwich

R. J. M. Corrall, Department of Medicine, Bristol Royal Infirmary, Bristol

P. Dieppe, Department of Medicine, Bristol Royal Infirmary, Bristol

W. M. Edmunds, British Geological Survey, Wallingford, Oxon

J. M. Evans, Department of Medicine, Bristol Royal Infirmary, Bristol

A. B. Hawkins, Department of Geology, University of Bristol, Wills Memorial Building, Queens Road, Bristol

A. Heywood, Research Fellow, Department of Renal Medicine, Southmead Hospital, Bristol

G. A. Kellaway, 14 Cranedown, Lewes, Sussex

S. Kilvington, Public Health Laboratory, Royal United Hospital, Bath

M. McNulty, Royal United Hospital, Bath

P. G. Mann, The Old Vicarage, Southstoke, Bath

D. L. Miles, British Geological Survey, Wallingford, Oxon

N. D. Millar, Department of Medicine, Bristol Royal Infirmary, Bristol

J. P. O'Hare, Royal United Hospital, Bath

R. Rolls, 35 Great Pulteney Street, Bath

W. I. Stanton, National Rivers Authority Wessex Region, Rivers House, East Quay, Bridgwater, Somerset

D. C. Warhurst, Amoebiasis Unit & PHLS Malaria Reference Laboratory, London School of Hygiene & Tropical Diseases, Keppel St, London

List of plates

Aerial photographs

Erratum. Plates 9 and 13 have been transposed in the colour sections and thus appear with the wrong caption. We apologise for this error.

Foreword

Rt. Hon. Lord Kearton of Whitchurch

Ever since Roman times Bath has been famous for its hot springs and for the treatment or alleviation of rheumatism and other medical conditions. Queen Elizabeth I placed the hot springs under the care of the Civic Authority of Bath in 1590 and, from that time until its complete closure in 1977, Bath was an important medical and social centre which attained the peak of its fame in the 18th Century. In 1977, the discovery of pathogenic amoebae in the water resulted in the closure of the City's spa water bathing pools and the pipeline supplying the water for treatment purposes to the Royal Mineral Water Hospital. For the first time since the beginning of Roman occupation, the thermal springs were not in use by man.

In response to this disaster, Bath City Council mounted a geological investigation in order to restore a supply of biologically 'clean' thermal water. This work was carried out on the advice of Dr Geoffrey Kellaway with the assistance of the Wessex Water Authority and the co-operation of scientists from the British Geological Survey, the Universities of Bristol and Bath and the Hospital for Tropical Diseases in London. Much of the microbiological work was carried out in the Pathological Laboratory of the Royal United Hospital in Bath, and the other medical investigations were based on the Royal Infirmary at Bristol.

As has happened before at Bath, the loss or reduction in flow of the thermal water has prompted investigations which have revealed scientific results of great interest. Aspects of the history of science and medicine have also been revealed by the present work which has thrown much light on the possible origin of the thermal water and on comparatively youthful earth movements connected with the appearance of the hot springs.

The investigations carried out at Bath since 1977 provide an outstanding example of integrated multi-disciplinary research. The basis of the success in drilling for amoeba-free thermal water lies in the correct assessment of the geological structure, in preventing oxidations of the thermal water before its arrival at the surface and in the isolation of this supply from any oxidized water capable of sustaining the growth of *Naegleria fowleri.*

Biologically clean thermal water now issues from the fountain in the famous Pump Room at Bath and restoration of the Spa is in sight.

The City of Bath, with which I am closely associated as Chancellor of the University of Bath, has been fortunate to have a scientist of Dr Kellaway's distinction and commitment to lead this important work. I congratulate him, and all those who have been associated with this major project, in bringing it to a successful conclusion; and I also congratulate him on his initiative in promoting the symposium at the Royal Society which has resulted in this book. It is essential that the symposium's papers, from a range of disciplines, should be brought together as a permanent record and valuable source of reference. Each paper is important in itself; but so too are the connections across the disciplines so well exemplified in this book. I commend this book most warmly to the scholar and interested layman alike.

Introduction

Councillor Elgar Jenkins, OBE, BA, Leader of Bath City Council (1984–1988)

For all Man's great skills and knowledge, sometimes our very existence depends on matters outside human control. It is a fortunate accident of nature that rain fell tens of thousands of years ago on hills surrounding the Bristol and Somerset Coalfields and then, after descending deep into the earth, emerged as hot springs in Bath. It is the existence of these hot springs that has made Bath the great city it is today.

Bath has always valued the importance of these waters and I welcome the contributions in this publication. They were first presented at the meetings at the Royal Society sponsored by the Royal Society of Arts, Royal Society of Medicine and the Geological Society of London. The aim has been to bring to bear human minds, skill and scientific knowledge to help achieve a greater understanding of the nature of Bath thermal water and, thus, to contribute to the safeguarding of its continued existence. It is for this reason, with the help of Dr Geoffrey Kellaway and encouragement from our Member of Parliament, Chris Patten, that the reports of our contributors are published in full.

Until the 16th Century it was first the Romans and then the Church who had the waters in their care. It was the drive of the City Council in the days of Queen Elizabeth I that ensured civic control. On 4 September 1590, in the Charter of Incorporation of the City, it was 'granted and confirmed to the aforesaid Mayor, Aldermen and Citizens and to their successors all and singular such and the same waters . . . which the Mayor and Citizens of the said City now have, or of right ought to have, for ever'.

From around this period people began increasingly to come to the City to take the waters. Tourism replaced cloth making as a major local industry. Various writers and Doctors helped with their works in praise of Bath waters. It was these, together with the problems of getting to Continental Spas because of wars on the mainland of Europe, which helped to spread the name of Bath in the 16th and 17th Centuries. It was noted that not only did the popularity of Bath waters increase, but the Baths became somewhat cleaner as a result of civic involvement. The growth and use of the Spa in the 18th, 19th and early 20th Centuries is known to us all.

What of recent years? Between 1978 and 1990 the Council has undertaken works and investigations, particularly concerned to monitor the flow of water at various sites, produce a pure supply and protect the water supply. The total cost of these works has come to over £400,000. Wessex Water Authority have been, may I record, of considerable help to the Council in many practical ways with this work, including the solution of technical problems. The Council is very appreciative of their support. Since reorganization of the water industry, the responsibility for groundwater resources has been assumed by the National Rivers Authority with whom we maintain close contact.

In addition to the monitoring work, the Council intends to make Spa water available free in the City Centre for the benefit of residents and visitors alike. It has been available for a small payment inside the Pump Room since 1986 when a pure water supply was re-established there.

Bath cannot hope reasonably to have as large a Spa development now as it once had in the past. Spas have continued to exist on the Continent, although a number of these have their problems. Largely because of the existence of a free Health Service in Britain, Spas have virtually ceased to exist here. In Bath the health treatment for people who would previously have gone to the Treatment Centre is now provided for at the Royal United

Hospital or at the Royal National Hospital for Rheumatic Diseases. Our aim as a Council now is, with financing from the private sector, to seek to provide an opportunity for people to bathe again in Bath Spa water with, we hope, local medical voluntary support.

Much of the material in the book is necessarily specialized and designed for the attention of scientific and medical opinion. It also has a substantial content which is of general and historical interest and which should enable non-technical members of the public to appreciate the state of knowledge regarding the hot springs and some of the problems which are as yet unsolved.

As a Council we are trustees for the waters of Bath. They are part of the heritage of our Country. We are conscious of our trusteeship and we are working effectively to fulfil it. The Council has ensured the provision of pure Spa water, has carried out work to protect and examine the sources of the hot springs and is doing all it can to ensure the return of a spa facility.

Preface

G. A. Kellaway

Bath and its residents have figured so widely in well known literary works from the time of Chaucer onwards that the later history of the Spa and its social life have become familiar to most of us. Visitors to Bath may carry away indelible memories of its Georgian crescents, of the interior of the Pump Room or of the view of Pulteney Bridge from the banks of the River Avon at Bathwick Weir, but those who descend below street level to the King's Spring and Great Roman Bath (Plate 6) find themselves transported back two millennia to the early days of the Roman occupation of Britain. It is below ground where the King's Spring rises in its Roman reservoir that public questioning about the source and origin of the hot springs is at its keenest and where the need for more and better information about the thermal water has long been felt.

Since the winter of 1977–1978 and the closure of the Spa a concerted effort has been made to achieve its restoration and some of the medical treatments formerly in use have been reassessed. As so often happens in scientific research, the solution of one problem discloses another. New information is often found to have a critical bearing on problems which would originally have been considered to be quite unrelated to the primary objectives of the work.

In this way the detailed consideration of the date of formation of the hot springs which is important in environmental assessment, reveals that the springs must have emerged at some time when appreciable dissection of the Mesozoic cover rocks (Triassic, Jurassic and Cretaceous strata) had taken place. The solution of this question involved a study of the Pleistocene succession and drainage patterns, resulting in conclusions which are now supported by recent work at the Open University (Chapter 14). This confirms the transport of the 'foreign stones' of Stonehenge by natural means and not by human transport. Geological considerations indicate that the series of profound changes in the geomorphology which culminated in the formation of the hot springs may be considered to have started in late Oligocene and early Miocene times with the establishment of a major drainage route from the Welsh borders to Southampton Water. This system was modified by tectonic movement and by the onset of glaciation in late Pliocene times, a major glaciation at 2.47 Ma BP (million years before present) being well placed to accord with the stage reached by erosion of the Mesozoic and Tertiary rocks and the deposition of Welsh rocks on Salisbury Plain. Water continued to drain eastwards across the site of the Severn estuary for about 1.5 Ma, but in mid-Pleistocene times, when much of the British Isles was covered by ice, the drainage pattern underwent a drastic change with the rapid evolution of the Lower Severn and the formation of the Bristol Avon and its gorges (Chapter 13).

A general synthesis of the geological record from the period following the mid-Tertiary earth movements (*c.* 23 Ma BP) and preceding the last (Devensian) cold period of the Pleistocene (*c.* 100,000 years BP) is given in Chapters 13 and 14. This brings together many different kinds of evidence and relates them to a modern timescale. Some of the conclusions are necessarily speculative, but a model is badly needed if only to direct attention to those aspects which require further consideration. In reading these and other contributions to the geology and hydrogeology it will be found helpful to refer to the maps of the Geological Survey (listed on p 23) on which much of the work is based.

Thermal water springs are known to occur in the Avon Gorge at Hotwells and Clifton, Bristol as well as at Bath. Though the daily yield of the main Hotwells Spring (24°C) is only about 30 per cent of that of Bath, there are a number of other minor springs and thermal water sources in the area and some of these, long forgotten, have only recently

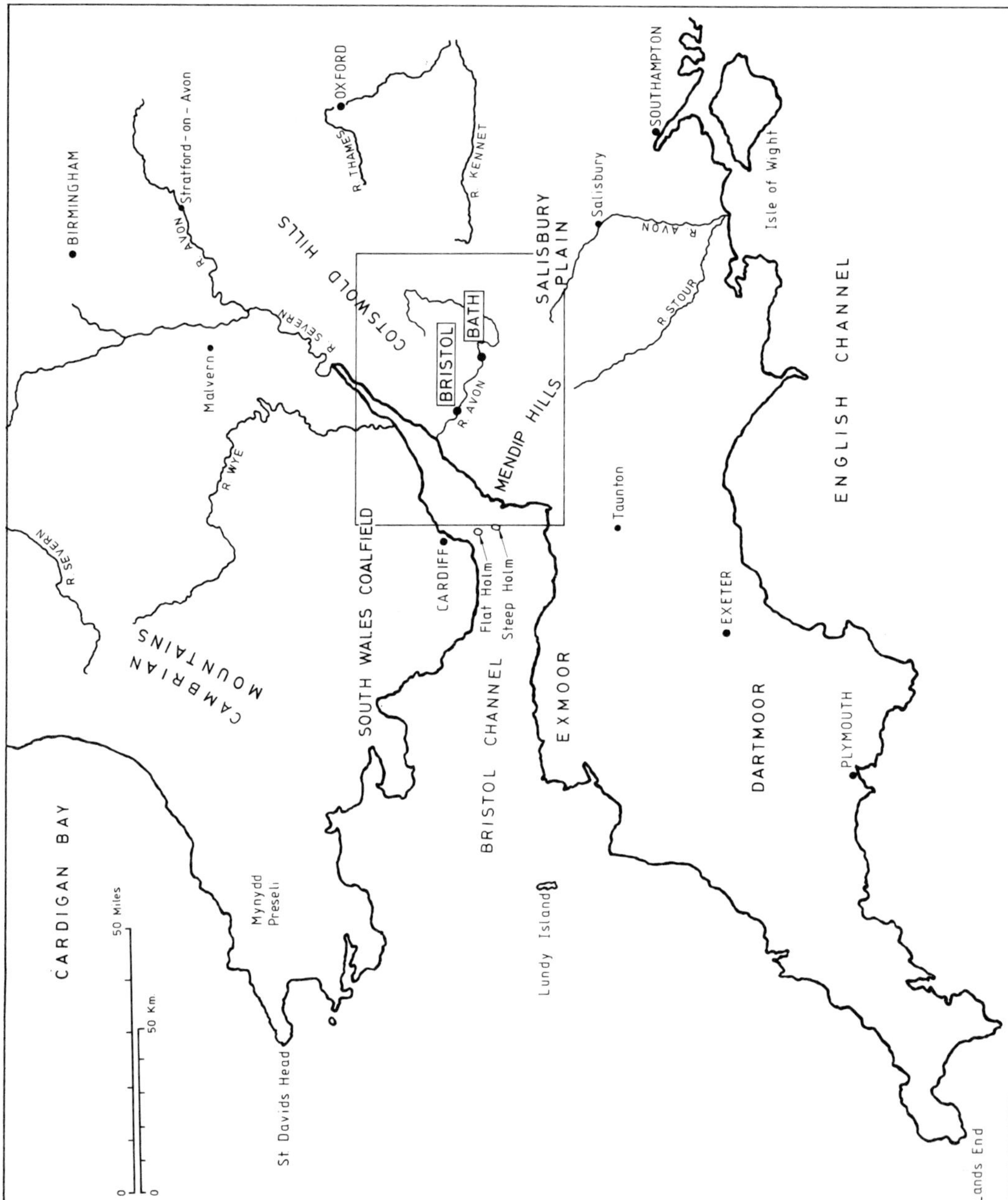

Fig. Pf1. Key map showing the position of Bath and the River Avon.

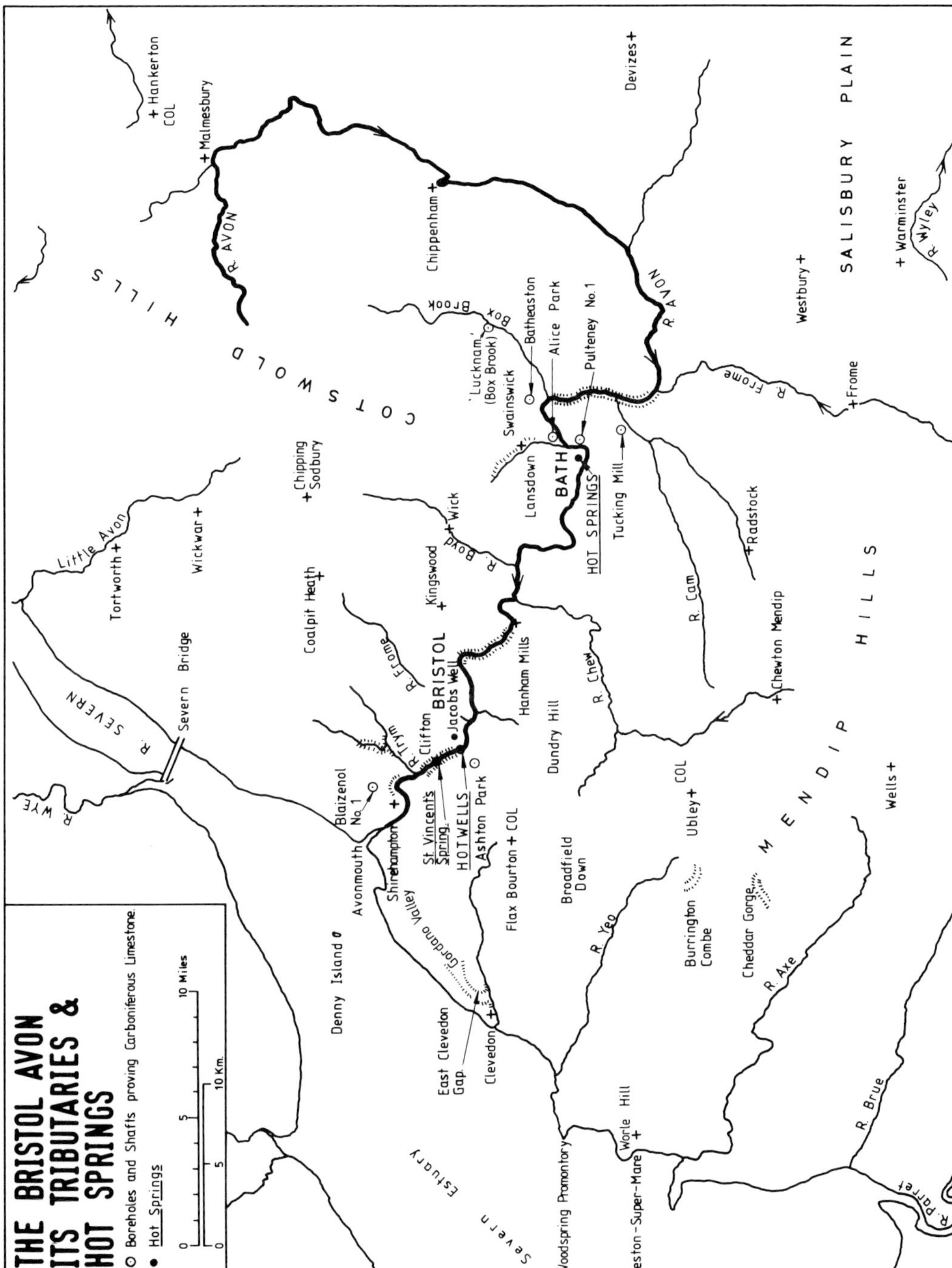

Fig. Pf2. The Bristol Avon: its tributaries and hot springs.

been rediscovered (Chapter 12). Whether any other springs or thermal water seepages remain to be found is therefore a legitimate question.

Nevertheless, Bath remains the only Spa in Great Britain where water with a temperature in excess of 40°C rises naturally at the surface, the total yield being about 275,000 gallons (approximately 1.25×10^6 litres) of water per day with a temperature of 41–47°C. Having regard to all the known sources it is likely that the total yield of the mineral water springs and seepages in the Avon valley is of the order of 400,000 gallons per day (1.82×10^6 litres) though a proportion of this (possibly 20–25%) is cold water of shallow origin which is mixed with the hot thermal water.

In the account which follows we present the results of interdisciplinary work devoted partly to the solution of the problems of contamination which have been known to affect the hot springs of Bath since 1977–1978 and partly to the investigation of the composition and medical use of the thermal water. The studies include microbiological and medical research which now form the basis of preventative measures which have been taken to protect users of the Spa facilities. Much of the work carried on at Bath has been done in close collaboration with the City Council's Engineers and has been specifically designed to ensure a supply of thermal water which can safely be used by the public. This has involved the solution of complex geological and engineering problems in the face of constraints imposed by working in densely built-up areas and by stringent Public Health requirements for microbiological cleanliness of the thermal water.

Archaeology

During the course of recent investigations at the King's Spring at Bath by Professor Barry Cunliffe and his colleagues, archaeological evidence has been found of human activity in pre-Roman times including Mesolithic implements as well as Iron Age coins. No direct evidence of the participation of Neolithic or Bronze Age man has been discovered, though the existence of the springs is hardly likely to have escaped their notice. The Iron Age people who inhabited the areas adjacent to the springs left behind them coins which may have been votive offerings cast into the water as is customary at the present day in Bath or in Rome. Many of the objects found in the springs during the course of this work are exhibited in the Museum alongside the Temple of the Goddess Sulis Minerva, the Roman Reservoir of the King's Spring and the Great Roman Bath. Some of the information revealed by these excavations was helpful in dealing subsequently with practical engineering problems. Historical and archival material preserved in Bath has also yielded information which is both of immediate practical value and of more general interest, especially in the field of the history of medicine and of science.

The presence of a Mesolithic site alongside the King's Spring suggests that the water was rising from the spring pipe at the time of the occupation (*c.* 7,000 years BP). As the spring pipe is filled with mid- to late Quaternary gravel it dates from before the Mesolithic site and could be at least as old as 80,000 years BP. The most likely interpretation is that the flow of the springs has diminished since mid-Quaternary times when the hot springs originated on tectonic rifts formed during the replacement of an easterly directed drainage system by the westward draining River Avon (Chapter 13).

The early use of the hot springs

Construction of the *thermae* at Bath commenced at an early date in the Roman occupation of the British Isles, possibly about the time of the Emperor Claudius. The main centre of interest was the King's Spring which was enclosed by the wall of the Roman reservoir and flanked by the Temple of the Goddess Sulis Minerva. From the reservoir, water flowed first into the Great Bath and thence to the Roman Drain. The two smaller hot springs situated about 100 m west of the King's Spring, namely the Cross Bath Spring and the Hetling Spring (see Fig. 1.2), were also used by the Romans. Most of the Roman installations could accommodate a larger flow than exists at the present time but in the absence of any record, comparison of the total yield in Roman times with that of the present day is impossible. On the other hand the archaeological and historical evidence shows that the flow of the springs was well maintained throughout the Roman occupation.

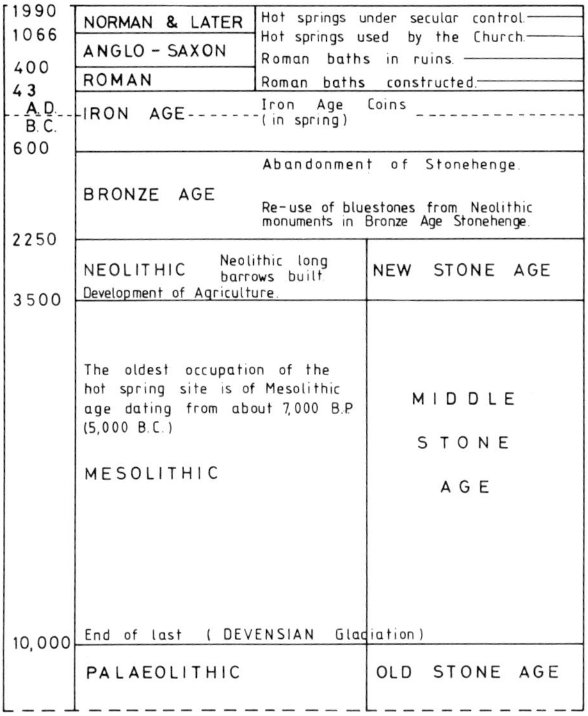

Fig. Pf3. Principal cultural changes in post-Glacial times in the regions around Bath.

Aquae Sulis remained in British hands until the defeat of the Romano-British armies by the Saxon invaders at the Battle of Deorham (Dyrham) in 577 AD. A Saxon poem ('The Ruin') records that thermal water issued freely from the remains of the abandoned Roman buildings (Cunliffe 1983). During the Dark Ages, Bath was known as Akemanceaster. However, in a deed of 676 AD, Osric of Mercia referred to it as *Hat Bathu*, implying that the hot springs were in use at that time. The next recorded change of any significance took place in Norman times when John de Villula, a physician from Tours, bought the town of Bath from King John for 500 silver marks in 1091. He moved the See from Wells to Bath, converting the abbey church at Bath into a cathedral. The monastery being a Benedictine foundation, the thermal water would have been much used by the monks under his jurisdiction for the care and treatment of the sick. Two baths are said to have been constructed by John de Villula, enclosing and conserving the thermal water springs. By the later part of the Middle Ages the baths were more widely used and in the mid-15th Century they were visited by King Henry VI whose modesty was somewhat affronted by the sight of naked men engaged in communal bathing in the hot baths.

In the Dark Ages the collapsed Roman buildings with their blocked pipes and drains were largely buried under debris into which floodwater from the River Avon introduced silt and stones. With the passage of time the thermal water migrated through the rubble and debris covering the Roman remains giving rise to numerous distributaries which were often mistaken for separate springs. Shortly before the suppression of the monasteries by King Henry VIII, Bath was visited by John Leland who described the springs and their use in his famous *Itinerary*. In this he tells us that there are three springs of which the water of the Hot Bath (or Hetling) Spring was the hottest though the Cross Bath was the largest bath. This does not imply that the Cross Bath Spring necessarily had the largest *flow*. The Cross Bath has long been cooler than the other springs and therefore more pleasant to use. The King's Spring (45–46°C) and the Hot Bath or Hetling Spring (47°C) are too hot to permit continuous exercise without exhaustion or overheating of the body. The Cross Bath Spring (41°C) gives an average bath temperature of about 100°F (37.7°C). This is more comfortable and it was the temperature of the bath water rather than the flow of the spring, that accounted for its size and popularity in Tudor and Stuart times.

Investigations of the nature and source of the hot springs

Following suppression of the monastery at Bath in 1539 the hot springs passed out of the control of the Church and their use may have declined. A revival of interest seems to have followed Queen Elizabeth's accession and this was accompanied by the appearance of accounts by Dr William Turner in 1568 and Dr John Jones in 1572. In 1590, Elizabeth granted the City of Bath its second Royal Charter and from this time forward the Corporation (later the City Council) of Bath became trustees responsible for the maintenance of the springs. The internal evidence of Sonnets 153 and 154 suggests that William Shakespeare knew of and perhaps may have visited the springs and in 1613 and 1615 the baths were used by Anne of Denmark, King James the first's Queen. This led to increased public interest, though improvement of the baths was delayed by the Civil War. A feature of the accounts immediately preceding the Civil War is the growing interest shown in the character and origin of the thermal springs as exemplifed by the work of Dr Edward Jorden in whose care Queen Anne of Denmark had been placed.

Edward Jorden (1569–1652) was educated at Oxford and subsequently at Padua where he took his degree of MD. While at Padua he would have learned of the researches of Galileo (1564–1642) whose work had become famous throughout Italy and who was appointed to the Chair of Mathematics at Padua in 1592. This almost certainly accounts for Jorden's interest in hydrostatics and for the attention which he paid to the origin of the thermal water at Bath. In this he was closely followed by J. Glanvill (1669).

Jorden returned to England sometime before 1595 and in 1610 took up residence in Bath where he later produced his *Discourse of Natural Bathes and Mineral Waters* of which successive editions appeared in 1631, 1632 and 1633. This work was popular and after the Civil War two more editions were produced in 1669 and 1673.

Jorden is generally described as a physician and chemist but, as John Aubrey observed at a later date, chemistry had not improved greatly with the abandonment of the search for the Philosopher's Stone, for, as he said, 'now they only look after medicines'. In his account of the Iatrochemistry of the Bath waters, Jorden reflects contemporary professional standards and attitudes, but he was also interested in the origin of the hot springs and the application of the principles of hydrostatics. He was the first to attribute the temperature of the thermal water at Bath to the passage of groundwater through hot rocks in the interior of the earth and to explain its reappearance as being due to hydrostatic pressure. The 1632 edition of Jorden's *Discourse* came to the notice of John Aubrey some time before 1691 as mentioned in his *Natural History of Wiltshire* (ed. Britton, 1847). At this time he was in close contact with Robert Hooke and other leading scientists at the

Royal Society, recognition that the whole interior of the earth is hot having become, in Aubrey's words, 'the most commonly received opinion'. Being situated near ancient coal, lead and iron mines where the rise in temperature with depth of cover would have been familiar to miners, Bath hot springs were well placed to prompt consideration of the possible link between the elevation of the temperature of the spring waters and the depth of their circulation in the rocks. In some academic circles, however, theories such as those of Werner (1749–1817), who believed that volcanoes were due to the subterranean combustion of coal, and others relating it specifically to chemical reactions or latent volcanicity, continued to appear until well into the 19th Century.

Working in Bath at the close of the 18th Century, William Smith (1769–1839) may well have encountered some of these arguments, though he seems not to have become embroiled in them. This may have been due in part to his humble origin and consequent lack of formal education and social status and partly by reason of his inherently pragmatic and observational approach to nearly all such problems. While he may have known of Jorden's works it is unlikely that he would have seen John Aubrey's writings on the subject since they remained in manuscript in the Bodleian Library at Oxford until their publication in 1847. Such records as we have of Smith's views on the hot springs indicate that he believed the thermal water to be groundwater heated by its passage through the rocks. The effects of hydrostatic pressure were familiar to him from his work underground and on the canals around Bath and he thought it most likely that the thermal water is derived from the Palaeozoic and Triassic rocks forming high ground near the head of the Swainswick valley north of the city (Plate 20). At no time does he give credence to any source of heat other than that stored generally in the interior of the earth.

The on-going stimulus given to geological investigation by the mere presence of the mysterious hot springs runs like a continuous thread through the accounts written in the 17th and 18th Centuries. By introducing the concepts of stratigraphy Smith clarified the relationship of the rocks and sediments from which the thermal water issues at Bath and thus prepared the way for further consideration of its source (Chapter 1).

Geothermal gradient

It has proved very difficult to obtain reliable data as to the geothermal gradient at Bath. Measurements made in boreholes up which thermal water is flowing are of little value and those made in local coalmines are liable to error. Temperatures in the rising thermal water remain constant down to the deepest levels proved to date. A borehole sunk at Tuckingmill near Midford (7635 6163) in 1988 has a very low water yield and may eventually provide useful thermal data. At the time of writing this hole is closed to enable the ground temperature to recover from drilling. The figure given by E. Wethered (Chapter 7) for Kingswood is 1°F for 76 ft (24°C for 1 km), but neither this nor any other estimated gradients which have been used, have been confirmed by reliable in situ measurements. Accepting Wethered's figure and assuming a mean annual temperature at sea level of 50°F (10°C), the thermal water would have circulated to depths of at least 5000 ft (1524 m) in order to acquire its temperature.

Chemistry and hydrogeology

There are three springs at Bath situated on the alluvial plain of the River Avon at an altitude of about 17–18 m A.O.D. The water is artesian and has a piezometric level of about 30 m A.O.D. at the King's Spring, though the present level at Stall Street is about

27 m. So the artesian pressure at natural ground level is approximately one atmosphere at the King's Spring. Exsolved gases include very small amounts of helium and radon resulting from the decay of radioactive minerals while hydrogen and oxygen isotope analyses indicate that the water is of atmospheric origin and that the proportions of the various atmospheric gases have undergone substantial changes as a result of solution and passage of the water through the rocks. The average composition of air at sea level in southern England (excluding water vapour and the rare gases) is nitrogen 77.32%, oxygen 20.8% and carbon dioxide 0.03%. The exsolved gases from the hot springs gave nitrogen 96% and carbon dioxide 3.5% while oxygen is absent, or present in infinitesimally small amounts. The absence of oxygen is due to the abstraction of that gas by rock-forming minerals, notably marcasite, pyrite and siderite. The increase in the percentage of carbon dioxide from 0.03% to 3.5% and the presence of hydrocarbons (1.4%) in the exsolved gases are due to additions from underground sources.

Iron sulphides and carbonates are abundant in slates, mudstones and shales and, to a lesser extent in impure limestone and sandstone. The sulphides marcasite and pyrite (FeS_2) react with oxygenated water to produce sulphuric acid which attacks calcium carbonate (calcite) and other minerals. The change in the composition of the exsolved gases is therefore explicable in terms of the chemical processes which take place as the water travels through the rocks under hydrostatic pressure. When the hot mineralized water emerges at the surface, oxidation produces the orange and rust-coloured ferruginous stains seen on almost everything with which it comes in contact (Plate 9).

Analysis of the waters of the King's Spring and Stall Street inclined borehole include the isotopic composition of the water (Chapter 9). Earlier analyses gave comparable figures for the amount of dissolved solids which average about 2.2 g/litre and of this amount 65% is in the form of sulphates of sodium, calcium and magnesium. Since the sulphate oxygen and ^{34}S both point to a pre-Permian source for the SO_4 there is a high degree of probability that the SO_4 is derived primarily from the oxidation of ferric sulphide in the Palaeozoic wall rocks. The isotope analysis also suggests that non-marine aquifers are unlikely to have contributed to the carbonate content. This would exclude the Lower Old Red Sandstone in which there are cornstones (impure limestone) of non-marine origin.

The geochemical results could be explained if the main flow of thermal water travelled from North to South at depth in fractured Palaeozoic rocks alongside the reactivated faults of the Malvern Fault Zone and thence to the Kingswood Anticline. At Tortworth, the Lower Old Red Sandstone is missing and the Upper Old Red Sandstone (mainly sandstone and Quartz Conglomerate) rests unconformably on Silurian shales and limestone overlying Tremadoc shales (Chapter 13). With the exception of the Upper Old Red Sandstone almost the entire Palaeozoic sequence between the Tremadoc and the Namurian (Millstone Grit) is composed of marine strata. The removal of the Mesozoic clays and marls which acted as a seal or cover may not have become effective until mid-Quaternary times.

On the other hand it is abundantly clear that the thermal water has spent a substantial proportion of its travelling time in Carboniferous Limestone and it emerges from that formation both at Hotwells and at Bath. The groundwater in the Carboniferous Limestone of the north–central Mendips may play a part in balancing or controlling the deep groundwater movement, but the very strong flows and high piezometric levels in the Wick–Batheaston area cannot be omitted from this equation. At Batheaston (Chapter 1) very strong flows of water have been recorded in the Blue Lias and the middle of the White Lias. When the trial shaft for coal at Batheaston first penetrated the Dolomitic Conglomerate, it appeared to be dry and this led William Smith to issue an optimistic Report in 1808 which was contradicted shortly afterwards by a sudden inburst of water ('The Cataract'). It is highly probable that the water in the White Lias and the Dolomitic Conglomerate was encountered in fissures and under considerable pressure, and that the abstraction of water at Batheaston affected the hot springs at Bath.

In measuring piezometric levels at the King's Spring and the boreholes in Bath, it has been found that when the White and Blue Lias have been lined out the standing water levels tend to be lower than those where the borehole is not lined or the spring pipe is 'open'. In the centre of Bath this appears to have contributed to a lowering of the standing water level in the Stall Street inclined borehole by about 3 m.

Enough has been said to show that the question of the age and origin of the thermal water raises complex issues which are further discussed in relation to radioactivity and hydrology (Chapters 8 and 10). If it is assumed that the Carboniferous Limestone beneath the Bristol and Somerset coalfield has a highly developed karstic system of connected cavities and channels with a large capacity for the storage of thermal water, then conclusions as to the age of the water would be affected. Unfortunately there is only a limited amount of evidence bearing on this. Such deep boreholes and tunnels as have been made in the Carboniferous Limestone e.g. Blaize No. 1 Borehole near Lawrence Weston, the Ashton Park Borehole, Bristol, the Northern Storm Water Interceptor Tunnel at Bristol and the Tuckingmill Borehole near Midford have given little evidence of large-scale karstic development at depth. In uplands such as the Mendips there is abundant evidence for the development of karstic phenomena due to solution of the limestone under cold aerobic conditions in Quaternary (Devensian) times. Complex cave systems were developed above sea level in free draining limestone uplands when the solubility of CO_2 (and therefore of limestone) was maximized by low temperatures. The possible effect of lowered sea level during the last glaciation must be taken into account but in general Karst phenomena tend to die out at depth.

The implications for geomorphology and archaeology are considerable as there is substantial evidence for the former presence of an eastward flowing Solent (or Southampton) River first envisaged in 1921 by Varney. In early Pleistocene times the Mesozoic–Tertiary strata (of which part remains) acted as a cover on which the Stonehenge erratics travelled from Wales and also as a restraint forcing water draining off the Welsh uplands to accumulate in the older rocks beneath the younger strata. The possible effect of hydrodynamic processes beneath a thick Mesozoic cover north and east of Bath has a bearing on deep groundwater movement and has been given special consideration (Chapter 11). It seems reasonable to suppose that the early to Mid-Quaternary warping and fissuring which produced the Avon Gorge made possible the emission of the thermal water and this became cooler as the Carboniferous Limestone of the Mendips and Severn estuary region was progressively exposed by removal of the Mesozoic cover.

Recent calculations of the age of the thermal water at Bath based on radioactivity and chemical evidence (Chapters 9 and 10) range from 'a few hundreds' to 8,000 years BP. The difficulties which arise in age determination stem partly from the fact that the thermal water is a mixture. Nevertheless the geological evidence indicates a much greater age for the orifices from which the water is now issuing and this is not altogether surprising in view of the history of the physiographical evolution of the region. The source of the thermal water is not only of academic interest as it has environmental implications for the maintenance of the hot springs. Although a start has been made (Chapter 8) much still remains to be done by monitoring the temperature and flow of the thermal water from the springs and boreholes at Bath. Thermal yields have been found to be more informative than simple flow measurements in assessing the relative importance of hot springs.

Medical and microbiological researches

Unlike Hotwells (Bristol) where the thermal water was used primarily for drinking, the hot springs of Bath specialized in medical treatments based on immersion (Chapter 4). No evidence has yet been found to show that the chemistry of the water has any particular quality which is of outstanding medicinal value and the early claims for its high level of radioactivity (then supposed to be beneficial) have turned out to be unsubstantiated (Chapter 3). There is, however, clear evidence that the former use of the thermal water for the treatment of certain diseases, notably painters' colic (lead poisoning) and gout, was efficacious and that other diseases related to rheumatism were greatly relieved by treatment with the thermal water (Chapter 5). For many visitors the general regime may have been beneficial as it commonly involved exercise both in and out of the baths, a moderated diet, fresh air and rest with the intake of a quantity of water that may, for some, have reduced their consumption of alcohol.

Closure of the Spa in 1978 followed the discovery of a pathogenic Amoeba in the thermal water (Chapter 6). Until its discovery in Bath in 1977 few people in Britain apart from specialists, had heard of *Naegleria fowleri*. Yet it is widely distributed around the world, though restricted to temperate and tropical climates. In many Icelandic and Alaskan hot springs the emergent temperature of the water is too high and winter air temperatures are much too low for *Naegleria* to survive in the open, but it is relatively common in temperate New Zealand where the Maori people have long known that it is dangerous to lower one's head beneath the surface of natural hot pools. Notices to this effect are posted up at many pools, but should accidental immersion occur, the advice is to seek medical help immediately as treatment can be effective if given promptly. Many sparsely populated areas of the world may contain colonies of *Naegleria* that have not yet been identified. One well known thermal region where it has recently been found includes the Grand Teton and Yellowstone National Parks, USA. Dr William O'Dell of the University of Nebraska discovered *Naegleria fowleri* in Huckleberry Hot Springs and the search was also extended to the Yellowstone Springs (*The Arizona Republic*, September 12, 1987).

Being an aerobic organism, *Naegleria* requires oxygen to live. Its control has been achieved by means of specially designed boreholes which convey unoxidized thermal water to the installations where it is used. Following the completion of the Stall Street inclined Borehole (Chapter 7), the thermal water was connected directly to the Pump Room fountain in 1987 (Plate 5). The constant flow of water, which is regularly tested, provides no time for oxidation to take place and is a permanent safeguard. When used for communal bathing however, chlorination or similar treatment of the thermal water will be required in accordance with Public Health regulations. As to the popularity of the warm water baths and the importance of Bath as a social and medical centre from the mid-17th Century onwards, there can be no doubt (Chapter 2); renewal of the thermal water bathing facilities therefore stands high in the list of priorities for restoration of the Spa.

Reference

Cunliffe, B. 1983. 'Earth's grip holds them' *In*: B. Hartley & J. Weller (eds) *Rome and Her Northern Provinces*. Allan Sutton, Gloucester, 67–83.

Editorial notes and acknowledgements

For reasons beyond control, the publication of the papers presented at the Symposium in 1987 was delayed and it was then decided to present the contents as chapters of a book rather than individual scientific papers accompanied by a discussion. To those who submitted discussion material we offer our apology for its non-appearance. Individual papers have, where necessary, been updated, a joint Preface has been provided and the links between chapters strengthened by cross-referencing, additional illustrations and a common index. Professor Barry Cunliffe's contribution has not been included as its subject matter is already available in a well illustrated guide book, *The Roman Baths and Museum* published by the Bath Archaeological Trust (1990).

To Councillor Laurie Coombs, Chairman of the Spa and Recreation Committee (1979–1988) and subsequently to Councillor Elgar Jenkins, we owe much encouragement in dealing with the difficult problems which Bath has faced since 1977. In the early stages of this work the interest of Mr David Beeton, then Chief Executive, and later, his successor Mr Clive Abbott gave much needed support. None of the investigations of the thermal water would have been practicable without the help of the Council's engineering and museum staff, notably by Mr David Vine (formerly City Engineer), Mr M. Froggatt, Mr T. Bletso, Mr Stephen Bird and their colleagues. Mrs N. Driver has made many of the drawings which illustrate this book and Miss Linda Kettletey and Miss Josie Smith have done most of the typing and word processor work on the text.

The editor thanks all those involved with the production of this book, including Bath Typesetting, Woodwards of Bath (colour printing), Barnabus Design and Print, Truro (cover artwork and printing), Moira Greenhalgh (index), The Alden Press, Oxford (text printing) and Cedric Chivers, Bristol (binding). David Ogden of the Geological Society Publishing House, Bath, is thanked for his management of this project through the production process.

All colour photographs not otherwise attributed were taken by the author; aerial photographs not otherwise attributed are from the Cambridge University Collection of Air Photographs and are reproduced by permission of the Curator and the Committee for Aerial Photography. We are indebted to West Air Photography for permission to reproduce Plate G. The photograph of the Boles Barrow stone, now exhibited in Salisbury Museum, was taken by Colin Bowen with the permission of the Curator.

Professor Troy L. Péwe of Arizona State University and Professor and Mrs W. Benninghoff of the University of Michigan, Ann Arbor, have provided me with much information and advice.

During the course of the investigations, help and advice has been given by the officers of the British Geological Survey and the former Wessex Water Authority and other research organizations. Much of this is incorporated in the papers now presented. Other help is acknowledged by individual contributors. On a personal note the editor may be permitted to express his gratitude to his wife Bronwyn for her constant and unfailing support.

List of maps

Ordnance Survey. 1:50,000

Sheet 162 (Gloucester and Forest of Dean)
172 (Bristol and Bath)
173 (Swindon and Devizes)
182 (Weston-super-Mare and Bridgwater)
183 (Yeovil and Frome)
184 (Salisbury and the Plain)
1:250,000 (Quarter-Inch Map)
Sheets 7, 8 & 9 Southern England

British Geological Survey. One-inch and 1:50,000

Sheet Nos 233 (Monmouth), 234 (Gloucester),
250 (Chepstow), 251 (Malmesbury), 263 (Cardiff),
264 (Bristol), 265 (Bath), 266 (Devizes),
279 (Weston-super-Mare),. 280 (Wells), 281 (Frome).
Geological Survey 'Ten Mile' map (South Sheet) 1:625,000
Third Edit. (Solid) 1979

Engraving of 1829: John Wood's Hot Bath (1775–1778).

From Kellaway, G. A. (ed.) *Hot Springs of Bath*, pp 25–54

1. The work of William Smith at Bath (1799–1813)

G. A. Kellaway

Bath attracted the attention of a succession of eminent 17th Century diarists including John Evelyn (1654) and Samuel Pepys (1688), but the most informative account is that of Celia Fiennes, an indomitable lady traveller with remarkable powers of observation. Writing during or before 1687 she lists five baths then in use. "the Hot Bath has the most hot springs, its but small and built all round which makes it the hotter: out of it runs water into a bath called the Lepours". The third or Cross Bath she describes as "something bigger then the former and not so hot". After commenting on the strength of the springs in general she says "the springs bubbles up so fast and so strong and are so hot up against the bottom of ones feete, especially in that they call the Kitching" (i.e. the central area) "in the K(ing's) bath". The fourth bath (i.e. the Queen's Bath) was filled from the King's Bath and had no separate sources, as Guidott's plan confirms (Guidott 1691). Celia Fiennes commented that the water of the King's Bath is "very hot and tastes like the water that boyles eggs". She observed that after the baths are emptied "I have seen all the springs bubble up as thicke out of the ground when the baths have been empty, the bottom is gravell . . .". So the floors of the spring basins were not paved, and the rising streams of hot water and gas seen at the surface could be felt under foot by the bathers.

Celia Fiennes wrote up her account in 1702 the year in which Bath was visited by Queen Anne whose presence set the final seal on Bath's popularity. As time went by the pressure on its resources grew greater and by the middle of the century improvements were long overdue. The flow of the Hetling Spring had for some time been dwindling and with past experience in mind it was suspected that much of the water was leaking away.

In 1772 a new stone cylinder was constructed to house the Hetling Spring and the following year the old Hot Bath was pulled down. On the wise advice of John Wood the new Hot Bath was built on the east side of the old bath while Hetling House on the west side was developed as a Pump Room. The stone cylinder which supplied the new Hot Bath is now situated beneath the highway, roughly midway between Hetling House and the front of the 18th Century "Old Royal Bath" or Hot Bath (Plate 2).

Wood's account (1778) of the stone cylinder states that it is 8 ft in diameter and extends "full fifteen feet below the level of the street, surrounds the whole spring (*sic*) and is so well secured on the outside with cement and masonry, as to prevent either the hot water of the spring from escaping, or any cold water mixing with it". Wood's measurement of the flow after the construction of the cylinder (1778) gave 140 tons of water in 24 hours (31,360 gallons per day). He pointed out that "in Dr Guidott's time", i.e. about 1681, "by the waters rising through rubbish and loose earth" the spring produced only 108.25 tons in 24 hours (24,304 gallons per day).

Thus in 1778 the yield of the Hetling Spring appeared to have increased to about 30% above that of 1681. Yet by the early part of the 19th Century the problems that have so often affected the Hot Bath and the Cross Bath were once again causing disquiet to the Town Council.

Not all of those who took up residence in 18th Century Bath were patients requiring treatment or gamblers looking for an alternative venue to London. Many residents had wealth and influence combined with agricultural, medical, or scientific interests and many Clubs and Societies devoted to such activities came into being. It was inevitable that with the Baths and Grand Pump Room forming a centre of medical and social life in the very heart of the City, questions as to the permanence of the hot springs should be asked. In view of their importance, the continuity of the life of the springs became a source of anxiety to the Town Council whenever there was any evidence of diminution of flow. Anyone who could throw light on these problems or suggest a solution was likely therefore to attract interest and support.

From the 17th Century onwards an ever-increasing stream of publications dealing with the hot springs of Bristol and Bath began to appear (Whittaker *in* Richardson 1928, pp 263–270). Among these were accounts of the rocks from which the thermal water issues at Clifton and Hotwells in the Avon Gorge and the strata in the floor of the valley at Bath. One work by J. Glanvill (1669) includes the first description of the White Lias and Blue Lias, collectively called "Lyas" by Strachey (1719, 1725) and other 18th Century authors. These now-familiar names appear with others in William Smith's *Table of Strata* (1799) and it is no accident that the hot springs of Bath should be at the centre of William Smith's first geological map also published in 1799. The visitors to Bath demanded Guide books accompanied by maps and it was one of these that Smith used as the base for his circular map covering a radius of five miles around the centre of Bath. Through the steadily increasing interest of influential and wealthy residents and the medium of the Bath Agricultural Society, William Smith's work became known nationally. This brought him in contact with British scientists of international repute, notably Sir Joseph Banks, who introduced his work and ideas to the geologists of France (Eyles 1985).

The year 1799 was an important one in the life of William Smith (1769–1839), being marked by the production of his first geological map (Sheppard 1920; Eyles 1969) and the recording of the "Order of the Strata" in the house of the Rev. J. Townshend in Great Pulteney Street (Phillips 1844; Bather 1926; Cox 1948). For some of the citizens of Bath, however, it was a year of great tribulation marked by landslides and mudflows on steep slopes saturated by the persistent heavy rainfall (Phillips 1844; Kellaway & Taylor 1968). In this same year, due to a disagreement with the Somerset Coal Canal Company, Smith's association with the Company was terminated (Torrens 1975). This left him free to work on other problems and it was not long before his services were in demand to deal with the steep unstable hillsides of Bath (Plate 3).

One of his most outstanding achievements was to stabilize the landslips which were threatening to destroy Combe Grove House in 1799 and 1800. Smith drove a tunnel into the hillside and tapped the water issuing from the base of the Great Oolite Limestone overlying Fullers Earth Clay. In this way he drained the slipped mass of limestone and clay on the steep slope below Combe Down, bringing the slips to rest (Phillips 1844).

Batheaston Shaft and Boring (1805–1813)

By 1804 Smith had become well known as the expert in all forms of mining and engineering geology. In this year proposals were put forward by one Thos. Walters, for the sinking of a shaft (or shafts) in search of coal at Batheaston. William Smith is not thought to have originated this proposal, but he was certainly consulted at an early stage as the Monthly Magazine of 1st June, 1804 (Vol. 17) states that one of the Company's shares is "to be presented to Mr William Smith, mineralogist for his able assistance and advice."

Batheaston is situated about 2.5 miles (4 km) east of Bath Abbey, the shaft being located on the steep western slope of Bannerdown north of the main road (A4) from Bath to London and some 5 miles (8 km) east of the nearest coal seams then in work at Harrington's Coal Works, Newton St Loe. The selection of such a site can hardly have been Smith's, since Walters apparently owned the mineral rights at Batheaston before the Company was floated. Moreover, Smith would have been aware of all the canal developments then in progress or impending and an alternative site alongside the Avon Navigation or the canals already built (or then building) would have had many practical advantages as well as being nearer areas where Coal Measures were known to exist.

The evidence, therefore, points to Walters' ownership of the land and mineral rights as being the determinative factor and it is noticeable that in his later criticism of Smith, Walters does not once mention the siting of the shaft, though this was the cause of its failure to prove coal. Nor, in spite of fulsome praise of Smith at the outset of the work, did the Company always follow William Smith's advice. Thus his preference for a powerful pumping engine designed by Trevithick rather than one made by Boulton and Watt was not heeded (Eyles 1970–1971), though subsequent events suggest that Smith gave sound advice based on his knowledge of water problems in the coalfield.

Information as to the site and altitude of the Batheaston Shafts has been obtained from a variety of sources including a personal inspection of the site of the shafts by Dr W. Bullerwell and the writer in 1947 during the course of the primary six-inch Geological Survey. The site shown on the published one-inch to one-mile (or 1:50,000) Geological Survey map (Sheet 265) is believed to be substantially correct (Nat. Grid Ref. ST 7818 6774). The surface level of the ground recorded by Dr R. H. Rastall (Report dated 1924) is approximately 193 ft (58 m) A.O.D. This level was recorded before the construction of the Elmhurst Estate which now occupies much of the hillside.

Two 9 ft external diameter shafts were constructed in Batheaston. According to the Bath Herald (10th March, 1809) both were walled round to a depth of at least 50 fathoms. A third shaft was projected but not constructed. The two shafts known to have been built were lined with masonry, one being the Engine Pit, the other the By Pit.

No drawings showing the details of the shafts and drainage level have been found, though two sections illustrating the geological succession proved in the shafts and boreholes are extant. Copies of these are preserved in the Municipal Reference Library at Bath [B662.182]. The first is dated 7th March, 1808. It includes a verbatim statement by William Smith with a log of the strata terminating in Dolomitic Conglomerate (or Millstone) at a depth of 337 ft 3 in. (102.8 m) below the surface (see Appendix 1.2). This section is almost identical to the one presented to the Geological Society of London by Mr Meade on 2nd February, 1810 and is reproduced in summarized form by H. B. Woodward (1893, pp 135–136) and L. Richardson (1928, p 49). Smith's accompanying statement was drawn up after the difficulties caused by the influx of water from the Blue Lias and White Lias had taken place and before the second and much greater influx when the Dolomitic Conglomerate was penetrated. It gives no indication that the drainage level had been constructed at this time.

The second important document held by the Municipal Reference Library at Bath is the broadsheet by Thomas Walters dated 3rd March, 1813 (referred to above) and addressed to the Committee of Management of the Batheaston Coal Company (B 622.182). Walter's statement is accompanied by a section of the strata proved in the shaft with additional MS notes in an unknown hand describing the rocks proved by the trial borings made in the bottom of the shafts up to the time of abandonment in 1813. This record was lost sight of after the closure of the shafts and was eventually re-discovered by Dr Victor Eyles who drew the writer's attention to it. The upper part of the log matches William Smith's record of 1807, the bottom part from 421 ft (128.3 m) to 683.25 ft (208.25 m) showing the Dolomitic Conglomerate extending to a depth of 511 ft (155.7 m). Between 511 ft (115.7 m) and 683.25 ft (208.25 m) the strata are thought to be Palaeozoic rocks, but their dip is not stated, a curious omission unless the dip was indeterminate owing to the nature of the samples recovered by drilling (Fig. 1.1).

One important item briefly mentioned in Walter's account but not in Smith's Report is the drainage level, which carried water from the upper part of the shaft to the valley of St

Catherines Brook. According to Walters' account the greatest influx of water proved in the shafts was encountered at 421 ft (128.3 m) below pit top and 361 ft (110 m) below the drainage level. This implies that the inset of the level in the shafts was 60 ft (18 m) below ground level, i.e. 130 ft (or 39.6 m) A.O.D.

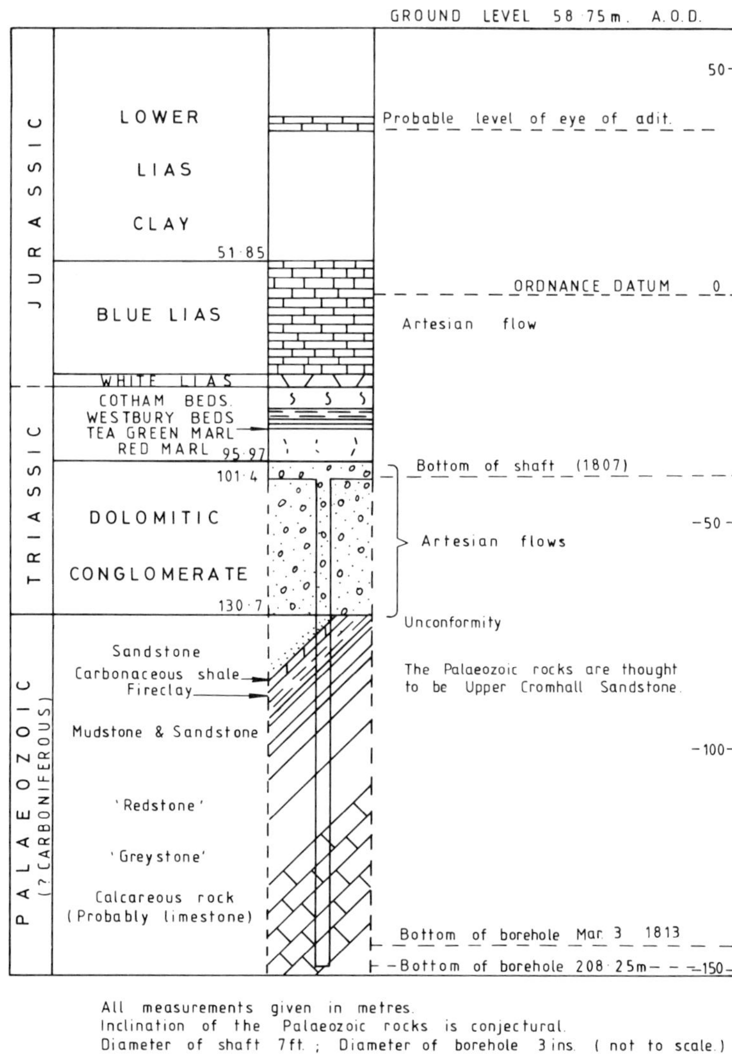

Fig. 1.1. Section through the Batheaston Shaft and borehole (1804–1813).

Evidence from several sources indicates that the mouth of the adit was situated about 1,100 ft (335 m) SE of Batheaston Church (National Grid Reference ST 780677). The altitude of the level mouth would, therefore, be about 27 m A.O.D. and the length of the adit about 600 ft (183 m) with an average inclination from the horizontal of about 1:11 (or five degrees). The construction of the adit must have been carried out mainly in clay or mudstone and shale and it would, therefore, have required a lining as well as a properly constructed channel to convey the water. When visited by Robert Etheridge in 1888 (75 years after closure and infilling of the pit), water was still issuing from the adit at a temperature of 61°F (16°C) at a rate of 57,600 gallons per day. In his Report to Bath Town Council of 9th March, 1888, Robert Etheridge pointed out the importance of determining the depth in the shaft at which the adit level is placed. If the depth of the opening or eye of this level from pit top is 60 ft (18.2 m) as suggested above, then this implies that the opening was made in the bands of harder rock (possibly the Jamesoni–Valdani Limestone) proved at depths of 19.1 to 23.56 m in the shafts.

The answer to this is provided in a MS note on Walters' section (1813) indicating that the greatest emission of water took place in 1811 when the boring in the bottom of the shaft penetrated to a depth (ex. surface) of 421 ft (128.3 m). At this time the water rose 60 fathoms, i.e. 360 ft (109 m) or 61 ft (18.5 m) below pit top level which approximates to the calculated inset level of the adit given above. Therefore, the drainage level was probably constructed sometime after the influx of water from the Lias and the upper part of the Dolomitic Conglomerate had been excluded by walling the shaft, or reduced it sufficiently by pumping to enable construction of the adit to take place. However, the final flow of water from the Dolomitic Conglomerate completely overwhelmed the steam pump and, to quote Walters, "has increased within the last six months . . ." (i.e. October 1812–March 1813) "producing a stream issuing from the Level equal in appearance at least to the discharge when the Steam Engine was working."

The Batheaston shafts therefore proved large volumes of groundwater with a slightly elevated temperature and a hydrostatic head of more than 58 m A.O.D. The chemical composition of the water can now be shown to be very similar to that of the Pulteney No. 1 (Sports Centre) Borehole made in Bath in 1982. William Smith's notes show that by 1811 after the influx of what he described as "The Cataract", which issued from the Dolomitic Conglomerate at a depth of 128 m (Fig. 1.1), he was already considering the possibility that the abstraction of water at Batheaston was affecting the flow of the hot springs at Bath. This was referred to by Wilkinson (1811, p 16) and subsequently by William Hill in a letter published in the Bath Chronicle (6th July, 1869) in which it was stated that the "pit engine [of Batheaston] was sufficient to empty the pit without working on Sundays which gave Mr Smith the opportunity of observing during six weeks its effect on the Cross Bath which took two hours more to fill on Saturdays than it did on Mondays from the engine's rest on the Sabbath." This effect was noticed more strongly at the Cross Bath than at the other springs as its bed was the highest. According to Wilkinson (in Warner, 1811) the time of filling fluctuated from 16 to 28 hours but William Smith's notes indicate that 16 to 18 hours is correct. Unfortunately Smith's notes on this subject are incomplete and Phillips' (1844) account is somewhat generalized and does not go into details. The situation in which William Smith found himself in 1811 was nevertheless a difficult one. He was acting on behalf of Bath Corporation in order to restore or preserve the hot springs while at the same time advising the would-be colliery owners to sink a shaft which was now known to affect the yield of the springs. The dilemma did not last long as by 1812 the influx of water at Batheaston had become such a drain on the resources of the company as to threaten the continuance of the operation. In the document circulated to the management committee on 3rd March 1813 Thomas Walters urged abandonment of the enterprise. At this time the boring sunk in the pit had attained a total depth from the bottom of the shaft of about 115 yds (105 m).

In his explanation to the shareholders, Walters attempted to transfer much of the blame for failure to William Smith. Thus he quoted Townshend's verdict to the effect that they had been sinking in "Mountain Limestone at the depth of Fifty Fathoms", implying that William Smith had allowed them to proceed from 1807 to 1813 when he must have been aware that it was futile to do so. In fact the shaft was sunk in Triassic rocks at 50 fathoms (100 yds or 91.4 m) and it is probable that Townshend's identification was of a specimen obtained from Dolomitic Conglomerate in which pieces of Carboniferous Limestone probably constituted the dominant rock. The base of the Dolomitic Conglomerate was encountered at a depth of about 143 yards so there is no possibility that Smith was either mistaken or neglectful of his responsibilities.

Once the boring entered the Palaeozoic rocks new problems of identification arose and Walters enlisted the help of a number of other professed "experts" including the familiar figure of Dr Charles Hemmings Wilkinson (1763–1850), a certain Dr Davis and Mr H. Smithers, a colliery surveyor of Bristol. Dr Davis gave it as his opinion that the presence of lead ore and "Marble" (Calcite) was an indication that the Mountain Limestone had been reached at 50 fathoms. From this it is clear that his knowledge of lead mining left much to be desired, the Dolomitic Conglomerate being a common host rock for galena. Dr Wilkinson agreed in principle with his judgement but needed more time to make a further "analysis". Mr Smithers was "unable to discover any reason for proceeding" but was so busy preparing a "Report on the Strata of Great Britain", that he was prevented from giving an opinion before "revisiting the Mine". Such was the "evidence" presented by Walters in his attack on the professional conduct of the Father of English Geology. This event gives an impression of one of the many dangers to which Smith was exposed and from which his friends were so concerned to protect him (Phillips 1844, pp 27–29, 43).

The identity of the rocks proved beneath the Dolomitic Conglomerate is not known but the general character of the succession suggests Upper Cromhall Sandstone. It is unlikely that either William Smith or anyone else involved at Batheaston had any previous experience of drilling this formation. The upper part of the succession proved beneath the base of the Dolomitic Conglomerate may well have resembled Coal Measures, but the downward passage into calcareous rocks would have told William Smith that the strata were older. His comment to William Hill in September 1812 that the calcareous rock then being drilled "was very discouraging" was probably his response to the first reliable evidence he had seen of the presence of limestone in the Palaeozoic rocks.

In 1807 the works foreman is reported to have said that the Pumping Engine then being erected "would melt 1000 guineas a year." Since that time the pumping costs must have increased still further. The proof of rocks older than Coal Measures was the final straw and boring ceased in 1813. From this time forward we hear nothing more of Thomas Walters whose obvious disappointment at the outcome of what was a daring but somewhat risky enterprise would have been regarded sympathetically had it not been marred by an unworthy attempt to shift the responsibility on to William Smith.

In his broadsheet (1813) Walters referred to Townshend's comment in his book (1813, p 312) in which the latter remarked truthfully but not too tactfully that "should the Kennet and Avon Company take a hint from the experience of these unfortunate coal adventurers and, in Claverton, sink through the blue marl into the lyas, and then into the mill-stone [Dolomitic Conglomerate] there is a probability that they will have the same springs, for a never failing supply of water in the construction of their canal between Bath and Bristol." Unpalatable though this may have been to Walters, it illustrates the progress in understanding the principles of hydrogeology which had been brought about, primarily by William Smith's work between 1799 and 1813.

The Batheaston Shaft and boring did not prove Coal Measures because there was no means of determining the structure of the concealed Palaeozoic rocks, a problem which even with the help of modern drilling techniques assisted by palaeontology and geophysics, is still soluble only with great difficulty when the structure of the concealed Palaeozoic rocks is as complex as it is at Bath.

That such an important event in Smith's life, linked as it was with his investigation of the hot springs, should have been so little understood is due in large measure to the disappearance of nearly all the original records including the journal of the sinking and the plans and sections of the shafts and drainage levels of the Batheaston shafts and boring. This may have arisen through deliberate destruction of the evidence, or to unwillingness to allow it to be published at the time, followed by loss or destruction of papers regarded as worthless at a later date. Closure of the pit, which is thought to have been back-filled on William Smith's instruction in order to avoid damaging Bath hot springs, was not however the end of the Batheaston saga.

In the years following the closure of the shaft, some water continued to flow from the adit mouth and was observed by Robert Etheridge in 1888 as mentioned above. Etheridge's investigation arose as a result of the need for an additional source of public water supply for Bath as set out in the Report of the Special Cold Water Committee presented to the Town Council in March 1888. This recommended a proposal put forward by Councillor Mr Handel Cosham, M.P., whose experience in mining in the Bristol Coalfield was considerable. Cosham suggested that a shaft in the Batheaston area might provide an additional source of water. He seems to have been unaware of the effect produced on the hot springs by pumping at Batheaston, in itself a surprising fact, but not less surprising than the apparent absence of any immediate firm opposition by other members of the Council.

Worse was to come however, for Robert Etheridge, having consented to advise the Council, produced a forecast section for a proposed borehole. This was not sited at Batheaston on the slopes of Bannerdown, but in Chilcombe Bottom between Charmy Down and Solsbury Hill. Before his Report could be printed, however, Etheridge was shown by the Council's Engineer, Mr W. S. Mitchell, details of the strata proved in Batheaston Pit "constructed in 1812". Subsequently he received from Major C. E. Davis a copy of the original section "by Dr W. Smith". Etheridge, who clearly had no idea that these sections existed, added a codicil and described Davis' document as "very important" saying that "it should be secured by the Corporation." No copy can be traced in the Council's archives, however, and it is not known whether this advice was taken.

Etheridge's forecast section which accompanies his Report is only broadly similar to that of the Batheaston Shaft and borehole section. Because it starts at a much higher stratigraphical level the upper part of his section is not comparable with that of the Batheaston Shaft. Etheridge envisaged drawing water from the Blue Lias and White Lias and he seems to have considered that this would pose no danger to the hot springs. However, his section shows the proposed well extending down through the New Red Sandstone into the Carboniferous Limestone and he appears not to have realized the problems with which he would have been faced. It is possible that he may have been shown William Smith's Report and section reproduced by Cruttwell (1808) or the later version by Walters (1813) without its MS notes of the deeper levels proved by the borings in the bottom of the shaft.

The surprising feature of the whole affair is the apparent lack of general knowledge of the events which took place prior to 1813. Yet William Hill (Jnr) sent a letter to the Bath Chronicle in 1869 indicating that the memory of these earlier events had not been lost. Etheridge's record of the temperature of the water issuing from Batheaston Mine (61°F) in 1888, and Heriot's analysis of 1874 (published 1875) showing the relatively high level of sodium chloride, should have provided some indication of the nature of the water at Batheaston. To this warning signal may be added the information in the possession of Mitchell and Davis which may eventually have reached the Council. In the long run realization of the true situation must have dawned, for Etheridge's Report was not acted on and the proposal to drill for a public water supply in or near Batheaston was quietly dropped.

In spite of these adverse discoveries, the extent of the Triassic Dolomitic Conglomerate in the area east of the hot springs is a matter of great importance for Bath since this formation contains a large volume of low-temperature thermal water. The western limit of the conglomerate lies between the Sports Centre and the King's Spring, its eastern limit must be situated between the Batheaston Shaft and the so-called "Lucknam Borehole" (Whittaker & Edmunds 1925, p 55) in the valley floor at Box (Fig. Pf 2). The Triassic rocks proved in the Westbury Ironworks Borehole at Westbury, Wilts. (Pringle 1922), consist almost entirely of mudstone (Mercia Mudstone) but the Dolomitic Conglomerate was proved to have a thickness of 221 ft (67 m) farther west in the Buckland Dinham Shaft [7446 5079] NW of Frome (Cruttwell 1881). The margin of the coarse Dolomitic Conglomerate is seen at Mells and Whatley, where it is resting on the Carboniferous Limestone of the Mendips and also farther west between Stratton on the Fosse and Old Down whence it extends northwards and northeastwards towards Radstock and Camerton north of the Mendips. The greatest recorded thickness of the Dolomitic Conglomerate in the Avon valley is 34.7 m at Batheaston but the general distribution and coarseness of the conglomerate as a whole indicates derivation from several sources including the Mendips and the area around Wick and Hamswell. If, as the "Lucknam" Borehole seems to indicate, there is another extensive tract of Carboniferous Limestone concealed by Triassic rocks east of Box, then this also may have contributed to the formation of this concealed body of Triassic conglomerate.

For Bath, however, it is the presence of a large store of groundwater and the relationship between this and the thermal water of the hot springs which is the critical problem. In 1979 the Bath Act of 1924 was replaced by the new Avon Act and the opportunity was taken to extend the area of control of borings and sinkings to cover Batheaston. This does not, however, give complete protection, since the abstraction of water from the area situated between Wick and Hamswell in the north and Mells in the south, and between Dunkerton in the west and Box in the east, might have an adverse effect. Even if quarrying did not cause any significant pressure changes in the deep fissure system in the Carboniferous Limestone, it could still have an effect on the flow of the hot springs if it caused a local disturbance of the balance between the pressure of water in the Dolomitic Conglomerate and the thermal water rising in the deeply penetrating fissure system at Bath.

In addition to monitoring groundwater levels in the Carboniferous Limestone of the eastern Mendips, Wick and other places, it is therefore essential to obtain more information about the water in the Blue and White Lias as well as the Triassic rocks east of Bath and, if possible, to establish monitoring stations. With this object in view, it is planned to sink two further boreholes, one midway between Batheaston and the King's Spring and the other at Tucking Mill at the southern end of Horsecombe Vale in order to carry out hydrogeological and geothermal investigations. The Tucking Mill borehole which, with the co-operation of the Wessex Water Authority, is sited at their research station, is within a short distance of William Smith's house where he resided for a time during the construction of the Somerset Coal Canal.

The Hot Springs (1808–1813)

It is probably true to say that William Smith's remedial work on the hot springs of Bath attracted the attention of a wider and more numerous section of the public than did many of his other contemporary activities. This arose from the substantial social, cultural and

Plate 1. William Smith (1769–1839), born in the Oxfordshire village of Churchill, was a founder of the science of Geology. His original observations made in and near Bath as a mine surveyor and canal engineer led to the development of stratigraphy and geological cartography and to the concept of a connected account of the history of the earth based on the succession of the stratified rocks and their contained fossils. The portrait, by H. Fourau, was painted in 1838 and is reproduced by permission of the Council of the Geological Society of London.

Plate 2. The stone cylinder surrounding the Hetling Spring (exposed by excavation in May 1986) is seen in the road fronting the Hot Bath (p. 116). Photo – Stephen Bird.

medical status of the Spa at the close of the 18th and commencement of the 19th Century. Nevertheless the results of his work on the thermal water also added much to scientific progress. So fragmentary is the existing record of this work that its reassessment has only become feasible in the light of recent discoveries made at Bath. The details of all Smith's operations at Bath may never be fully known, but we can now understand the nature of the practical problems which he faced and the reason why he was successful while others were not.

In the first place there is a distinction to be made between the remedial work (p 27) at the Hetling Spring, which was undertaken specifically to restore the dwindling supplies of thermal water, and the activities at the King's Bath. The Hetling Spring had been giving trouble for many years as described by John Wood (1778) while the inadequate flow at the Cross Bath was also a never-ending source of complaint. The flow at the King's Bath was less troublesome and the work done between 1809 and 1811 was designed to improve conditions for users rather than to increase the supply.

William Smith's work from 1808 onwards is described in a Report compiled by the Committee appointed by the Common Council of the City of Bath to manage the Baths (28th November, 1811). Because of its importance, part of this document is reproduced in Appendix 1.3. The report of 1811 draws attention to the failure of Abel Fisher's attempt to improve conditions in the King's Spring as a result of which William Smith, who was then approached for help, suggested "a different mode of procedure" should it be necessary to have recourse to it at a later date. It is obvious from this report that Smith had already identified beds of redeposited Lias Clay interbedded with other sediments in the Hetling Spring and that Abel Fisher, the Council's engineer at that time, had failed to appreciate the difference between Lias Clay in situ and redeposited or slumped clay layers in the gravel filling of the spring pipe of the King's Spring. Several clay layers of this kind were encountered in 1980 during trial drilling of the basin of the King's Spring and the uppermost one so closely resembled undisturbed Lower Lias Clay that it was not until the samples were washed that the presence of sand and tiny pebbles was recognized.

Additional light is thrown on Smith's activities at the King's Spring by MS notes made at the time of the affair of Pinch's Well (1836) presumably in preparation for the lawsuit in the case of the Corporation v. Pinch as described by Armstrong (1838). Among the items of evidence in Bath City Archives, Corporation v. Pinch (1836), recorded in handwritten notes, are the statements of George Jervis, who worked on the puddling of the Hot Bath Spring for two months with "Mr Smith an Engineer". Whether this was William Smith or his brother is not certain but William Smith was in Bath for most of this time and was in personal control of much of the work.

George Jervis's statement goes on to describe subsequent work at the King's Spring which involved opening the cylinder in which the spring rose. Jervis, who was in control of the pump, stated: "It was dug down several feet and a good foundation of stone laid loosely . . . Abel Fisher worked there and Mr Smith the Engineer, superintended it. He was sent for—I heard of no scarcity of water there . . . I don't recollect Smith being employed subsequently on any other occasion at Bath . . . there was a large body of sand in the King's Bath . . . which fell in on the Spring as the water was pumped out and a quantity of the brass railing fell in after . . . Railing was fixed in a stone which rested on the sand and as the sand gave way the railing fell in. There was no stone work at the bottom but as we got down we found two oak Beams and a large Lead cistern seven or so feet from the surface."

Other details follow, the work having been done according to Jervis "to cleanse the spring".

Taken with the Baths Committee Report of 1811 this is therefore fairly conclusive evidence that William Smith, having dealt with the Hetling Spring and learned about the structure of its spring pipe, then turned his attention to Fisher's work at the King's Spring at the Council's request, where he was able to detect the cause of the trouble. We do not know by what method Smith proposed to deal with the problem, though it may have involved anchoring the structure to the Lower Lias Clay (or to buildings founded on the clay) surrounding the spring pipe, but there being no shortage of water it was not carried out. In the case of the Cross Bath, Smith realized that he could not tackle the problem effectively without temporary removal of at least part of the building. The Committee seem to have been desirous of doing this, but nothing further was put in hand until 1885 when Major Davis carried out his first investigations at the Cross Bath.

In addition to these activities and his work at the Batheaston Shaft and borehole sinking Wiliam Smith found time to record the great flood (Appendix 1.4) which took place on 25th January, 1809. From his notes (Smith MS) we read that he spent the day, "examining the strata of the extraordinary flood marking the height etc." The following day (January 26th) was spent "Writing on flood etc." from this it appears that Smith compiled an account or report on the flooding which had caused water to invade many of the baths and installations. The original document does not appear to have survived but there is reason to believe that the measurements published by the Rev. R. Warner in his "New Guide through Bath" (1811, pp 26–27) may be based on those made by William Smith. Warner attributes most of his information to Dr Wilkinson, "philosophical lecturer and occupant of the Abbey Baths". Wilkinson seems to have persuaded Warner that the Abbey Baths (or Kingston Baths as they became known) were fed by independent springs and not, as indicated by Leland in his Itinerary, by water draining from the King's Spring. The Kingston Baths, of which Wilkinson was Lessee, were unpopular with the Council which suspected, not without good reason, that these baths were entirely supplied by water leaking away or draining from the King's Spring and several attempts were made, without success, to cut off this supply. William Smith must also have supported the Council in its conviction that there were only three hot springs in Bath for at no time does he give any indication of belief in the existence of independent springs on the old Priory site.

Major C. E. Davis, whose archaeological investigations have not always met with unqualified approval, was finally instrumental in closing down the Kingston Baths. In 1871 he showed as a result of his excavations that the Kingston Baths were indeed fed by waste thermal water from the King's Spring and during the succeeding period (1871–1880) revealed and preserved a large part of the Roman thermae while effecting substantial improvements in the drainage of the central area of the City (Davis 1880). The closure of the Kingston Baths (which were almost certainly using contaminated water) and the introduction of improved sewerage were two highly beneficial results of Major Davis' activities on behalf of the Council.

At various stages in their history all of the baths fed directly by the three hot springs have been accompanied by pump rooms. The most famous is the Grand Pump Room attached to the King's Bath. This beautiful and historic building, constructed between 1789 and 1799, overlies the site of the Roman Temple of Sulis Minerva at the very centre of Aquae Sulis and modern Bath. At the present day it is usual to refer to the Pump Room of the King's Bath simply as "The Pump Room", the smaller pump rooms attached to the other baths having long since ceased to operate. Thus John Phillips' (1844, p. 64) reference to William Smith's restoration of the water to the Baths and the Pump Room has been widely misinterpreted. In William Smith's day the 'Pump Room' attached to the Hetling Spring was in use and highly regarded. Thus in 1799, the year in which Smith dictated the 'Order of the Strata' in Great Pulteney Street, Jane Austen wrote (2nd June) of her brother Edmund Austen Knight who was ailing and in need of treatment "He drinks at the Hetling Pump, is to bathe tomorrow, and try Electricity on Tuesday." With regard to the new-fangled electricity she added "He proposed the latter himself, but I fancy we are all unanimous in expecting no advantages from it."

In his account of William Smith's work Phillips says that Smith was "hastily sent for" to restore the flow of the "Hot-bath spring" (Hetling Spring). This statement is not entirely correct as there are indications that the need for Smith's assistance was recognized for some time before he finally undertook the work. Thus an intimation that all was not well at Bath appears in the Minutes of Bath City Council for 2nd October, 1808. This states "Committee appointed to consider Mr Smith's report and letter relative to the hot and cold spring water in and around the city." On 24th November William Smith wrote to John Gurney, Esq. (Smith MS) saying that he had been busy working on the hot springs at Bath for some days and that this will prevent him coming to Norfolk. Thus by January 1809 Smith must have been involved for about six months in dealing with the Hetling Spring problem. He was in Bath almost continually during January and February 1809 where he compiled notes and reports on his findings. On 25th January the River Avon was in flood and, as has happened before and since, the flood water penetrated the area of the Roman Baths which are situated on the original flood plain of the river. Work on the Hetling Spring went forward again in February and on the 15th he "attended the Mayor and others at the Town Hall and delivered to them my Report." On 20th, 21st and 22nd February the men were "searching and cutting through the leakage under the Lepers Bath." On the 23rd he explained to the Mayor and Town Clerk what had been done and received payment of "£100 on account." On 25th February he notes in his Diary "Had the Copper Pipe drawn out and the hole choked and filled with clay." This, presumably was a bypass which he had been using to drain off the water while work was in progress, so its withdrawal marked the completion of the main task. Mrs Joan Eyles has suggested to me (pers. comm.) that he had entrusted this work to his brother, who was left to complete the job. Smith arrived back in London by coach on 26th February and did not return to Bath until 13th March when he attended a meeting of the Batheaston Coal Company whose twin shafts, then sinking, were soon to pose problems of a different kind.

Spring pipes and origin of the hot springs

In his reminiscences noted down while resident in Yorkshire in 1832 (Phillips 1844, p 65), William Smith observed that "the Bath hot-springs have a deep source, and rise as through a natural borehole in the blue marl to the surface." This observation is based primarily on his study of the Hetling Spring supplemented by inspection of the excavations (referred to above) which were made in the King's Spring during the course of Fisher's unfortunate attempt to found a structure on top of the infilling of the spring pipe. We may take it that William Smith's refusal to deal with the Cross Bath Spring without temporary removal of the building indicates that he regarded the condition at all three springs as being similar, differing only in the size of the orifice. There are considerable differences in the infilling of the King's Spring and Cross Bath Spring pipes but the nature of the infilling of the Hetling Spring is, as yet, uncertain, though it does contain some gravel.

Partly because of the situation of the hot springs on the north side of the River Avon and partly because of its proximity, William Smith looked to the area around Hamswell and Wick as the probable source of the thermal water. There is no question that the hydrostatic head of the water in the Dolomitic Conglomerate between Hamswell and Batheaston is considerable as the level of the crop at Hamswell is about 91 m A.O.D. while that of the Dolomitic Conglomerate in Batheaston Shaft (taking the highest and lower levels at which a strong influx of water was recorded) is about −45 to −65 m O.D. In the Blue Lias, which also crops out at Hamswell, the main artesian flow entered the shaft at about −5 m OD and, it is said, overflowed at the top at about 59 m A.O.D. If the old records are accurate the standing water level of the water from the Blue Lias and from the upper part of the Dolomitic Conglomerate is likely to be at least 60 m above O.D. at Batheaston. The artesian pressure in the "Cataract" may be even greater but cannot be determined as the drainage level had certainly been constructed by 1811 when the influx occurred.

All these artesian pressures are greater than those recorded in the Hot Springs at Bath and Smith may have suspected this, even though he is hardly likely to have been able to measure the full standing water levels of the hot springs.

William Smith's analogy comparing the thermal water spring pipes to boreholes shows however that he attributed their formation to water rising under pressure from some fault or fissure in the older rocks and literally drilling its way upward through the overlying clays. This is the mechanism which gives rise to some cylindrical pipe or well-like structures in other formations composed of materials such as sand or gravel, or pulverized and broken chalk cut by linear fissure systems. The significance of these non-karstic structures was not fully appreciated until investigations commenced at Bath in 1978.

By 1813 when Smith's work at Bath was drawing to a close he was resident in London and increasingly involved in other work, not least in relation to the production of his 'Map of the strata of England and Wales' which was published on 1st August 1815 (Eyles 1969). From this time onwards, memory of William Smith's work at Bath, though not lost, faded to little more than a collection of anecdotes, a situation which Phillips' biography of 1844 did something to improve, though, as will be seen from the account given above, his record contains a number of ambiguities and leaves out much of historical and scientific interest.

Acknowledgements. Victor and Joan Eyles, whose passing I have much cause to lament, assisted me in the early stages of preparing this account, and latterly, Mr Johnstone (Bath City Council Archivist) and Ms E. A. Holland. Members of the staff of Bath Reference Library have given much help in tracing documents relating to William Smith. To them and to all those who have assisted me I offer my gratitude.

References

Anon 1804. 'Batheaston Coal Company'. Note in *Monthly Magazine*, 17, June 1st, 1804.

Anon 1909. 'Batheaston Colliery'. Letter in *Bath Herald*, March 10th, 1909.

Armstrong, W. 1838. 'An Account of Tapping and Closing the Springs of Hot Water at Mr. Pinch's Brewery, Bath'. Bristol.

Bather, F. A. 1926. William Smith 'The Father of English Geology' *Royal Literary and Scientific Institution*. Bath.

Cox, L. R. 1948. 'William Smith and the Birth of Stratigraphy' *International Geological Congress, London*.

Cruttwell, A. C. 1881. 'On the Geology of Frome and its neighbourhood'.

Cruttwell, T. M. 1808. Batheaston Coal Company: call for additional support including a Statement by William Smith accompanied by a section of the shaft to a depth of 337 feet. Bath Reference Library R67/1587; B622.182.

Davis, C. E. 1880. Report to the Baths and Pump Room Committee of Bath City Council, December 1880 (relates to clearance of the Roman Drain).

Etheridge, R. 1888. Report of Mr. Etheridge FRS presented to the Special Cold Water Committee of the Town Council, March 9th, 1888. Bath City Archives, enclosed with Council Minutes of 27th March, 1888.

Eyles, J. M. 1969. 'William Smith (1769–1839). A bibliography of his Published Writings, Maps and Geological Sections, Printed and Lithographed'. *Journal of the Society of Bibliography and Natural History*. (1969) **5**, 87–109.

—— 1970–71. 'William Smith, Richard Trevithick and Samuel Homfray: Their correspondence on Steam Engines, 1804–1806'. *Transactions of the Newcomen Society*, **43**, 137–161.

—— 1985. 'William Smith, Sir Joseph Banks and the French geologists' in 'From Jeremiah to Darwin'; commentaries on the history of biology and geology, *Society for the History of Natural History*. London, pp 37–50.
Fiennes, Celia 1687. 'The Journeys of Celia Fiennes'. Ed. Christopher Morris with a Foreword by G. M. Trevelyan (1947). Cresset Press, London.
Glanvill J. 1669. Observations concerning the Bath Springs. *Philosophical Transactions,* **49**, 977–982.
Guidott, T. 1725. 'A Collection of Treatises relating to the City and waters of Bath'. Bath.
Heriot, M. 1875. The Mineral Spring at Batheaston *Proceedings of the Bath Natural History and Antiquarian Field Club,* **3**, 171–177.
Hill, William (Jnr) 1869. 'Batheaston Pit and the Bath Waters' *Bath Chronicle* 3, No. 5776, July 8th, 1869.
Jones, J. 1572. 'Bathes of Bathe's Ayde'. London.
Kellaway, G. A. & J. H. Taylor 1968. 'The Influence of Landslipping on the Development of the City of Bath, England' *23rd International Geological Congress,* **12**, 65–76.
Leland, J. 1907. 'Itinerary'. Edited by Toulmin Smith.
Phillips, J. 1844. 'Memoirs of William Smith, LLD.' John Murray, London.
Pringle, J. 1922. On a boring for coal at Westbury, Wiltshire. Summary of Progress of the Geological Survey (1921), 147–153.
Richardson, L. 1928. *Wells and Springs of Somerset*, with a Bibliography of the Bath Thermal Waters by W. W. Whittaker. Memoirs of the Geological Survey of Great Britain.
Sheppard, T. 1920. 'William Smith: his Maps and Memoirs'. Hull.
Smith, William 1819. 'Section of the Strata Through Hampshire and Wiltshire to Bath, on the road from Bath to Salisbury'. John Cary, London, May 1st, 1819.
Strachey, J. 1719. 'A Curious Description of the Strata observed in the Coal-Mines of Mendip in Somersetshire'. *Philosophical Transactions*. **30(360)**, 698.
—— 1725 or 1726. 'An Account of the Strata in Coal Mines etc.' *Philosophical Transactions*. **33(391)**, 395.
Torrens, H. 1975. The Somersetshire Coal Canal Caisson Lock. *BIAS Journal* **8**, 4–10.
Townshend, J. (Rev.) 1813. 'The Character of Moses established for veracity as an Historian, recording events from the Creation to the Deluge.' Bath and London.
Walters, T. 1813. Batheaston Coal Company: Letter addressed to the Management Committee by Thomas Walters, March 3, 1813. Printed by Richard Cruttwell, Bath. Bath Reference Library B622.182.
Warner, R. (Rev.) 1811. 'A New Guide through Bath and its Environs'. Bath: Cruttwell.
Whittaker W. & F. H. Edmunds 1925. 'The Water Supply of Wiltshire'. Memoirs of the Geological Survey of England and Wales.
Wilkinson, C. H. (Dr.) 1811. 'Analytical researches into the properties of Bath waters'. Bath.
Wood, John (Jnr) 1778. 'Description of the Hot Bath at Bath'. London.

Appendix 1.1 The Lias of William Smith

White Lias and Blue Lias are among the oldest stratigraphical terms in use in Western Europe, and, like their Norman equivalent, *liais*, are of Celtic origin (Donovan & Kellaway 1984, p 1). The flat platy or slab-like appearance of the limestones which constitute the Lias is well seen in many cliff sections, such as those of West Dorset or Glamorgan. White Lias and Blue Lias limestones were employed for building, paving and other purposes at least from Roman times onwards and their traditional names are still widely used at the present day.

In the cliff section at Lyme Regis, Dorset, the White Lias and Blue Lias are seen in sequence and in inland areas the two formations were commonly worked together in the same quarry. The men who worked them for building stone, flag stones, lime or cement were as familiar with the stratigraphical succession in the Lias as the contemporary coal miners were with the sequence of the coal seams. From the south coast of England to Warwickshire and beyond, almost every limestone was known by name to the older quarrymen (Woodward 1893). Some of these terms are plainly descriptive or indicate the uses to which the limestone could be put. Most are extremely old, a few are known to be of Celtic origin.

The Lyas in Tudor times

It is arguable that with the possible exception of Coal which is essentially a mineral or rock name rather than a stratigraphical unit, the Chalk (Leland, *Itinerary c.* 1538) and the Lias, or Lyas as it was spelled in the 16th Century, are the only British strata which are clearly identified in writing before the 17th Century. In 1572, J. Jones published a book under the title of *Bathes of Bathe's Ayde* in which he describes the hot springs of Bath and refers to the presence of hot water issuing from the Lyas at Dunkerton.

From his description it seems likely that he was referring to an area situated near the disused Dunkerton Colliery and subsequently covered by the colliery tip, where a slipped mass of Lower Lias Clay, Blue Lias and White Lias appears in the valley of the Cam Brook [6990 6880]. No confirmatory evidence of the presence of a thermal water flow has been obtained at this place and most later authors have rejected the claims as founded on hearsay. However it is possible that water issuing from the slipped rocks may have been heated by the limekilns or that the spring water may have come in contact with quicklime. Whatever the explanation of the curious record it suggests that Jones accepted without question the possibility that thermal water might be found issuing from the Lyas.

It is probable that this association of thermal water with the Lyas has been known for a long time in the Avon valley at Bath. The relationship of the Lower Lias Clay to the Blue Lias and White Lias would have been familiar to the quarrymen working Lyas at Twerton, Weston, and other localities west of Bath and shallow wells sunk through the Lower Lias Clay into the Blue and White Lias on the western side of the city would have encountered thermal water. In the early 19th Century investigations at Pinch's well (Armstrong 1838) in Kingsmead Street showed that all the water in this well, which was originally sunk to the top of the Blue Lias, was of thermal origin. Shallow trial bores made for engineering purposes have since confirmed this.

Research in the 17th and 18th Centuries

Most of the accounts of the Lias[1] in the 16th to the mid-17th Century relate to the presence of fossils or are purely descriptive in character (Childrey 1661). One of the earliest records of any stratigraphical significance is that of J. Glanvill (1669) whose paper on 'Observations concerning the Bath springs' was published in the *Philosophical Transactions of the Royal Society* (Vol. IV, No. 99, pp 977–978). Glanvill was writing at a time when increasing use of the thermal water for medicinal purposes was stimulating enquiry about the origin and source of the hot springs. His specific reference to the White Lias and Blue Lias is of interest as both formations crop out at Twerton, Weston and Newbridge on the west side of Bath, where the Lias may have been quarried from very early times. They were being quarried in 1822 (Sutcliffe 1822, p 23) and throughout the remainder of the 19th Century. William Smith would therefore have been familiar with them.

The Lias seen in the quarries at Newbridge dips eastwards towards Bath, where the thermal water rises from fissures which cut the White and Blue Lias before it penetrates the overlying Lower Lias Clay, emerging as hot springs in the floor of the Avon valley. Glanvill was therefore drawing attention to rocks which act as channels for the thermal water.

Substantial advances were next made in 1719 and 1725 when two papers by John Strachey appeared in the *Philosophical Transactions*, these were followed by a pamphlet printed for the author in 1727. All three accounts were based on investigations of the local coal mines which were then working the Farrington seams, either at crop or under shallow Mesozoic cover at the western margin of the Radstock basin. Among the features illustrated by Strachey was the unconformity between the Coal Measures and the overlying Mesozoic rocks including the 'red ground' of the miners (now described as Mercia Mudstone) shown on the published Geological Survey maps as Keuper Marl and sandstone. On these red rocks lie the deposits described by Strachey as 'yellowish spongy earth', recognizable as a description of weathered Tea Green Marl. The latter is overlain by strata described by Strachey as 'Marle' i.e. the Westbury Beds and Cotham Beds (Rhaetic); upon this rests the Lyas.

From the point of view of basic stratigraphy and their influence on geomorphology the combined White Lias and Blue Lias form prominent limestone plateaux between Stowey and Clutton and elsewhere in the Radstock basin and the Avon valley and this is how Strachey presented them in his section of 1719. In his later accounts he recognized the presence of two divisions of the 'Lyas' the lower one (dominating the heights above Stowey) being the White Lias.

[1] The rocks at Pucklechurch referred to by Childrey (1661, p 82) are Blue Lias not Coal Measures as stated by S. Smith (1943, p 412).

Strachey's 1719 section was only four miles in length and this accounts for his exclusion of any rocks above the 'Lyas'. In 1725 and 1727 he extended his original section to make it 20 miles in length with the Chalk of Salisbury Plain at one end and the Carboniferous Limestone of Broadfield Down at the other. He was unable to observe the relationship of the Carboniferous Limestone to the Coal Measures, the contact being obscured by the presence of the 'red ground' or 'red earth'. Nevertheless he placed the 'lead ore', which occurs in the Carboniferous Limestone, as being beneath the 'Coal' and, it will be observed, put the Freestone (or Oolite) between the Lias limestone and the Chalk.

At a later date (about 1788) the Rev. John Michell, who became Woodwardian Professor at Cambridge and was Rector of Thornhill near Dewsbury, worked out the succession between the Coal Measures and the Chalk (including the Lyas) on his journeys between Yorkshire and Cambridge. Although the list was not published until after his death it appears that he may have supplied the eminent engineer John Smeaton with a copy of his table (Phillips 1844).

None of Michell's findings would have been known to William Smith whose first opportunity of checking his own conclusions on a long traverse across England took place when he made a journey from Bath to Newcastle after his appointment as engineer to the Somerset Coal Canal Company in 1794. Both the White Lias and Blue Lias are listed as components of the Lias in the original 'Table of the Strata' dictated by William Smith and written down in the house of the Rev. J. Townshend at 29 Great Pulteney Street, Bath in 1799 (Fig. 1.A1).

In this account Smith described the strata beneath the White Lias as 'Mudstone, Indigo and Black Marls' but in later lists these items are omitted, as for example in the explanation of his map of 1815. This is due to the fact that the strata in question are too thin to show as separate items, the problem of scale (referred to below) having raised its ugly head even at this early date. The same problem forced William Smith to simplify the geology and to omit very thin beds of rock in drawing his long sections across England and Wales. Thus his section from Snowdon to London (1817) shows the Lias (i.e. the combined White Lias and Blue Lias) resting directly on Red Marl and Sandstone and overlain by Blue Marl (Lower Lias Clay).

The top of the White Lias as described by William Smith is quite clearly taken at the contact with the Blue Lias. From Smith's MS notes preserved in the Department of Geology at Oxford University, we can see that he regarded the 'Cotham Stone' as marking the base of the White Lias. This is a reference to the Landscape Marble or Cotham Marble of Cotham, Bristol, described by Edward Owen in 1754 which marks the base of the White Lias at Cotham House. In view of the misleading statements made about the White Lias by other geologists at a later date William Smith's personal memorandum is an important factor which has to be taken into account.

Research in the 19th Century

After 1815 Smith's attention was largely focussed on Yorkshire where he was accompanied on visits to the Yorkshire coast sections by his nephew and biographer John Phillips. Following the production of Smith's map of Yorkshire (1821), Phillips continued to work on the coast sections which he later described in his 'Illustrations of the Geology of Yorkshire Pt. 1. The Yorkshire Coast' (1829). A second edition appeared in 1835 and in these Phillips proposed the extension of the term Lias to cover all the rocks up to the base of the Dogger. He divided the Lias into three, the Lower Lias extending to the base of the 'Marlstone Series' (later Middle Lias), the remainder of the succession above the Marlstone Series being ascribed to the Upper Lias.

Some general grouping of the kind proposed by Phillips is essential if the relationship of the Blue Lias to the Lower Lias Clay is to be covered. Unfortunately, the section which he used at Robin Hood's Bay is unsatisfactory for this purpose. The Sinemurian rocks beneath the Bucklandi Zone are largely submerged while the presence of the basal beds of the Blue Lias including the Planorbis Zone (Hettangian) are only known from weed-transported blocks floated on to the beach (Kendall & Wroot 1924, pp 751–752). As an early reference to weed flotation this record is of some interest, but the absence of any precise definition of the base of the Lias and the use of a defective section far-removed from the type locality of the Lias was a violation of the elementary principles which should be employed in stratigraphical definition. To have absorbed William Smith's 'Blue Marl' in an extended Lias is one thing, to have omitted the greater part of the original Lias including its base is another and quite different proposition.

Yet a large part of the typical Blue Lias and all the White Lias of Dorset, Somerset and Avon are either invisible or absent in the Yorkshire section. The Lower Lias as defined by Phillips had no firm ascertainable base and its upper limit, drawn between the Lower Lias and Middle Lias (Marlstone

Series) proved to be one of the most difficult and troublesome boundaries in the Jurassic System. The Lower Lias could hardly have got off to a worse start!

In the West of England the implied change in the description of the base of the Lias was simply ignored. Thus in his Report on the Geology of Cornwall, Devon and West Somerset, De la Beche (1839) identified the White and Blue Lias at Lyme Regis as 'Lias limestones' in accordance with the historically correct use of the term Lias as used by William Smith (1799).

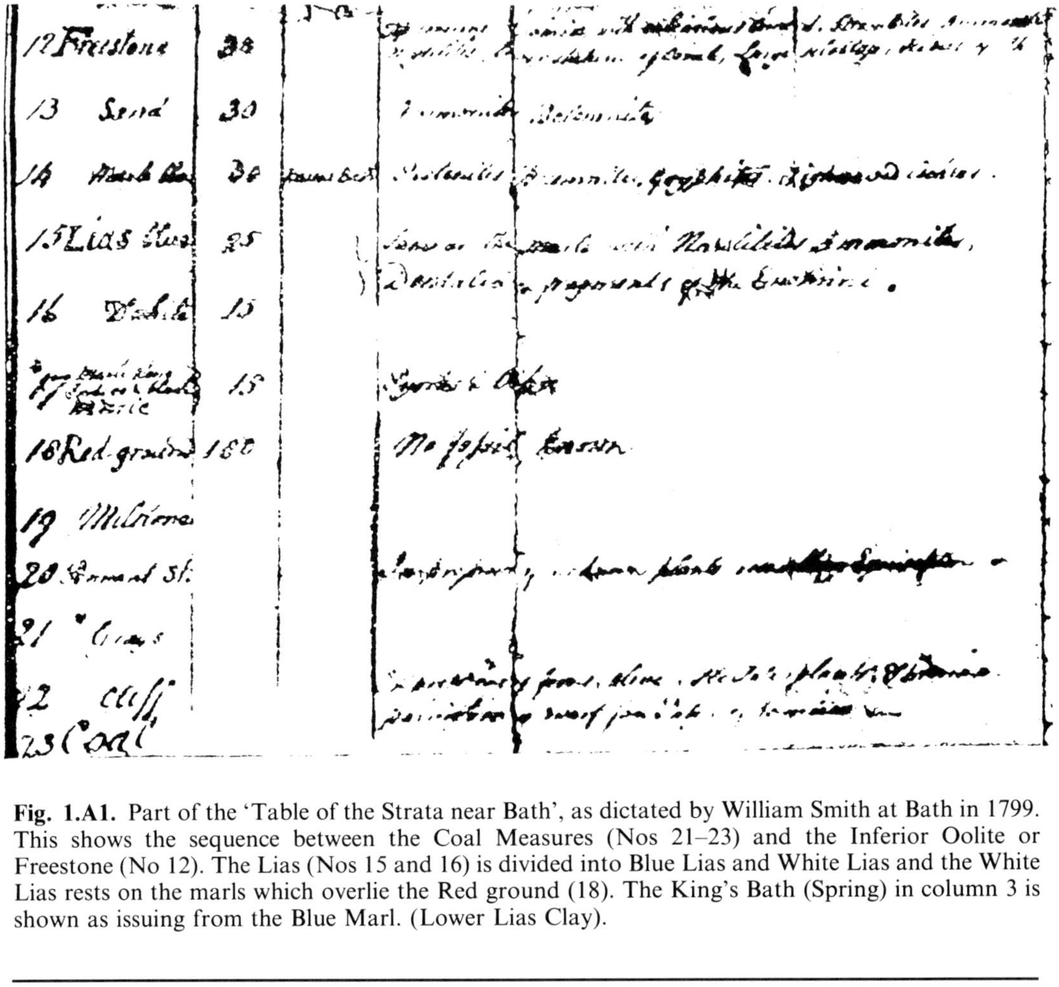

Fig. 1.A1. Part of the 'Table of the Strata near Bath', as dictated by William Smith at Bath in 1799. This shows the sequence between the Coal Measures (Nos 21–23) and the Inferior Oolite or Freestone (No 12). The Lias (Nos 15 and 16) is divided into Blue Lias and White Lias and the White Lias rests on the marls which overlie the Red ground (18). The King's Bath (Spring) in column 3 is shown as issuing from the Blue Marl. (Lower Lias Clay).

12	Inferior Oolite
13	Midford Sand
14	'Blue Marl' (Lower Lias Clay)
15	'Lias blue' } (The Lyas of Strachey and others)
16	'do white' }
17	'Mudstone, Indigo and Black Marls' (Cotham Beds and Westbury Beds)
18	Red-ground (red Triassic marls or Mercia Mudstone: The Tea Green marl is thin at Bath and is not specifically mentioned).
19	Millstone (Dolomitic Conglomerate)
20–23	Coal Measures

From this time onwards there developed conflicting views about the base of the Lias though most geologists working in the West of England adhered to those of De la Beche. In 1860, however, T. Wright published an account in which he described some limestones of an abnormally pale colour within the Blue Lias as 'White Lias'. Charles Moore (1861), corrected some of Wright's errors, but Moore himself was uncertain whether there was only one Cotham Marble or, if so, precisely where it was situated with respect to the base of the White Lias. The Cotham Marble is seldom seen in situ beneath the White Lias in inland areas as the quarrymen were in the habit of leaving the rather marly basal beds of the White Lias to act as a floor, covering the very soft sticky greenish grey clays of the Cotham Beds which accompany the Cotham Marble. Ignorance as to the detailed succession beneath the hard white limestones forming the upper part of the White Lias was therefore widespread and became a persistent source of fruitless controversy in later years.

In the mid-19th Century biostratigraphical evidence began to exert an ever increasing influence on the correlation of the Upper Triassic and Lower Jurassic rocks. Charles Moore (1861, 1867) was among those who recognized elements of a Rhaetic fauna in the black '*Avicula-contorta* Shales' or Westbury Beds, which rest on Tea Green Marls at the top of the Mercia Mudstones. These faunal similarities suggested a correlation with the Alpine Rhaetian succession, in itself a useful addition to knowledge. Not satisfied with this however, Moore proposed the 'removal' of the White Lias from the Lias and also renamed it the 'Rhaetic White Lias' on the grounds that all the rocks between the Tea Green Marl and Blue Lias have a Rhaetic fauna. This change was strongly opposed by geologists working in and around Bristol where the position of the Landscape Marble was first recognized by Owen in 1754 and the White Lias was regarded, quite correctly, as an integral part of the Lias by original definition and was used by Smith on his great map of 1815 (Sheppard 1920, p 119).

Moore's disastrous intervention was typical of many such changes proposed by stratigraphical palaeontologists in the second half of the 19th and early 20th Centuries. It involved 'adjusting', or even abolishing, perfectly sound, clearly defined and well understood lithostratigraphical boundaries in the interest of establishing biostratigraphical classifications which, had they been given an independent identity in the form of zones or stages, could have been of lasting value.[1]

Both the White Lias and the Blue Lias of William Smith, are lithostratigraphical units.[2] The White Lias has a conformable base and though owing to offlap the contact with the overlying Blue Lias is non-sequential at some localities, the angular disconformity is always so slight as to be unnoticeable. Unlike the Blue Lias, however, the White Lias is of almost uniform lithology, the base being more marly than the top, but the deposits are remarkably constant in character showing almost no lateral changes of facies except in limited areas such as the Mendips and Broadfield Down where they pass into massive shelly or conglomeratic deposits. Locally the conglomerative facies extends upwards into the basal Blue Lias as in Broadfield Down. In the area between the Avon valley at Bitton and the Mendips, the White and Blue Lias form an extremely important mass of limestone with well-marked geomorphological, hydrogeological and engineering characteristics.

The significance of the clays, mudstones and shales, now known as the Cotham and Westbury Beds, which form a very effective aquiclude underlying the Lias, did not escape William Smith. To lump together the upper part of the weak marly mudstones, shales and clays with the lower part of the overlying limestones of the Lias would have had no appeal whatever for him! The Cotham Beds are partly continental (including some sabkha-type evaporites), partly of fresh water, brackish and shallow marine origin, contrasting with the wholly marine White Lias. Smith would not have been aware of this, but his instinct, which was unerring in such matters, was to take the Cotham Stone as a marker band at the base of the White Lias, bracketing the White Lias and Blue Lias as Lias in his 'Table of the Strata near Bath' (Fig. 1.A1).

Even if the Rhaetian (Triassic) age of the White Lias could be conclusively proved this would have no bearing on the definition of the Lias or of the White and Blue Lias as described by William Smith and his predecessors. These and other traditional names incorporate centuries of observation and experience and their identity and stratigraphical order were recognized in the 17th and 18th Centuries, long before the Triassic and Jurassic Systems had even been envisaged.

Charles Moore's uncertainty about the position of the Cotham Marble may have strengthened his belief that all fossils found in the pale greenish-grey marls between the black '*Avicula-contorta* shales' (Westbury Beds) and the Blue Lias are Rhaetic. However, very few of the fossils characteristic of the Westbury Beds and basal Cotham Beds are present in the White Lias where the fauna includes large

[1] The alteration or suppression of long established lithological divisions of the Mountain (Carboniferous) Limestone was a feature of the Avonian classification of Vaughan (1905), now abandoned.

[2] This has recently been emphasised by Cope (1991) in a discussion of the Jurassic–Triassic boundary problem (*Journal of the Geological Society, London*, **148**, 420–422).

Modiola, as well as *Pleuromya* and *Plagiostoma*, forms which are quite common in the Blue Lias but exceedingly rare or unknown in the Westbury Beds. By 'lumping' these disparate elements Moore succeeded in proving to his own satisfaction that he was justified in changing William Smith's White Lias into 'Rhaetic White Lias'. Others were less impressed and the name was still-born.

In a chronological sense none of these fossils can have much diagnostic value and no dispassionate observer could justify the arbitrary exclusion of the White Lias from the original Lias on such trivial and irrelevant evidence. However, Moore was not the only geologist to misread the field evidence relating to the stratigraphical position of the base of the White Lias. Among others was H. B. Woodward, author of the Geological Survey Memoir covering 'East Somerset and the Bristol coalfields' published in 1876. The two old Series One-inch Geological Survey maps covering the Bristol and Somerset Coalfield, Sheets 19 and 35, were revised and published in 1873 and 1864 respectively. Before this was done H. W. Bristow and Robert Etheridge carried out an examination of critical sections in the West of England and South Wales with the intention of using a suitable place name from a British locality to replace the term Rhaetic.

As a result of their investigation Bristow proposed to use 'Penarth Beds' taking the name from a cliff section at Penarth near Cardiff which exhibits the local succession between the top of the red Triassic marls and the base of the Blue Lias: this does not include the Cotham Marble. Woodward's summary of the strata included in the 'Penarth Beds' (1876, p 69) shows that in so far as the definition was applied to the Bristol and Somerset Coalfield it was fatally flawed, as it contained the usual contemporary error regarding the identity and position of the Cotham Marble and the base of the White Lias of William Smith. A thin representative of the White Lias has been identified at Penarth (Whittaker 1978), but Woodward's comment about the Cotham Marble (1876, p 72) is worth quoting as it shows how the muddle and confusion about the position of the base of the White Lias produced such a baleful effect. He says 'The Cotham Stone generally (sic.) occurs at the base of the White Lias, at or near the top of the Black paper-shales.' The fact is that Woodward had failed to recognize the Cotham Beds as distinct from the true White Lias. The Cotham Marble has *never* been found in contact with the black shales of the Westbury Beds (unless faulted); neither has it ever been found in any other position than at or immediately beneath the base of the White Lias save (as at Aust) where the White Lias is missing. At Aust the impersistent Cotham Marble is overlain non-sequentially by the basal beds of the Blue Lias (Pre-Planorbis Beds).

In spite of these problems some form of general grouping of the thin formations between the Mercia Mudstone and the Lower Lias Clay was essential because of the small scale on which geological mapping was carried out in the mid-19th Century. The base map was the one-inch to one-mile (1 : 63,360) map which was then without spot heights or contours. It would have been quite impossible on this scale to map individually all the very thin lenticular strata between the red Keuper Marl and the Blue Lias. In the end, however, the 'Penarth Beds' turned out to be no more than a 'bucket' term designed to separate the Blue Lias from the White Lias and include the latter with all the other underlying beds down to the top of the red marls of the Mercia Mudstone. In short, it was a thinly veiled attempt to map a biostratigraphical 'boundary' separating Charles Moore's 'Rhaetic' from the defective 'Lower Lias' of Phillips. The term Jurassic was not in use by the Geological Survey at the relevant time, the rocks later classified as Jurassic being referred to as the 'Lias' and 'Oolitic Series'.

With hindsight it can be seen that the one-inch revision could never have achieved success, as an elementary mistake was made by not mapping the Lias limestones in the simple form that Strachey, Smith, De la Beche and others had originally defined them. The surveyors, who were presented with the unenviable task of drawing one boundary of the 'Penarth Beds' at the top of the red Mercia Mudstone (or Keuper Marl) and another in the middle of the Lias limestones in Somerset and Dorset using an uncontoured one-inch map, thus lost the help which would have been accorded them by mapping the natural features at the base and the top of John Strachey's Lyas. The resulting map did little to enhance the Survey's reputation. In practical terms, definition of the crop of the 'Penarth Beds' and 'lower Lias' was valueless for the study of geomorphology, water supply, resources of lime and building stone, agriculture, engineering or tectonics. Most geologists working in the West of England refused to acknowledge the term as having any scientific or economic value and the Geological Survey map was, in this respect, written off by them as useless.

The position following the publication in 1876 of the Geological Survey Memoir covering the Bristol & Somerset coalfield and its environs could hardly have been worse, and the classification remained in this lamentable condition until about 1890 when the construction of a deep cutting on the face of Totterdown at the approaches to Temple Meads Station, Bristol brought about a dramatic change of attitude.

Edward Wilson, who described the Pylle Hill cutting in 1891 had much experience in dealing with the Upper Triassic and Lower Jurassic rocks in Nottinghamshire and Leicestershire. While working in the Midlands he had, as he later explained, no idea as to the basis of the separation of the true White Lias from the underlying strata. Soon after his appointment as Curator of Bristol Museum, he started to study the problem of the Rhaetic rocks. This brought him in contact with geologists who had adhered strictly to William Smith's definition of the Lias as it had been handed down from the beginning of the 19th Century. Wilson's account of the Pylle Hill cutting which proved a fine section extending from near the base of the Mercia Mudstone to the Blue Lias was first published by the Geological Society of London in 1891. At the discussion of the paper some of the old misconceptions were brought out, notably by Winwood who had long supported Charles Moore's interpretations. In a second account published by the Bristol Naturalists Society in 1894, Wilson dealt specifically with these problems, including the position of the base of the true White Lias, pointing out that the Cotham Marble which marks the contact with the base of the White Lias, is closely associated with shales carrying *'Avicula' (Meleagrinella) decussata* and that these may extend slightly above the Cotham Marble at some locations.[1]

Wilson described the black shales (Westbury Beds) as Lower Rhaetic and the overlying pale marls and clays (Cotham Beds) as Upper Rhaetic, but pointed out that these divisions were not of more than regional significance.

Research between 1900 and 1939

Wilson's account of Pylle Hill (1894) was followed by the appearance of a short, but valuable paper by W. H. Wickes (1901) giving an account of a section at Redland, Bristol. This is of interest as it described the lower part of the Upper Rhaetic (Cotham Beds) including the rocks with *Estheria* and *Naiadites* and underlined the fact that the attenuation of the White Lias (here only $10\frac{1}{2}$ inches thick) is independent of changes in thickness of the Cotham Marble (8 ins) and the associated shales with *Pseudomonotis (Meleagrinella) decussata*. Wickes followed Wilson's example in dividing the strata between the White Lias and the Tea Green Marl into Lower and Upper Rhaetic. There followed an account by Tutcher (1908) of the railway cutting at Charlton, between Filton and Avonmouth, arguably the best section seen in these rocks north of Watchet, including a thin White Lias with the Cotham Marble at the base.

At about this time the railway cutting at Saltford, midway between Bristol and Bath was widened and deepened at its eastern end. This time Winwood (1910, pp 45–48) who had long supported Charles Moore's views, was obliged to concede that the Cotham Marble marks the base of the true White Lias as William Smith had recorded it in his notes almost a century earlier. The Saltford section was described by Richardson (1910, pp 49–50) who contributed a photograph showing the basal Blue Lias, the whole of the White Lias (8 ft thick) and the Cotham Marble *in situ* in the top of the greenish grey shaly clays of the Upper Rhaetic (or Cotham Beds). Richardson then proceeded to exclude the White Lias from the Lower Lias, on what evidence is not stated. Presumably this was a biostratigraphical interpretation, which, like so many others of this date, involved the false equation of the lithostratigraphic term Lias with Jurassic.

Two years later Richardson (1911) produced a classification of the Rhaetic, White Lias and basal Blue Lias. Having investigated sections at Langport, Watchet and Blue Anchor he re-named the Lower and Upper Rhaetic divisions as Westbury Beds and Cotham Beds respectively and added the name of another unit, the Watchet Beds (now included with the Blue Lias). The White Lias was renamed 'Langport Beds' an invalid synonym of White Lias which violated the priority of William Smith's name without any justification whatever. The terms Cotham Beds and Westbury Beds have since been widely used in place of Lower and Upper Rhaetic, though Richardson did not record a typical section at Cotham, Bristol where the Cotham Marble was first described.

Cotham Hill is a prominent outlier, now lying well within the City limits of Bristol. The lower part of the Blue Lias and the White Lias were formerly worked in quarries on the hill and a section through Cotham Quarry was published by W. W. Stoddart (1868, pp 203–204) and two sections were described by C. O. Groom-Napier in the same paper (pp 204–206). Both these accounts contain much of interest but, without local knowledge, are difficult to interpret. Stoddart's section of Cotham Quarry shows the White Lias 2 ft thick resting on 4 ins of Cotham Marble which is said to be resting

[1] This is only one of several features which can be used to define the base of the White Lias in the absence of the Cotham Marble.

on 'Keuper Marls'. As the black shales of the Westbury Beds overlying Tea Green Marl are known by the writer to be present beneath the Cotham Beds at Cotham House, the identification of 'Keuper Marl' beneath the Cotham Marble is clearly wrong. The upper part of the Cotham Beds consists of greenish grey mudstone and both Stoddart and Groom-Napier mistook these mudstones for Tea Green Marl (then regarded as the upper part of the Keuper marl). However, Groom-Napier recorded *Pecten (Valoniensis?)* and bones of *Ichthyosaurus* and *Plesiosaurus* from strata below the Cotham Marble at Cotham thus confirming this interpretation. One particularly interesting feature is the presence of celestite in the dark grey clays associated with the Cotham Marble, suggesting the development of a short-lived sabkha or intertidal flat immediately after the formation of the Cotham Marble and preceding desposition of the purely marine White Lias.

These records fail to give an account of the full thickness of the Cotham Beds which were later exposed in an excavation at Oakfield Road Pumping Station, Clifton (Tawney 1878). This section is at the southern end of the Cotham outlier but it provides the most complete record of the Cotham Beds and Westbury Beds in the vicinity of Cotham Hill and is therefore proposed as a stratotype.

The proposal to replace White Lias by Langport Beds (Richardson 1911) has, until recently, been tacitly rejected by most geologists as a breach of priority. It may have been intended to avoid the problems associated with the definition of the White Lias in part of West Somerset and Glamorgan where the Cotham Marble is very thin or has not been identified. In practice, however, the White Lias can, with experience, be recognized, either by the presence of the shale with *Pseudomonotis (Meleagrinella) decussata* or by the general character of the white porcellanous limestones and of the associated marls or clays. In the West Somerset coast sections it has been found that the White Lias can be identified even where it is only 1 to 1.5 m in thickness and the Cotham Marble is not present (Whittaker 1978)[1].

The reasons for the choice of the Langport section as stratotype of the Langport Beds (Richardson 1911) are obscure. The section in the White Lias of Langport does not bear comparison with that of Sparkford and Camel Hill north of Yeovil, first described by Charles Moore (1867, pp 461–464) and later re-examined by Kellaway & Wilson (1941, pp 198–240, fig. 11). Here the White Lias is 21 ft (6.4 m) thick and is superior to that of Langport in so many respects as to make it impossible to understand Richardson's unsuitable choice even if any valid reason had been given for substituting another name for William Smith's White Lias. Smith worked in the area between Bath and Radstock, where the White Lias is comparable in thickness with that of Sparkford. North of the Mendips the White Lias achieves its maximum thickness in a narrow belt extending through Clandown north of Radstock to Buckland Dinham (Donovan & Kellaway 1984, fig. 2) and it is here that the stratotype should, if possible, be located. Fortunately there is no difficulty in achieving this as Tutcher & Trueman (1925) provided the necessary information (Fig. 1.A2).

Shortly before the Second World War the results of Professor T. M. Harris' work on the flora of the British Rhaetic was published in 1938. This included a useful summary of the stratigraphy as known at that date. Primary six-inch mapping of the Bristol and Somerset Coalfield was commenced shortly after by the Geological Survey (1939) but this was broken off owing to the outbreak of war and not resumed until late 1942 and 1943.

The Geological Survey's primary six-inch programme (1942–1962)

At an early stage in the primary six-inch mapping of the Bristol District Special Sheet it was established that the Tea Green Marl, White Lias and the Blue Lias could be mapped as independent formations on a scale of 1 : 10,560. Only in those areas where the White Lias is on average less than 1.0 m thick was it necessary to map it in combination with the basal Blue Lias (Pre-Planorbis Beds) as a single unit. On the published one-inch to one-mile and 1:50,000 scales however, the White and Blue Lias may perforce be shown as undivided; locally the White Lias is thicker than the Blue Lias (Fig. 1.A2).

Unlike the top of the Tea Green Marl (or Blue Anchor Formation) the contact of the Westbury Beds and Cotham Beds is not marked by a mappable feature and the clays of both formations weather down to give material of low bearing-strength which forms extensive slumps. The Westbury Beds are transgressive and as they overstep the rising surface of the older Triassic and Palaeozoic rocks they tend to thin from the base upwards. When this is combined with mass attenuation of the

[1] In some parts of the Bristol coalfield the White Lias has been mapped in open country to the limits of about 0.25 m both with and without the presence of the Cotham Marble. In closely built-up areas the problem is more difficult and the limit is about 0.75 m. Whittaker (1978) has shown that in West Somerset and Glamorgan the problem is not as insoluble as it has been claimed to be and the White Lias can be identified in most sections.

Westbury Beds and the Cotham Beds, as it is in parts of west Bristol, the combined thickness can be reduced to 3 metres. Folding and faulting of the contact can only be identified with the aid of an intensive programme of augering or trial-holing and concealment by large urban areas poses almost insuperable cartographic problems.

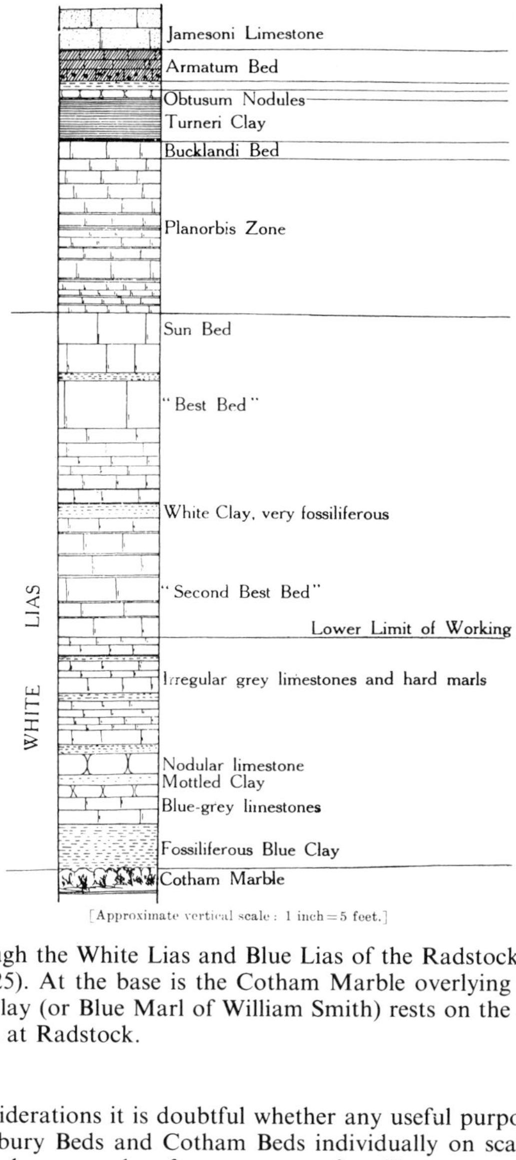

Fig. 1.A2. Section through the White Lias and Blue Lias of the Radstock Basin at Rockhill (from Tutcher & Trueman 1925). At the base is the Cotham Marble overlying the clays of the Cotham Beds. The Lower Lias Clay (or Blue Marl of William Smith) rests on the Jamesoni Limestone, the top bed of the Blue Lias at Radstock.

Apart from such considerations it is doubtful whether any useful purpose would be achieved by trying to map the Westbury Beds and Cotham Beds individually on scales smaller than 1:2,500. Treated as a single unit, however, they form a group of argillaceous strata which is mappable at 1:10,650. They have important engineering properties and also form an effective aquiclude. The name Rhaetic has been applied to them on one-inch and six-inch Geological Survey maps, but as the basis of identification on the map is essentially lithostratigraphic, the term Rhaetic is unsatisfactory and a practical replacement is suggested below (p 48).

In spite of their separation by a non-sequential contact the depositional pattern of the Westbury Beds and Cotham Beds is closely related and in Oxfordshire there is a passage between them. The area over which they were laid down remained as stable as did the parallelism of their bedding. Deposition of the Cotham Marble (an algal limestone marker band) signalled a pause in sedimentation over a remarkably wide area and can be identified as far away as Cheshire, but the absence of this marker in some areas, notably Glamorgan, is said to give rise to problems of correlation.

Though not worse than those experienced in the Bristol and Somerset coalfield, these are of a different kind. In the Bristol and Somerset coalfield the chief problem is to deal with extreme marginal attenuation of the White Lias. This has been solved by combining the attenuated White Lias with the basal Blue Lias limestone ('Pre-Planorbis Beds'). In Glamorgan the absence of the Cotham Marble and attenuation of the White Lias also gives rise to problems of classification. These can only be solved by local methods as at Bristol. Suppression of the term White Lias and the merging of a so-called 'Langport Member' with a 'Cotham member' in a 'Lilstock formation (Warrington *et al.* 1980) raises much larger problems than it solves.

The form of the White Lias sedimentary basin has little in common with that of the underlying Cotham Beds and combination with the so-called 'Langport Member' using the Langport section as a replacement stratotype for William Smith's White Lias is totally unacceptable.

The strata for which a suitable name is urgently needed are those described by Strachey (1719) as Marle, i.e. the combined Westbury Beds and Cotham Beds, or Rhaetic as shown on the published maps. The most obvious choice for a stratotype is Aust Cliff where a permanent and extensive section is visible from the M4 motorway and the Severn Bridge. Historically this is the most famous of the sections in the Severn Valley, known since the 17th Century and the source of remarkable collections of vertebrate fossils.

A number of accounts of the section have been published including the one given in the Geological Survey's Chepstow and Monmouth Memoir (Welch & Trotter 1961). The Cotham Marble is known to be present locally beneath the Blue Lias at Aust even though the White Lias is missing (Wilson 1897).

The combined Westbury Beds and Cotham Beds were referred to by Strachey (1719, 1725) as Marle and their range is taken as extending from the top of the Tea Green Marl to the base of the White Lias. They could therefore be described as the Aust Marls and this name could be used as a lithostratigraphical term replacing 'Rhaetic'.

Stratal units and their stratotypes

The rocks above the Mercia Mudstone include the Tea Green Marl and Grey Marls of central and north Somerset with Avon. Substitution of the term Blue Anchor Formation (Warrington *et al.* 1980) in the area south of the Mendips and particularly in West Somerset is clearly advantageous, as it would also be at Uphill in the western Mendips where the Grey Marl facies is present (Kellaway & Oakley 1934). North of a line drawn from Goldcliff to Locking, Wells and Shepton Mallet, however, the status of the Tea Green Marl is less clear. Locally up to half the thickness of the Tea Green Marl may be Mercia Mudstone which has undergone a secondary colour change to produce red and green mottlings and bandings which are not of depositional origin.

Table 1.A1. Stratotypes

Blue Lias	(William Smith 1799) Stratotype: Saltford Railway Cutting (Cope *et al.* 1980)
White Lias	(William Smith 1799) Stratotype: Rockhill, Clandown, Radstock (Tutcher & Trueman 1925, fig. 5)
Cotham Beds	(Richardson 1911) Stratotype: Oakfield Road Pumping Station, Clifton, Bristol (Tawney 1878).
Westbury Beds	(Wright 1860; Richardson 1911) Stratotype: Garden Cliff, Westbury-on-Severn.
Aust Marls	(This Volume, p 49) Stratotype: Aust Cliff, Gloucestershire (Welch & Trotter 1961, p 121)

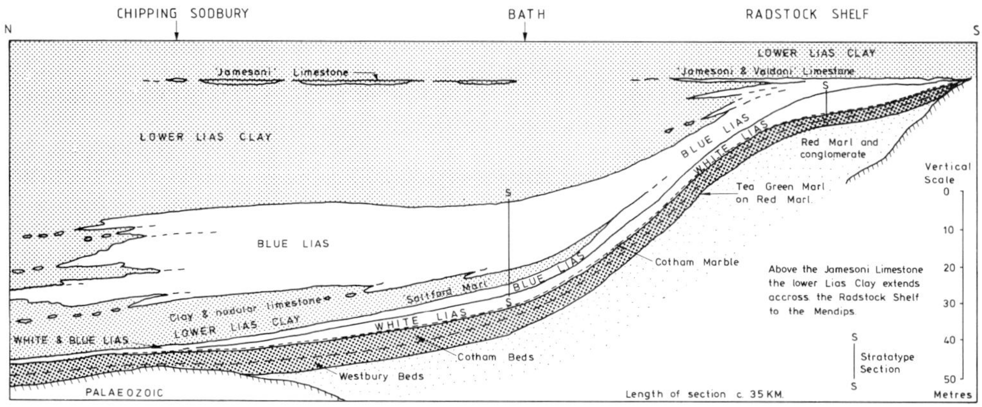

Fig. 1.A3. Facies changes in the Lias between the base of the White Lias and the 'Jamesoni' Limestone (top of Blue Lias) of the Radstock Shelf. The Blue Lias stratatype section is at Saltford, 7 km NW of Bath. North of Chipping Sodbury at Hawkesbury and Stinchcombe the main mass of the Blue Lias is represented by clay, only the thin combined basal Blue Lias and White Lias having survived the change.

The Blue Lias and Lower Lias Clay require regional stratotypes in accordance with regional changes (Table 1.A1). These would be related to major sections such as those of the Dorset, West Somerset and Yorkshire coasts, or to inland borehole data.

Facies changes controlling the relationship of the Blue Lias and Lower Lias Clay (Fig. 1.A3) have been recognized in principle since the 19th Century and are illustrated by Woodward (1893, fig. 1). They are shown in a striking fashion by the changes in the Blue Lias and Lower Lias Clay between the Mendips and the Severn valley near Berkeley and were originally elucidated in the Radstock area by the use of ammonite zonation by Tutcher & Trueman (1925) and in the Avon valley at Keynsham and Saltford by Tutcher (1923). North of the Avon valley on the eastern side of the Bristol and Somerset Coalfield the greater part of the Blue Lias passes laterally into clay, the change being almost complete in the area north of Wickwar (Cave 1977; Donovan & Kellaway 1984). The assessment of such changes shows that it is imperative to define basic lithostratigraphic units in a form which can be compared with independent zonal or other biostratigraphical divisions without prejudice to the limits of either of them. Within the main mass of the Lower Lias Clay however, lithostratigraphic units can seldom be established on surface evidence alone. Exceptionally this may be done in areas adjacent to major coastal sections but inland, additional borehole and geophysical data are required and problems are likely to arise in relating surface to subsurface classifications.

The use of over-formalized hierarchical classifications is, in the author's view, to be deprecated. In use they are unnecessarily restrictive, wearisome to write and boring to read, witness the conversion of White Lias which has been described in print as 'the Langport Member of the Lilstock Formation'. Shortening them by omitting the words 'Member' or 'Formation' appears to present a way of moderating repetition, but may produce such nonsensical effects as 'the plateau of Langport at Radstock' or 'the miner's houses at Radstock are built of Langport'. A more serious objection is the way in which a rigid hierarchical system cannot be adapted without great difficulty to deal with lateral change of facies. Because the grading of Group, Formation and Member is largely reliant on the thickness of the strata at the point of definition, we find absurd situations such as in the Radstock Basin where the White Lias (21 ft thick) is defined as the 'Langport Member' while in the same section the combined thickness of the Cotham Beds (Member), the Westbury Beds (Formation) and the Blue Anchor (Formation) is less than that of the Langport Member.

There is no way in which the disadvantages of such a rigid system can be rectified, indeed they are likely to increase with time. Most of the stratal units under discussion here are lenticular, thin, and of such contrasting facies that comparisons of thickness and sedimentation rates have little or no meaning. For most shallow water, marine, brackish and continental sediments a simple descriptive lithostratigraphical nomenclature is best. On a larger scale this is the reason for the survival to date of the remarkable national heritage bequeathed to us by William Smith's practical nomenclature e.g. Coal Measures, White Lias, Blue Lias, Cornbrash and Chalk.

The White Lias has been proved in a number of deep boreholes to the east and south east of the Bristol and Somerset Coalfield. It continues at depth as a key marker band, the dominant unit in the so-called 'Penarth Group', maintaining much the same thickness as it exhibits at Midford, Radstock and Westbury (Wilts) while extending at depth from Bruton (Holloway 1982) to Sparkford and thence eastwards beneath Salisbury Plain. The importance attached to it on the six-inch Geological Survey maps of the Bristol and Somerset Coalfield is therefore fully in accord with findings at depth farther east where the Cotham Beds and Westbury Beds become thin and inconspicuous. It may be possible to misinterpret the significance of the White Lias near the margin of its basin in West Somerset or Glamorgan and to award it a minor place in the hierarchy, but in West Dorset, at Sparkford, Radstock and Bath or beneath Salisbury Plain it is an important lithostratigraphical unit occupying an extensive independent sedimentary basin.

References to Appendix 1.1

Armstrong, W. 1838. 'An account of the tapping and closing of the springs of hot water at Mr Pinch's Brewery, Bath. Bristol.

Cave, R. 1977. Geology of the Malmesbury District. *Memoirs of the Geological Survey of Great Britain.*

Childrey, J. 1661. Britannia Baconica: Or, The Natural Rarities of England, Scotland and Wales. London (Printed for the author).

Cope, J. C. W., Getty, T. A., Howarth, M. K., Morton, N. and Torrens, H. S. 1980. *A correlation of Jurassic rocks in the British Isles Pt. 1. Introduction and Lower Jurassic.* Geological Society, London, Special Report, 14.

De la Beche, H. 1839. Report on the Geology of Cornwall, Devon and West Somerset. *Memoirs of the Geological Survey of Great Britain.*

Donovan, D. T. & Kellaway, G. A. 1984. Geology of the Bristol district: the Lower Jurassic rocks. *Memoirs of the British Geological Survey.*

Glanvill, J. 1669. Observations concerning the Bath springs. *Philosophical Transactions of the Royal Society,* **4** (No. 99), 977–978.

Groom-Napier, C. O. 1868. 'On the Lower Lias Beds occurring at Cotham, Bedminster and Keynsham near Bristol' (Abstract) *Quarterly Journal of the Geological Society, London,* **24**, 204–206.

Harris, T. M. 1938. The British Rhaetic Flora. British Museum (Natural History), Pl. 1–5.

Holloway, S. 1982. Bruton No. 1 Geological Well Completion Report. *Institute of Geological Sciences.*

Kellaway, G. A. & Oakley, K. P. 1934. Notes on the Keuper and Rhaetic exposed in a roadcutting at Uphill, Somerset. *Proceedings of the Bristol Naturalist Society*, 4th Series, **7**, 470–488.

—— & Wilson, V. 1941. An Outline of the Geology of Yeovil, Sherborne and Sparkford Vale. *Proceedings of the Geologists Association,* **52**(2), 131–174.

Kendall, P. F. & Wroot, H. 1924. *Geology of Yorkshire.* 2 Vols. (Printed for the author.)

Moore, Charles, 1861. On the Zones of the Lower Lias and the Avicula Contorta Zone. *Quarterly Journal of the Geological Society, London,* **17**, 483–516.

—— 1867. On abnormal conditions of secondary deposits when connected with the Somersetshire and South Wales Coal-Basin: and on the age of the Sutton and Southerndown Series. *Quarterly Journal of the Geological Society, London,* **23**, 449–568, Pl. 14–17.

Owen, Edward, 1754. Observations of the earths, rocks, stones and minerals, for some miles about Bristol, and on the nature of the Hot-well and the virtues of its water. London.

Phillips, J. 1829. Illustrations of the Geology of Yorkshire Pt. 1. The Yorkshire Coast.

—— 1844. Memoirs of William Smith LLD. London.

Richardson, L. 1910. Note on a White-Lias section at Saltford near Bath. *Proceedings of the Cotteswold Naturalist Field Club*, **17**, 49–50, Pl. 13.

—— 1911. The Rhaetic and contiguous deposits of west, mid- and part of East Somerset. *Quarterly Journal of the Geological Society, London,* **87**, 1–74.

Sheppard, T. 1920. William Smith his maps and memoirs. Hull.

Smith, Stanley, 1943. Seventeenth Century Observations on Rocks, Minerals, Fossils etc. of Gloucestershire and Somerset. *Proceedings of the Bristol Naturalist Society Fourth Series,* **9**(4), 406–426. Pl. 14.

Smith, William. 1817. Section of England and Wales from Snowdon to London. (Reproduced in the Philosophical Magazine, Jan. 1833 and by T. Sheppard (1920.)

Stoddart, W. W. 1868. 'Notes on the Lower Lias Beds of Bristol'. *Quarterly Journal of the Geological Society, London,* **24**(1), 200–204.

Strachey, J. 1719. A curious description of the strata observed in the coal-mines of Mendip in Somersetshire. *Philosophical Transactions* (No. 360), **33**, 968–973, Table II.

—— 1725. An account of the strata in coalmines. *Philosophical Transactions* (No. 391), **33**, 395–398, plates I, II.

—— 1727. Observations on the different strata of rocks and minerals (more particularly of such as are found in the coalmines of Great Britain). London. J. Walthoe.

Sutcliffe, J. 1822. *The Geology of the Avon.* Bristol.

Tawney, E. B. 1878. On an excavation at the Bristol Water Works pumping station, Clifton. *Proceedings of the Bristol Naturalist Society New Series,* **2**(2), 179–182.

Tutcher, J. W. 1923. Some recent exposure of the Lias (Sinemurian and Hettangian) and Rhaetic about Keynsham. *Proceedings of the Bristol Naturalist Society Series 4,* **5**, 268–278.

—— & Trueman, A. E. 1925. The Liassic rocks of the Radstock District (Somerset) *Quarterly Journal of the Geological Society, London,* **81**, 595–666.

Warrington, G., Audley-Charles, M. G., Elliot, R. E., Evans, W. B., Ivimey-Cook, H. C., Kent, P. E., Robinson, P. L., Shotton, F. W. & Taylor, F. M. 1980. *A correlation of Triassic rocks in the British Isles.* Geological Society, London, Special Report, **13**.

Welch, F. B. A. & Trotter, F. M. 1961. Geology of the Country around Monmouth and Chepstow. *Memoirs of the Geological Survey of Great Britain.*

Whittaker, A. 1978. The lithostratigraphical correlation of the uppermost Rhaetic and lowermost Liassic Strata of the West Somerset and Glamorgan areas. *Geological Magazine,* **115**(1), 63–67.

Wickes, W. H. 1901. A Rhaetic section at Redland. *Proceedings of the Bristol Naturalist Society New Series,* **9**, 99–103.

Winwood, H. H. 1910. Notes on a White Lias section at Saltford near Bath. *Proceedings of the Cotteswold Naturalist Field Club,* **17**(1), 45–48.

Wilson, E. 1891. On a section of the Rhaetic rocks at Pylle Hill (Totterdown) Bristol. *Quarterly Journal of the Geological Society, London,* **47**, 545–549.

—— 1894. The Rhaetic Rocks of Pylle Hill, Bristol. *Proceedings of the Bristol Naturalist Society,* **7**(3).

Woodward, H. B. 1876. Geology of east Somerset and the Bristol coal-fields. *Memoirs of the Geological Survey of Great Britain.*

—— 1893. The Jurassic Rocks of Britain Vol. III Lias of England and Wales (Yorkshire excepted) *Memoirs of the Geological Survey of Great Britain.*

Wright, T. 1860. On the Zone of *Avicula Contorta* and the Lower Lias of the South of England. *Quarterly Journal of the Geological Society, London,* **16**, 374–411.

Appendix 1.2. Section of pit sunk at Batheaston by the Batheaston Coal Company based on the Section drawn up by William Smith (1808) and published with additions by Thos. Walters, March 3, 1813. Bath Reference Library No B622.182.

Item No.	Ground Level 192.68 ft. A.O.D.	Thickness Ft Ins	Depth Ft Ins
	Lower Lias Clay (proved 171 ft 9 ins)		
1	Yellow Clay	8. 0	8. 0
2	Blue Marl	27. 0	35. 0
3	Stone [limestone]	0. 6	35. 6
4	Blue Marl	18. 0	53. 6
5	Stone [limestone]	0. 4	53.10
6	Blue Marl	9. 0	62.10
7	Blue Rock [argillaceous limestone]	12. 0	74.10
8	Marl	1. 0	75.10
9	Stone [argillaceous limestone]	1. 6	77. 4
10	Blue Marl	24. 0	101. 4
11	Rock [limestone]	3. 0	104. 4
12	Marl	6.10	111. 2
13	Stone [limestone]	0.10	112. 0

Appendix 1.2 (*continued*)

Item No.	Ground Level 192.68 ft. A.O.D.	Thickness Ft Ins	Depth Ft Ins
14	Blue Marl	41.10	154. 8
15	Blue Marl	7. 0	161. 8
16	Stone [limestone]	0. 9	162. 5
17	Blue Marl	6. 0	168. 5
18	Stone [limestone]	0. 4	168. 9
19	Marl	3. 0	171. 9
	Blue Lias (78 ft 11 ins)		
20	Marl and Stone [limestone]	11. 6	183. 3
21	Stone [limestone]	0. 9	184. 0
22	Marl	2. 0	186. 0
23	Stone [limestone]	3. 0	189. 0
24	Marl and Stone [limestone]	8. 0	197. 0
25	Marl and Stone [limestone]	12. 3	209. 3
26	Hard marl	4. 0	213. 3
27	Hard rock [argillaceous limestone]	18. 4	231. 7
28	Blue stone [argillaceous limestone]	5. 1	236. 8
29	Hard blue stone	12. 0	248. 8
30	Corn Grit Stone [a coarse calcarenite rock commonly developed at this level containing echinoderm and shell fragments]	2. 0	250. 8
	White Lias (10 ft)		
31(a)	White Lias rock	10. 0	260. 8
	Cotham Beds (12 ft 3 ins)		
31(b)	Cotham Stone [shown as base of White Lias, thickness not given]		
32	Blue Marl	6. 0	266. 8
33	Stone	0. 6	267. 2
34	Clay	0. 9	267.11
35	Rough Blue Marl	5. 0	272.11
	Westbury Beds (10 ft)		
36	Black [marl]	10. 0	282.11
	Tea Green Marl (Blue Anchor Beds) (2 ft)		
37	light blue marl with Fish Palates	2. 0	284.11
	'Red Ground' (Mercia Mudstone) (30 ft)		
38	Red Ground	30. 0	314.11
	Dolomitic Conglomerate (about 114 ft)		
39	'Millstone' (proved in 1807)	(24. 0)	
1*	'Millstone' (proved after 1811)	(90. 0)	428.11
	Lower Carboniferous		
2	Red Sandstone	7. 5	435. 4
3	Light Grey ditto	2. 5	437. 9
4	Black bass (= Carbonaceous shale)	0. 2	437.11
5	Stone and Earth (= Fireclay and sandstone)	3. 1	441. 0
6	Light grey stone	6. 8	447. 8

Appendix 1.2 (*continued*)

Item No.	Ground Level 192.68 ft. A.O.D.	Thickness Ft Ins	Depth Ft Ins
7	Marl Stone	14. 6	462. 2
8	Stone	18. 3	480. 5
9	Red Earth	0. 5	480.10
10	Grey Stone	5. 1	485.11
11	Red Marl	1. 6	487. 5
12	Red Stone	2. 9	490. 2
13	Marl	1. 3	491. 5
14	Red Stone	4. 5	495.10
15	Marl	0. 6	496. 2
16	Red Stone	6.10	503. 0
17	Marl	1. 4	504. 4
18	Red Stone	30. 7	534.11
19	Grey Stone	53. 0	587.11
20	Do. . darker ('calcareous rocks' at 592 ft.)	14. 5	602. 4
21	Do. Rock (probably limestone)	80.11	683. 3

* Second series of numbers (Dec. 1811 to 1813).

There has always been some doubt as to the correct thickness of the Dolomitic Conglomerate at Batheaston. This is due to the difficulty experienced in 1807 owing to the second inburst of water. The works were virtually abandoned until May 1811 when boring was resumed.

In 1807 the boring had reached a depth of 421 ft (128.3 m ex-surface) when it was abandoned. the hole was then plugged in order to reduce the total flow from the springs while the water from the Lias was controlled. Walters' account is quite clear in this respect. He says that "it was proposed to carry down a Wooden Pipe through the water to meet the top of the Boring hole withdrawing the plug from thence and recommencing the Boreing . . . and the Boreing has in this manner been proceeded in to the present day (March 3rd, 1813) to the depth of Six Hundred and Seventy Feet from the surface". He also states that the boring had proved limestone "similar to the Mountain Limestone . . . at the great depth of Five Hundred and Ninety Two Feet". This is probably Bed No 20 of the second series.

Appendix 1.3. The state of the Cross Bath in 1811

Extract from Report addressed to the Mayor, Aldermen and Common Council of the City of Bath, preserved in Bath City Archives, dated 28th Nov. 1811, signed by J. F. Davis, William Bowen and Chas Crook.

. . .The attention of the Committee appointed to manage the Baths has lately been called in a particular manner to the state of the Cross Bath . . . So great a length of time is now required to fill the Cross Bath that were it completely emptied after the Patients have bathed it could not be filled again by the following morning and therefore a small part only of the water is let off except on Saturday when the intervening Sabbath affords sufficient time for the Bath to be completely filled before it is again wanted. The want of a fresh Bath on the other five days is not the only inconvenience resulting from this alteration. The slow entrance of the Water and retention of so much of that used on the preceding day occasion a lower temperature of the Bath it being now at 91 'deg.'—of Fahrenheit instead of 97 . . . the Water in the Pump Room [i.e. the Pump Room of the Cross Bath] is also colder in a similar proportion.

Wood reports that the Cross Bath in 1693 was, according to Gilmore, filled in eleven hours and a half. In 1748 sixteen hours and a half were required. Previous to Mr Fisher's late attempt to secure the spring the time had increased to 26 hours. In the summer of 1810 after his undertaking it sank to 19 hours but at present thirty hours and upwards are necessary.

A comparative view of the Hot Bath [this bath was fed by the Hetling Spring] will it is hoped remove all apprehension as to the entire loss of the [Cross Bath] spring at a future period and direct us where to look for a remedy of the Evil as it now exists, if it should not afford an explanation of the increasing loss of water since the late measures were undertaken for its recovery. From Wood we learn also that in 1693 the Hot Bath was filled in the same time as the Cross Bath viz. eleven hours and a half. Previous to Mr Smith's attempt to secure the [Hetling] spring the time had increased to sixteen hours, whereas since that period no more than eight hours have been required and the time is at present by no means increasing.

It appears then to your Committee that from the success which attended Mr Smith's measures for the preservation of the Spring belonging to the Hot Bath [i.e. the Hetling Spring] his conjectures with regard to the escape of the Water were just and that his superior success is to be ascribed to a thorough knowledge of the Strata through which the water passes on its approach to the surface of the Earth. Mr Fisher adopted his mode of procedure [at the Cross Bath] as far as he was able but being unaccustomed to this kind of work it is probable that he was deceived by fallacious appearances in the Strata which one better versed in Geological pursuits would have guarded against. For instance he might have supposed he had reached the blue clay Stratum [Lower Lias Clay] when he had merely found a thin layer of this substance accidentally interposed between Beds of Gravel. His Puddle therefore being attached to this could not prevent the escape of water through the Gravel beneath, besides which there may have been some defect in the composition of his Puddle or the application of it arising entirely from his being unaccustomed to work with this material.[1] Mr Smith thinks it cannot be done effectively without a temporary removal of the Building.[2]

Your Committee have endeavoured to lessen the inconvenience with respect to Bathing at the Cross Bath as directing that a Temporary supply of Water be procured by means of a Pump from the Hot Bath but they beg leave to advise that at a convenient season Mr Smith be employed to increase the spring in the same effectual manner as he has done of the Hot Bath. It would not be proper to suffer the state of the Kings and Queens Bath to pass unnoticed although there is at present a supply of water more than sufficient for every purpose. From the authority already quoted it appears that in 1693 these Baths were filled in nine hours and forty minutes. In the Summer of 1810 thirteen hours being required and it was for this and other reasons I judged it proper to cleanse the Spring and endeavoured at the same time to ascertain and remove the cause of the leakage. It is not necessary to note the steps which were taken for this purpose as they are fresh in the recollection of every member of the Body. We shall merely remark that from the manner in which Mr Fisher commenced his operation nothing effectual could be done for it was found necessary to relinquish the undertaking through fear of injury to the foundations to the Pump Room and adjoining Buildings [i.e. the Kings and Queens baths]. In consequence of this failure Mr Smith was consulted and a different mode of procedure was suggested by him when it should be necessary to have recourse to it for the preservation of the spring. At present the same time (13 hours) is required to fill these Baths as before this attempt was made. The Evil is by no means as alarming as to demand any immediate steps for its removal for should the time required to fill these Baths amount to 16 hours instead of 13 there will still be sufficient water for every purpose.[3]

It might be asked whether any alteration in the areas of the Baths in consequence of architectural improvements can have occasioned their slower filling. The areas have however been diminished rather than increased by the surrounding accommodations, which circumstance ought to hasten the process of filling instead of retarding it.

In a treatise upon the Bath Waters lately published replete with misrepresentations[4] it is asserted that the comparative coldness of the Cross Bath Water is owing to cold springs mixed with the Warm and more effectively confined to the same Channel by Mr Fisher's operation than before. But no cold spring was observed during these operations and if what the author asserts were true the alteration of temperature which we have noticed should have taken place immediately after, whereas at that time the Bath was of a higher temperature than before the confinement and admixture of the supposed cold springs and it has gradually become colder as the supply of the water has diminished . . ..

Notes to Appendix 1.3

[1] This refers to the collapse of the structures built by Abel Fisher and founded on the deposits in the spring pipe; the events described by Geo. Jervis.

[2] William Smith discovered the spring pipe of the Hetling Spring during restoration of the water to the Hot Bath. He obviously suspected the presence of a similar spring pipe beneath the Cross Bath. In the light of contemporary engineering resources demolition of the Cross Bath building would probably have been necessary in order to anchor its foundation in the Lias clay surrounding the spring basin.

[3] In their opening remarks (not quoted above) the Councillors referred to the deep source of the hot springs and to the losses caused by the character of the alluvial matter resting on the Lower Lias Clay. It is clear from their comments throughout the Report that they had assimilated and understood the nature of William Smith's findings. By the end of the 19th Century all knowledge of this seems to have been lost and it was not until after work had started in 1978 that the various records were gradually recovered from the archives at the Guildhall and in the Public Reference Library at Bath.

[4] It may have been Dr Wilkinson who originated the idea about hot and cold springs mixing near the surface at the Cross Bath. This explanation of the temperature of the Cross Bath was supported by Townshend (1813) but recent investigations do not confirm it.

Appendix 1.4. The great flood of 25th January 1809

In his note on the flood Warner (1811, pp 26–27) states: 'The blue marl [Lower Lias Clay] is not visible in any of the reservoirs, except that of the Kingston Bath.[1] The beds of the other reservoirs have been raised by large quantities of alluvial matter brought from neighbouring districts . . . The relative elevation of the different beds of the Hot Bath (Hetling Spring), Cross Bath, Kings Bath and Kingston Baths were well ascertained at the great flood of January 25, 1809: the flood line 7 $\frac{3}{10}$ inches above the bottom of the Kings and Queens Baths; 7 inches above the Cross Bath and 8 feet 2 $\frac{3}{10}$ inches above the bed of the Kingston Baths.[2]

Notes to Appendix 1.4

[1] The absence of any alluvium or gravel in the floor of the Kingston Bath is a clear indication that there is no spring pipe of any appreciable magnitude at this place.

[2] The much greater depth to the floor of the Kingston Bath and the absence of alluvial deposits suggests that the bath has at some time been dug through the Alluvium into the underlying Lower Lias Clay.

From Kellaway, G. A. (ed.) *Hot Springs of Bath*, pp 57–63

2. Quest for the quintessence

R. Rolls

"For what can be more necessary and therefore useful than to know the principles of any waters we recommend to our patients." Thomas Guidott 1676.

At the end of the last century, the author of a spa guide book (Kerr 1891) wrote:

"The Bath Waters are so powerful in their action that a course of treatment by them must be entered upon with caution and not without a certain amount of responsibility."

Such was the lure of these waters that, over the centuries, they attracted hundreds of ailing people to Bath to immerse themselves in the mineral water bubbling incessantly from the famous geothermal springs. They had come to be douched, pumped and sprayed with it, to inhale its vapour or simply savour what Charles Dickens once called "the taste of warm flat irons". They expected to be relieved of their illnesses by absorbing an etherial quintessence which was released as the waters emerged from the mysterious subterranean world of brimstone fires.

They were certainly relieved of their money by the bevy of medical gentlemen who followed in their wake, swooping like eagles on gullible and desperate invalids. A visitor in 1709 had remarked that the physicians were very numerous but good natured and that he was cured of more diseases in one week than he had suffered in the rest of his life (*Guardian*, 30th September, 1713).

But a lot of visitors arrived penniless and were so destitute that they could afford neither bed nor board in the city. Not all had genuine illnesses: faked convulsions and exaggerated limps were common tricks used by beggars to elicit sympathy and prey on the increasing numbers each year. There were probably far more who were genuine, but for whom the expense of accommodation and medical fees was just too much for them to bear: charity was their only hope.

Relief came in 1738 with the foundation of Britain's first national hospital (Plate 4), instituted for the reception of impoverished invalids with infirmities reputedly cured by a regular dip in the salutary springs. The waters were so much the *raison d'être* of the institution that it became known as the Royal Mineral Water Hospital. Nowadays, the thermal waters no longer flow into the hospital's baths and the institution makes no mention of them in its title. The power of the Bath waters seems to have run out.

Some people lament the demise of the mineral water treatment and firmly believe it offers something special, almost magical, and thereby more efficacious than the stuff from taps. The waters emerge from the ground at an elevated temperature of between 45° and 49°C. Geologically they are even more wonderous, having originated in part from rain which fell on the land around Bath many thousands of years ago. But does this extraordinary time lapse, which allows them to pick up a varied assortment of minerals, gases and radioactivity as they percolate through the rock, give them medicinal qualities? Do they really have curative properties?

Both the heat and the mineral content play an important role in the dogma underlying the therapeutic use of the water. Since the 16th Century, there have been numerous treatises written by various medical men extolling the virtues of "the Bathe". Most of these treatises embrace theories of how the waters work and their indications. Physicians tried to prove to lay observers that they had a superior understanding of the water's healing potential in order to convince people of the need to consult them; these patients could have otherwise drunk or bathed in the waters on their own initiative. In this respect it is interesting to note that it was not until 1976 with only a few months to go before the final closure of the spa facilities that treatment became available to the public without the need for a doctor's prescription.

Humours

The first definitive treatise on English spas was printed in 1562. Dr William Turner's publication, "*The Book of the Natures and Properties of the Baths of England*", did much to bring about the revival of interest in Bath as a spa. His travels in Europe gave him the opportunity of visiting Continental spas which were flourishing on a far grander scale than those in England. Indeed, Turner was so disgusted by the sad state of affairs in his own country that he wrote that he "was ashamed for any stranger who has seen the foreign baths to look upon our baths" (Turner 1562).

In his book, Turner lists over sixty disorders which could benefit from bathing including such diverse conditions as piles, migraine, sciatica, wormes in the belly, forgetfulness, dullness of smelling, palsy, cramp in the neck and failure of menstruation, to name but a few. In modern terms, there seems very little to link these conditions together. But physiology was very different in Turner's day.

Disease was thought to be caused by the imbalance of four essential body fluids known as humours. These were yellow bile, black bile, blood and phlegm which ebbed and flowed like the tides. Man was a microcosm of the universe, in tune with the physical, cosmic and spiritual world. This paradigm of Greek philosophy was still the cornerstone of medical teaching in the 17th Century. Each humour had physical properties corresponding to their degree of heat and moisture: for example, yellow bile was hot and dry like fire whereas phlegm was cold and moist. An excess of any of these humours precipitated corresponding illness. Fortunately, nature had an innate propensity to cure disease by redressing the humoural balance, which explained why people often recovered from their illnesses spontaneously. The job of physicians was merely to set up favourable conditions to aid recovery. For example, nature eliminated excess phlegm by driving it through the nostrils, voiding it in the faeces, urine and vaginal secretions or causing it to erupt through the skin as scabs and spots. For the physician, the practical consequences meant directing treatment in such a way as to assist the elimination of phlegm so that the proper balance of humours could be restored. Phlegm, being cold and wet, was best treated by hot, drying remedies. Hot mineral water was the perfect choice. It might seem contradictory to describe water as 'dry', but the term is used in the sense that wine is called dry and refers to its mineral qualities.

The holistic approach of this time meant that consideration had to be given to the way in which the waters were recommended. A full moon, because of its effect on the tides, was thought to increase the flow of humours to the head, making it dangerous to administer heating remedies. Likewise, too much sex caused the brain to overheat and was best avoided during bathing. Any disorder caused by an excess of hot humours, for example fever, was an absolute contraindication to warm bathing.

The majority of diseases in Turner's list would have been classified as cold-moist disorders. For example, because a paralysed limb was cold and motionless, the palsy was thought to be caused by an accumulation of phlegm in the affected parts. Phlegm in the ears caused tinnitus and deafness (as indeed it can), while in the legs and abdomen, it caused dropsy. An excess in the brain, where the humour was thought to originate, explained forgetfulness and too much in the womb led to sterility. The reputation of the Bath waters for curing the latter proved attractive to certain English monarchs and Royal visits helped to popularise the town as a resort. Queen Catherine, wife of Charles II visited in 1663 but bathing was of no avail. Twenty-four years later, King James II dispatched his wife to the city to try the effect of the waters. Happily for the royal couple, Queen Mary gave birth to a son though whether this was due to the brimstone or the beaux in the bath, we cannot be sure.

Baths of brimstone

Brimstone had been suggested as the water's main mineral constituent from an early age. Some years before Turner's work, an absurd theory was propounded that Bladud had buried two tons of burning brass and two glass vessels containing sulphur, salts and wildfire, beneath the springs to provide the heat and appropriate mineral content (Levis 1973). Certainly, Turner believed the waters contained sulphur. "Bathes of brimstone soften the sinews, suage the pain that a man hath in defying the going often unto the stool and when he cometh thither can do little or nothing. They scour and cleanse the skin wherefore they are good for the white and black morphews (i.e. pimples), for leprosy and scabs, scurf, for old sores and blotches and for the falling of humours into joints" (Turner 1562).

By the 17th Century, chemical analysis of the waters was beginning to be performed in a more scientific way. Dr Edward Jorden published his "*Discourse on Mineral Waters*" in 1631. It is a general treatise on mineral water therapy but deals specifically with Bath in the final four chapters. It was ironic that one of his sons was drowned in the very water about which Jorden expressed so much enthusiasm. Most of Jorden's book is devoted to the specific properties of various minerals. He classifies them into their degree of heat and dryness, and whether they are penetrating, astringent, opening, cleansing or softening. The Bath waters he recommends for heating, drying, softening and cleansing by virtue of their sulphur, nitre and bitumen content, but he was reluctant to recommend their inward use in view of the difficulty of drawing the water in a pure state from the spring (Jorden 1631).

The belief that sulphur, nitrite and bitumen were constituents of the waters was a misconception which was perpetuated by the Bath medical establishment well into the 18th Century. Dr Thomas Guidott (1638–1705) who removed from his London practice to Bath in 1666 to escape the ravages of the Great Plague championed this theory by backing it up with experimental evidence. Some of his analytical experiments may have been made on the thermophyllic algae which form a profuse scum floating on the surface of the baths in summer for Guidott mentions products from the evaporated water burning with a "blew flame" and having a sulphurous smell. The terms sulphur, nitre and bitumen were applied rather loosely in the 17th and 18th Centuries and should not be taken too literally. But rather like radioactivity in later years, the presence of these chemicals had powerful connotations. The relationship with the constituents of gunpowder gave mineral waters the ability to produce powerful therapeutic effects.

Guidott's principal challenger at this time was the brilliant Oxford physician, John Mayow (1643–1679) whose own analytical investigations proved (correctly) that the waters did not contain any of these substance (Mayow 1907). Mayow's work was summarily dismissed by virtually all physicians in Bath for the next hundred years with the exception of a visiting physician, Charles Lucas (1713–1771), whose criticism of the established dogma was one of the reasons why he fell out with the local practitioners whose livelihood was so dependent on the reputation of the waters. Both Dr William Oliver (1695–1764) and Dr Rice Charleton (1710–1789) refused to have anything to do with Lucas and Oliver became embroiled in a slanging match which was made public in a series of pamphlets ('Letters between Charles Lucas and William Oliver', published at Bath, 1757–1758). Lucas' publications not only cast doubts on the analytical abilities of the Bath physicians but called into question the very medicinal virtues of the waters.

Charleton, who was a keen chemical investigator of the waters, was so anxious to re-establish sulphur as a constituent that he redefined the word in a later publication to embrace oily substances in general: "If any such oily substance can be found in these springs, it has all the just right to be called the sulphurous principle howeversomuch it may differ from common brimstone" (Charleton 1774).

Inward controversies

Controversial theories involving the Bath waters seem to have polarized primarily between the established medical fraternity residing in the city and those physicians like Mayow and Lucas who were visitors to Bath only during the "season". Cynics will say that the local practitioners who were more dependant on the waters for their livelihood were anxious to preserve the myths which provided an explanation of why they worked. The mode of action of the waters is by no means the only controversy which appears in the literature.

In the 17th Century there was much disagreement about their internal use. Again this seems to polarize between non-resident and resident practitioners. John Jones, who had no particular axe to grind and wrote about several spas, recommended patients to drink as much as their stomachs could bear (Jones 1572). Edward Jorden, Samuel Bave and Dr Guidott, all Bath residents, discouraged unrestrained drinking, giving various reasons (letter dated 12 July 1677 from Sir Robert Browne to Dr Robert Peirce quoted in Peirce (1697)). In Guidott's opinion, the body's fibres and nutrative juices were "so diluted and softened with a deluge of waters taken inwardly that ..... the good effects which were wont to be produced by external application of the water to the part affected" were inhibited (Guidott (1705)). In practice, unrestrained drinking absolved the patient's need to seek medical advice, an undesirable state of affairs. At a more practical level, the waters were subject to pollution. Even in recent times, mineral water pumped to the Rheumatic Diseases Hospital roof tank has unwittingly received enhancements of its flavour by the addition of avian excrement and drowned rodents. Notwithstanding the pollution, many people found the taste of the waters unpalatable and the prospect of drinking between three and six pints a day (the amounts usually prescribed in the eighteenth Century) was daunting. One solution was to mix it with tinctures of aromatic herbs like cinnamon and cardomoms, or boil it with asses milk and administer the whey after separating the curds, thus "reconciling the stomachs and palates of many who would not else be able to bear it" (Falconer 1772). Dr William Falconer disapproved of this practice as he considered it dissipated the "aerial and sulphurous impregnations" which were of prime benefit (Falconer 1772; see also Oliver 1764).

More than tap water

By the mid-18th Century, excess of phlegm was thought to be far too simple an explanation of disease. Humours there might be, but they were thought of as being acrid and corrosive rather than hot and moist. Logic could easily be twisted, and the waters were now considered relevant for diluting and washing out malicious humours and noxious effluvia which built up in various parts of a person's body causing disease in them. This theory introduced the possibility that ordinary water might work as well as the mineralized variety, an assertion which has continued to provoke controversy into the present day.

> "Without the interposition of any unintelligable influence, I can easily conceive how extraordinary cures may be performed by the mechanical effects of simple water upon the human body; and I fully believe that in the use of bathing and pumping, efficacy is

often ascribed to the mineral particles which belongs to the (water) itself, exclusive of any foreign substance." (Smollett 1752).

Smollett, like Lucas, was also a visiting medical practitioner and his opinions were not popular with physicians at Bath. But even Lucas was not prepared to propound such heresy, preferring to accept that the Bath waters, though they might not contain sulphur, nitre and bitumen, did at least contain something special. But as the 18th Century wore on, more doubts were voiced. William Saunders (1743–1817), a respected London physician, wrote in 1800 that the Bath waters differed in no respect from common water heated to the same temperature since any active ingredients were too minute to exercise a stimulant effect on the skin (Saunders 1805). Once again, established Bath physicians like William Falconer maintained there was a difference. If the solid constituents were losing credibility as the important principle in the water's therapeutic potential, the search for the quintessence was still a major preoccupation with physicians who were now turning their attention to the dissolved gases in the water. The final publication in this particular direction, by Dr Charles Hunnings Wilkinson (1763–1850), appeared in 1811 after which the medical profession, even in Bath, appears to have become disenchanted by their inability to demonstrate the presence of any component having therapeutic effect. The waters did not, after all, contain free sulphur, and the amount of iron was far smaller than had first been suggested. Secondly, until 1770, the established physicians at Bath were almost all graduates of Oxford or Cambridge. After this date, the proportion of Oxbridge graduates fell dramatically and the Scottish Universities, particularly Edinburgh, were chiefly represented. It was William Cullen, the great medical teacher of the Edinburgh school, who wrote in 1769 that those who cannot afford to go to Bath need not despair as he had known cases where the assiduous use of the common hot bath for a similar space of time had been equally serviceable in chronic rheumatism (Copeman 1964). Thirdly, the physicians who were protagonists of Bath water treatment in the latter part of the 18th and early 19th Centuries, Falconer, Gibbes and Wilkinson, were chemists.

The main antagonist, Caleb Parry, was an experimental biologist. Parry was the most successful physician of his time in Bath and his opinions were no doubt held in high esteem by his professional colleagues. Certainly, an anonymous letter in the *Journal of Science and Arts* (No. 25, p 30) attributes the neglect of the waters to the influence of Parry who never recommended their use. It is significant that the decline in publications on the waters coincides with a definitive upsurge in medical publications dealing with pathology and experimental physiology, and with the establishment of a scientific community in Bath in which medical practitioners played a significant part.

As a result, the spa sank into a decline such that a medical writer commented in 1841 that his medical brethren in Bath had nearly forgotten the waters were in existence (Granville 1841). There were also, of course, economic and social reasons for this decline. The Napoleonic wars caused a recession and general loss of confidence in schemes requiring the sort of capital expenditure to bring the standard of Bath's bathing facilities in line with those of continental spas, and Bath's heyday as the resort of fashionable society was long past. When the economy improved and the spa facilities were modernized in the 1880s, it was the physiological effects of bathing which interested medical men the most. They busied themselves measuring heart and pulse rates and changes in the chemistry of people's bodies rather than worrying too much about the chemistry of the waters in which they were immersed (Freeman 1897).

Since that time, the consensus opinion within Bath medical circles has tended to favour Smollett's hydromechanical theory. As methods of administering the waters became more complex with a mind-boggling selection of hydrotherapeutic gadgets from which to choose, the methods themselves began to adopt more importance than the media. In 1921, the medical board of the Mineral Water Hospital wrote: "Modern investigation tends to show that the beneficial result of the external application of natural waters depends as much on the means their application affords for the conveyance of heat, pressure, percussion, and other purely physical effects, as on any specific healing properties of the waters themselves" (Turner 1921).

Two Oxford scientists, Drs Peskett and Raiment, were invited to the hospital in 1925 to compare the effects of drinking tap and mineral water on healthy volunteers. They failed to show any essential difference in their action (Peskett & Raiment 1925), though their experiments were criticized at the time because the mineral water was given with meals and not, as had always been the custom, on an empty stomach. More recently, as we have heard, Drs Paul O'Hare and Audrey Heywood have come to similar conclusions on comparing the effects of immersion in the two media. It is ironic that these observations were conducted twelve miles away in Bristol as no suitable baths could be made available for use in Bath.

The counter-argument in Smollett's lifetime asserted that the Bath waters only worked if taken fresh from the spring. Removing them to Bristol, or anywhere else for that matter would cause the "aerial impregnations" to be dissipated as the water cooled down. According to Dr Oliver, even corks could not confine this volatile part (Nuttall 1979). What hope have browsers?

The idea that the waters lost a volatile curative principle on cooling was a fundamental premise put forward by physicians from the time of the earliest publications. This runs parallel with the notion that there was some vital difference between the geothermal heat of the Bath waters and the heat of ordinary tap water raised to the same temperature by artificial means. Dr Robert Peirce (1622–1710), an author often quoted by later writers, described his observations on their slow rate of cooling. He suggested that this was due to fermentation of particles which "until they are wholly evaporated continue the water more or less warm" (Peirce 1697). Cynics might say that a belief in a labile and elusive component which readily disappears on cooling was propounded of necessity by the local medical establishment as a means of getting their fee-paying patients to Bath for their treatment rather than receiving it in bottles at home.

In fact the water does lose some of its components as it surfaces. The first of these to be recognized was the dissolved gases, identified in the late eighteenth Century by Dr William Falconer, honorary physician at the Mineral Water Hospital 1784–1819. Falconer believed that the "fixed air" (carbon dioxide), which he discovered in samples he tested, gave them an antiseptic quality which might explain their efficacy (Falconer 1772). Other gases discovered later are discussed by Dr McNulty in Chapter 3. Here, for a short time, was the evidence that everyone had suspected for so long: the etherial quintessence radiating healing properties which rapidly escaped into the air. Radiation certainly had dramatic biological effects which is more than could be said for the chemical components of the water.

Other constituents of the waters have received attention from time to time but have later lost favour. In the 1920s, mineral waters were given a rating known as their "zymosthenic index", the value of which depended on their ability to enhance fermentation by the body's digestive juices (Waterhouse 1929). This effect is due to the presence of trace elements in the waters, and similar catalytic actions were thought to be facilitating cures within patients. Others thought that the mysteries of balneology could be explained by "the interrelationship between radioactive substance, ions and colloids" (Waterhouse 1929).

Ions have come back into vogue in recent years. An excess of negative ions in the atmosphere is associated with an increased sense of well-being. Ionizers in factories are supposed to increase the productivity of those who toil in them and mountain air owes its bracing qualities to the abundance of negatively charged particles at high altitudes. Conversely, stuffy rooms and thunder storms are producers of positive ionization, sapping vital energy and producing a generally unhealthy environment. Is the air immediately around the Bath mineral springs charged with negative ions and perhaps another reason for taking the cure at source?

It is tempting to dismiss spa water therapy as merely an effective placebo. There can be no doubt that mind-over-matter has played a large part in the success of the waters at Bath, but we must conclude that, for certain diseases at least, the water's efficacy was not just due to a placebo effect. We have evidence from Dr O'Hare's immersion studies and Dr Heywood's research of its physical benefits in plumbism. I might also quote the work of Schamberg, an American worker who, in 1927, demonstrated that syphilitic infection in rabbits could be prevented by 11 daily hot baths, each of 15 minutes duration (Schamberg & Rule 1927). In treating humans, he found improvements in the cutaneous manifestations of this disease. Evidence at this time suggested the hot bathing produced effects on the behaviour of white cells and enhanced immunity to infection (Colebrook *et al.* 1924). Such work was subsequently eclipsed by the arrival of antibiotics, but we should perhaps return to examine these more subtle ways of enhancing resistance to infection, particularly where we are dealing with agents unresponsive to antibiotics.

Both the local and generalized effects of immersion and skin stimulation, combined with the beneficial global effect of treatment in the environment of a spa with its emphasis on therapeutic handling and promotion of a positive attitude to health, are good reasons for reviving a medical use for the Bath waters even if their quintessence has lost much of its former credibility.

References

Charleton, R. 1774. *Three Tracts on the Bath Waters.* Bath, p 20.
Colebrook, Eidinow & Hill. 1924. *Brit. J. Exp. Path.*
Copeman, W. S. 1964. *A Short History of Gout.* University of California Press, p 195.
Falconer, W. 1772. *An Essay on the Bath Waters.* Bath, p 431.
Freeman, H. 1897. *New Methods of Cure at the Hot Springs.* Bath
Granville, A. B. 1841. *Invalids and Visitors Handbook to the Hot Springs.* p 76.
Guidott, T. 1705. *An Apology for the Bath.* London, p 88.
Kerr, J. G. D. 1891. *Popular Guide to the use of the Bath Waters.* Bath, p 47.
Jones, J. 1572. *The Bathes of Bathes Ayde, etc.* London, p 28.
Jorden, E. 1631. *Discourse of Natural Bathes and Mineral Waters.* London.
Levis, H. C. 1973. *Bladud of Bath.* West Country Editions, Bath, pp 92–93.
Mayow, J. 1907. *Tractus Quinque*, Alembic Club Reprints 17, Edinburgh.
Nuttall 1979. Correspondence of Phillip Dodderidge, Northants County Library, p 1196.
Oliver, W. 1764. *A Practical Essay on the Use and Abuse of Warm Bathing in Gouty Cases.* London, p 75.
Peirce, R. 1697. *Bath Memoirs.* Bath.
Peskett & Raiment. 1925. *Arch. Med. Hydrol.*, Aug.
Saunders, W. 1805. *A Treatise on the Chemical History and Medical Powers of Mineral Waters.* London, 178–179.
Schamberg & Rule. 1927. *JAMA*, April.
Smollett, T. 1752. *An Essay on external use of Water.*
Turner, P. V. 1921. *A Brief History of the Mineral Water Hospital.* Bath, p 20.
Turner, W. 1562. *A Book on the Natures and Properties of the Baths of England.* Arnold Birkman, Cologne.
Waterhouse, R. 1929. The spa treatment of disease. *British Medical Journal.* 12 Jan.

From Kellaway, G. A. (ed.) *Hot Springs of Bath*, pp 65–70

3. The radium waters of Bath

M. McNulty

In the 18th Century Bath was of course the heyday of style and fashion but tastes changed and public bathing was no longer attractive. The growth of the Turnpike Trusts made for better roads and improved coaching services and most of all the Prince of Wales moved the centre for the fashionable elite to Brighton.

In Bath, public assemblies, concerts, balls and the ritual drinking of the waters continued but the Golden Age had gone long before with the death of Beau Nash. Increasingly Bath became a backwater where pensioned officers relived their campaigns fortified with memories and regular doses of mineral water.

Visitors continued but their primary purpose was medicinal rather than social. Many earnest little books on balneology were published by the local physicians, stressing the beneficial effect of drinking the waters and usually including a chapter on the climate and *always* a chemical analysis of the waters.

The Victorians were now into chemistry and were determined to prove that the number of grains per gallon of sulphur, iron, calcium, etc., in the water from the hot springs was greater than anything their British and Continental rivals could produce—or if not that the composition of the Bath waters was more *digestible*. The "in" phrase was "mineral" and the Bath Hospital, founded much earlier by Dr Oliver, suddenly became the Bath *Mineral* Water Hospital.

In 1878, Major Davies, the City Engineer, undertook extensive excavations around the spring and discovered the Roman Reservoir. Bath now became a "Roman" Bath—a much better image than the faded glory of Georgian times.

A good selling point for medicine and the rechristened mineral waters was still lacking. Fresh chemical analyses regularly appeared but unfortunately the fashionable patients were now travelling to the continental spas.

Meanwhile in the world outside, the last two decades of the 19th Century had brought major advances in fundamental scientific knowledge. In 1896 Roentgen published his discovery of X-rays and in 1898 the Curies produced radium. It was Bequerel in Paris who had encourged them and it was he also who demonstrated the physiological effect of the new discovery. The Curies gave him a sample of their refined product which he carried in his waistcoat pocket and developed a burn on the skin of his abdomen beneath the surface of his waistcoat.

Sir William Ramsey had discovered helium in 1895 and in 1898 Lord Rayleigh found significant quantities of the gas in the vapours of the Bath Waters. This was of great interest to the scientific world but it was quickly pointed out to the City Council that although this inert gas was expensive to produce, it was not in sufficient quantities in the spring to make collection financially worthwhile.

Following the work of Ramsey and Rayleigh, Professor Sir James Dewar who had been researching the effects of pressure and temperature on gases, conducted an analysis of the gas rising from the spring in the King's Bath.

There was a short report in the *Daily Telegraph* in 1901 that the helium would cost £200 per cubic foot to collect—more than a million times the cost of coal gas. This must have belatedly struck a cord in the Council Chambers for in 1903 the Hot Mineral Bath Committee received a letter from Dewar "confessing he could see no reason to anticipate any money in Bath Gas".

In the same year Sir William Ramsay reported that radium evolved helium by spontaneous change. Professor Dewar collected some of the salt and lime deposits from the spring and Strutt, who was a son of Lord Rayleigh, found that the deposits contained radium but not in commercial quantities. He wrote to the Council that "the presence of radium in the deposit was very much less than in the ores from which it is at present obtained. The reason why radium is so easily detected is that the tests are exceedingly sensitive." The *Bath Herald* reported that "Mr Strutt did not wish to raise hopes in the minds of anyone that an additional source of revenue had been discovered. Notwithstanding they might pride themselves that this rare element existed in the Hot Springs of Bath and in no other springs in the United Kingdom". One got the feeling that the Council had hoped for something more substantial than pride and it seems that they did not pursue the matter of radium any further.

Bath's discovery was marked by a cartoon in Punch Magazine in which Beau Nash regarded the parvenue upstart Monsieur Radium with distaste.

It would appear that the Bath radium was a first for Britain, it was later found in lower concentration in Buxton; several of the Continental Spas were *later* shown to contain radium, some with a significantly higher concentration—notably at Gastein, Baden-Baden and Carlsbad.

It seems to me that at some time in the first decade of this century, logical thought regarding Spa treatment was turned upside down. It was, so to speak, a known fact, that bathing in Spa centres, including drinking the waters, was beneficial; everyone knew that, especially those involved in Spas which at that time were an important and profitable industry. Radium was a newly discovered and very active element and it was perhaps natural to assume that it was the new and previously hidden factor "X" which was responsible for the beneficial effect of the waters.

It would also seem that after Rayleigh, Dewar and Strutt had found traces of radium in the Bath Waters, scientific and medical opinion both in England and on the continent of Europe accepted that the beneficial and healing of Spa waters was due to its radium content.

In 1911 Professor Sobermann from Berlin addressed the British Medical Association; the proceedings of this meeting were subsequently published in the Archives of the Roentgen Ray. The results of his research were as follows.

1. The radium emanation in not too large doses promotes growth of healthy cells while morbid cells decay.
2. That it enhances urinary excretion.
3. That it stimulates activity of the digestive tract.
4. That it enhances excretion of uric acid.
5. There is a nerve soothing effect which relieves insomnia.
6. That radium increases sexual activity, which of course explains the reputation of some continental spas as "rejuvenating" waters.
7. It was also claimed that radium diminished blood viscosity and lowered blood pressure.

The first of Professor Sobermann's points—that radium emanation in not too large doses promotes growth of healthy cells, while morbid cells decay—is indeed partly true and was a principle laid down by Marie Curie; it is the basis of successful cancer treatment by X-rays and radium. Cancer cells are more primitive than the normal healthy tissue cells and consequently these primitive cells are more susceptible to the physical injury produced by radium treatment. If the radiation dosage is correctly calculated many of the abnormal or "morbid" cells will be destroyed while most of the healthy normal cells will survive the radiation injury. However, the contrary is *not* true and there is *no* evidence that exposure to

Plate 3. The City Centre of Bath seen from Beechen Cliff. The Pump Room and Roman Bath lie to the left of Bath Abbey. The background is formed by the steep wooded slope of Beacon Hill and Hedgemead. Photo – Stephen Bird.

Plate 4. The facade of the Royal Mineral Water Hospital looking down Old Bond Street.

Plate 5. The fountain in the Pump Room, supplied by artesian thermal water rising up the Stall Street inclined borehole. Photo – Bath City Council.

Plate 6. The Great Roman Bath, filled by thermal water from the King's Spring which rises in the Roman Reservoir and still flows through Roman pipes and channels to the River Avon.

minute amounts of radium will *stimulate* growth or healing, or that the beneficial effects of immersion in Bath Spa water is due to minute amounts of radium or that it has any beneficial effect on the growth of normal cells. There *are* undoubted benefits resulting from immersion in hot water but they have nothing to do with radium.

In March 1912 a meeting was held in Bath, to which all the press were invited, to announce plans for the expansion and complete re-development of the Spa Treatment Centre. The gathering was addressed by Sir William Ramsey himself and stage managed—if that is not too strong a phrase—by a newly arrived Spa Director, John Hatton.

Sir William had spent the previous Christmas investigating the Bath Waters, confirmed that they contained radium and now for the first time gave accurate measurements. He noted that there were small amounts of radium in the water from the King's Bath but a very considerable radium emanation in the gases. He later went on to say that the water was "*saturated* with the radium emanation". It was unfortunately impossible to compare his measurements with statements of radioactivity of foreign waters for the latter were stated in uncertain units; this was presumably a backhanded reference to Napoleon's metric system.

John Hatton, the new director of the Baths told the assembled press that the waters cured in the most wonderful way. Invalids crippled with rheumatism had come to them helpless and in their immobile state had been lowered into the baths by slings or hydraulic chairs and yet three to six weeks after their arrival they had been able to go for real country walks.

Even at the Mineral Water Hospital which received the poorest and therefore the worst cases the number of patients cured and relieved was as high as 90%. Clearly Sir William had given him a good script and he made the most of it. "All this we already knew" said Hatton, "but what we did not quite understand was the reason why the waters of Bath had such a marvellous effect". Now the finding of radium saturating the water had explained their healing powers and Sir William Ramsey's discovery provided a basis for important investigation. Much of Professor Sobermann's British Medical Association lecture of 1911 on the subject of radium waters and their beneficial effects was repeated at this meeting.

Over the next two or three decades, John Hatton maintained a steady output of pamphlets for the medical profession and for Spa visitors, repeating Sir William's remarks on radium saturation and the original measurements made in 1912. Hatton also had the good sense to say that radium was only one part of the healing character of Bath water and he also stressed the advantages of the "telluric heat" of the water in the Bath springs. Other Spas had to heat up their water and "if you heat a water artificially you at once destroy its character", he said. This, of course, was a sly dig at the rival Spas of Droitwich and Harrogate which are cold water springs.

Many people of course joined the radium band wagon. The Sparklett's Soda Water Company offered home treatment with a Spa radium bulb and the Baker's of Cheap Street offered radium bread. We tend to forget that radium and radiation was then regarded as beneficial and linked in a vague way to the health-giving properties of the sun. There was also a new fashion for sunbathing, and ultra-violet light. This went on well into the 1930s and early 1940s. I can remember, as a child, being giving doses of "radio-malt" and there were several other applications of the word "radium"—we still use the term radiant health.

Such it seems was the continuing conviction among the medical profession that radium was the active principle in Spa treatment that we find a highly reputable medical textbook, "*Recent Advances in Rheumatic Diseases*", still recommended small intravenous doses of radium as recently as 1931, two decades after Sobermann's original claims.

If we pause to remember that radium when injected, behaves in a similar manner to calcium and is deposited in bone where it can continue to irradiate bone marrow, and in time, erode bone itself, we can appreciate how much the profession had been misled by the radium waters.

One of the unique features of the new Spa arrangements was the collection of the "radium vapours" as they rose from the surface of the hot spring and their application to individual limbs of the affected patient. Unique also was the radium inhalitorium. You will see from the list of treatments that either the Bath waters or the vapours from the spring were applied to the nasal passages and for some cases the water taken directly from the spring was sprayed in jets into the throat and nasal passages of the patient. Being now aware of the behaviour of *Naegleria fowleri*, it is difficult to avoid speculating whether any of these patients developed fatal meningitis![1]

The end of the chapter for radium in Bath came in 1959. As you may remember, in the late 1950s there were reports of the radiation leakage from Windscale, which incidentally is now leaking under the new name of Sellafield, and we were all warned that the uncontrolled X-ray units in shoe shops created a hazard to the population. Alerted by reports of this kind, the staff of the Spa Treatment Centre and the guides in Roman Baths, who in particular knew all about the radiation levels in which they worked, because it was part of their speil for the tourists, expressed concern at the health risks to which they were exposed in the course of their employment. The Medical Officer for Health at that time, Dr Astley Weston, immediately called in the Radiological Protection Service. They in turn made careful measurements, noted that the radiation was one tenth of the figure reached by Sir William in his original measurements in 1912 and submitted a lengthy report, the conclusion of which was that the measurements of the radium and radon content of the Bath Spa Waters and of the radon concentration in the air of one of the treatment rooms show there is little likelihoood of this radioactivity causing any harm by drinking or bathing in the waters.

[1] It is probable that this water was taken directly from the central "fountain" above the exit of the King's Spring. If so, it may have been free of *Naegleria*.

From Kellaway, G. A. (ed.) *Hot Springs of Bath*, pp 71–76

4. Physiology of immersion in thermal waters

J. P. O'Hare, A. Heywood, N. D. Millar, J. M. Evans, R. J. M. Corrall & P. Dieppe

Man has immersed himself in water as a therapy for thousands of years and every major civilization has used water immersion as a form of therapy. Medical writings from ancient Chinese and Hindu manuscrips attest to the powers of water immersion and reference can be found also in Persian, Hebrew and Greek literature. Indeed the Babylonian word for "physician" signified "one who knows water".

Our research group were interested in the physiology of immersion in water at thermoneutral temperatures and developed the hypothesis that these physiological changes could be the scientific explanation for the emergence of the spas and the use of immersion as a means of therapy.

Many of these observations are not new and as early as 1764 Sutherland in his thesis (Sutherland 1764) clearly observes the diuretic effect in a subject with liver cirrhosis.

> "A scotchman in an ascites, was cured. By his girdle which I saw he fell six inches in five days, pissing freely all the time".

We have investigated the changes in eight normal subjects immersed in Bath spa water at 35°C and compared these to immersion in tap water. In a further series of experiments in seven subjects we have explored the effect of changes in temperature over the range 33–39°C. The results of the experiments using Bath water have been previously reported (O'Hare *et al.* 1985).

Patients and methods

Eight normal subjects, four male and four female, age range 20–56, were studied. Subjects were maintained on their normal diet but the same diet was followed for two days before each of the immersion or control studies.

Arising at 7.30 am each subject ingested 400 ml of water. At 8.30 am they passed urine and a further 200 ml was taken. Subjects then sat outside the bath for a control pre-immersion hour. At 9.30 urine was passed, a further 200 ml of water taken and subjects entered the spa bath and remained seated immersed to the suprasternal notch for two hours. At the end of each hour 20 ml of blood was taken, the subjects briefly emerged from the tank and passed urine and a further 200 ml water was taken. A further control hour after immersion was spent sitting outside the spa bath (Fig. 4.1).

Urine and plasma sodium and potassium concentrations were measured using flame photometry with an internal lithium standard (Corning model 450, Corning Instruments Ltd). Creatinine was measured using a Technicon auto-analyser based on the picric acid method of M. Jaffe. Red cells were counted using a Coulter counter.

Immersion was carried out in a modified spa bath (Aquatech Ltd., Newbury, Fig. 4.1) in water constantly recirculated and maintained at the set temperature. Bath City Council provided 500 gallons of spa water derived from the new borehole and transport was arranged courtesy of the Blakeney Hotel group.

A pulsed wave doppler blood velocity meter (Pedof) was used to obtain signals to measure ascending aortic blood velocity. Cardiac output was calculated as the product of mean aortic velocity and aortic cross sectional area. As a constant aortic diameter was assumed the results are expressed as changes in cardiac index (Evans *et al.* 1986). Statistical analysis was performed using analysis of variance.

Fig. 4.1.

Results

Cardiovascular studies

Immersion in spa water at 35°C resulted in a highly significant increase in cardiac index from 4.6 l/min to 7.4 l/min ($p < 0.001$). This increased cardiac output was accompanied by an increase in stroke volume and a reduction in peripheral resistance (Fig. 4.2). When studied in tap water at 35°C there was no significant difference in the increase in cardiac index compared to the results in spa water (see Fig. 4.6).

When studied at 33, 35, 37, and 39°C a progressive increase in cardiac index was achieved which reached a maximum of 11.7 l/min at 39°C. The increases in cardiac output expressed as percentage increments of resting values were 30%, 34%, 80% and 121% respectively for the four temperatures. Heart rate fell at 33°C to 70 beats per minute but rose to 87 bpm at 37°C and 109 ± 4 bpm at 39°C ($p < 0.001$).

Haemodilution

A highly significant fall in haemoglobin and red cell count and a fall in viscosity occurred during the first hour of immersion ($p < 0.001$) (Fig. 4.3).

There was no difference in the haemodilution in experiments using tap water. When the temperature was varied from 33 to 37°C it can be seen that the haemodilution was greatest and most prolonged with the higher temperatures (Fig. 4.4).

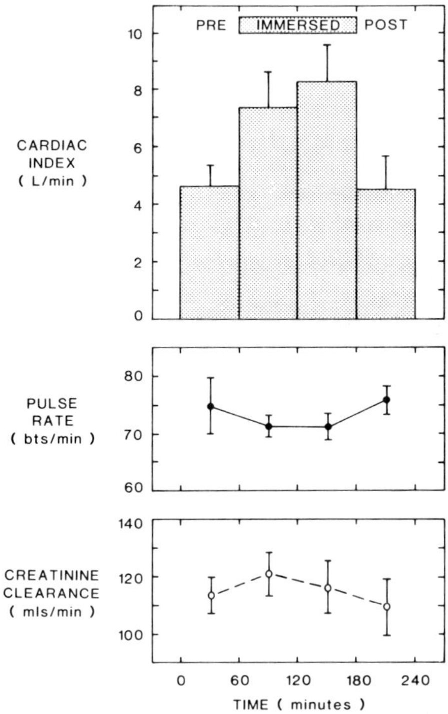

Fig. 4.2.

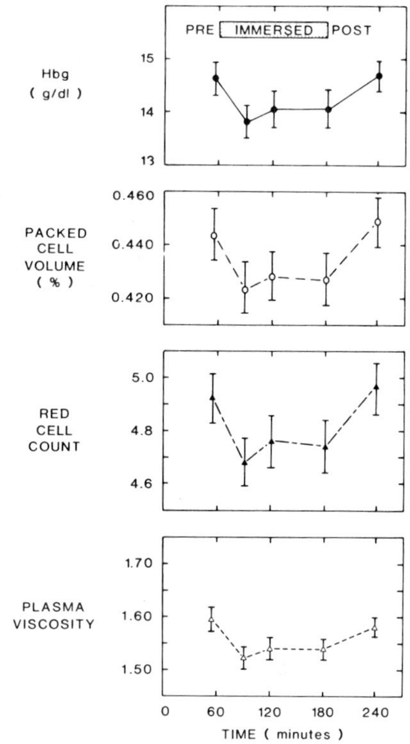

Fig. 4.3.

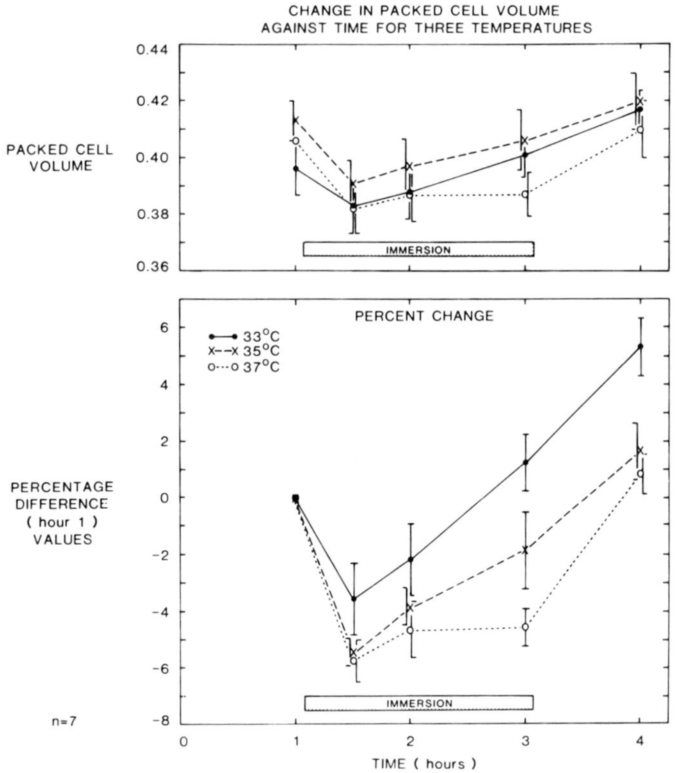

Fig. 4.4.

Renal changes

Immersion resulted in a highly significant increase in urine excretion, sodium excretion and potassium excretion (Fig. 4.5). A mean weight loss of 0.53 kg occurred. No significant difference was seen in these renal changes between spa water and tap water (Fig. 4.6). When immersion was carried out for two hours at 33, 35, and 37°C it can be seen that the diuresis was attenuated and natriuresis reduced at 37°C compared to the lower temperatures (Fig. 4.7).

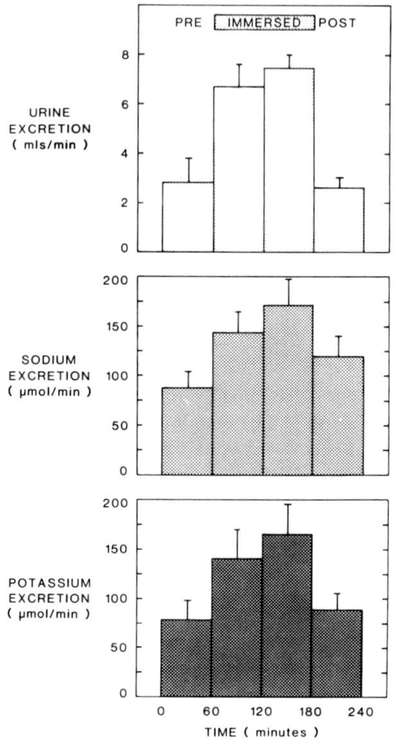

Fig. 4.5.

Discussion

It is perhaps ironic that it has only been as a result of mankind's need to conquer space that serious scientific investigation into the physiology of man during water immersion began. Immersion produces a hydrostatic pressure gradient which acts on the immersed limbs to squeeze blood back to the heart. This augmented venous return is then sensed and in turn leads to important renal changes and cardiovascular effects (Epstein 1978; Arborelius *et al.* 1972).

Immersion in spa water at 35°C produced a significant diuresis, natriuresis and kaliuresis. These renal changes are accompanied by many hormonal changes including a stimulation of atrial natriuretic peptide (Anderson *et al.* 1986), suppression of anti-diuretic hormone (O'Hare *et al.* 1985), plasma renin and aldosterone (Epstein 1978; O'Hare *et al.* 1986) and plasma catecholamines (O'Hare *et al.* 1986).

Significant and substantial increases in cardiac output and a haemodilution were also observed in the subjects immersed in spa water at 35°C. All these changes occurred also in

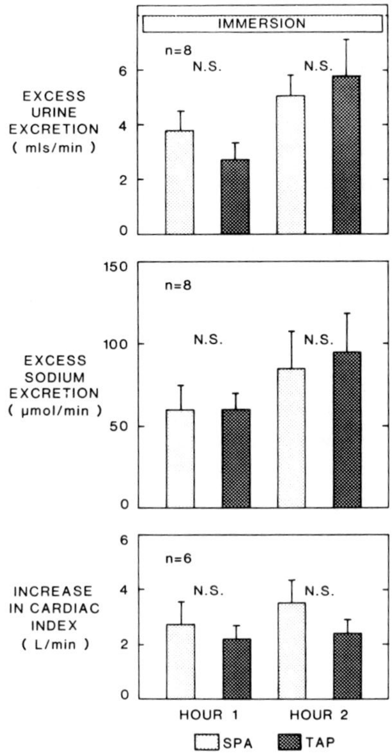
IMMERSION
n=8
N.S.
N.S.
EXCESS URINE EXCRETION (mls/min)
6
4
2
0
150
n=8
N.S.
N.S.
100
EXCESS SODIUM EXCRETION (µmol/min)
50
0
6
n=6
N.S.
N.S.
INCREASE IN CARDIAC INDEX (L/min)
4
2
0
HOUR 1
HOUR 2
SPA
TAP

Fig. 4.6.

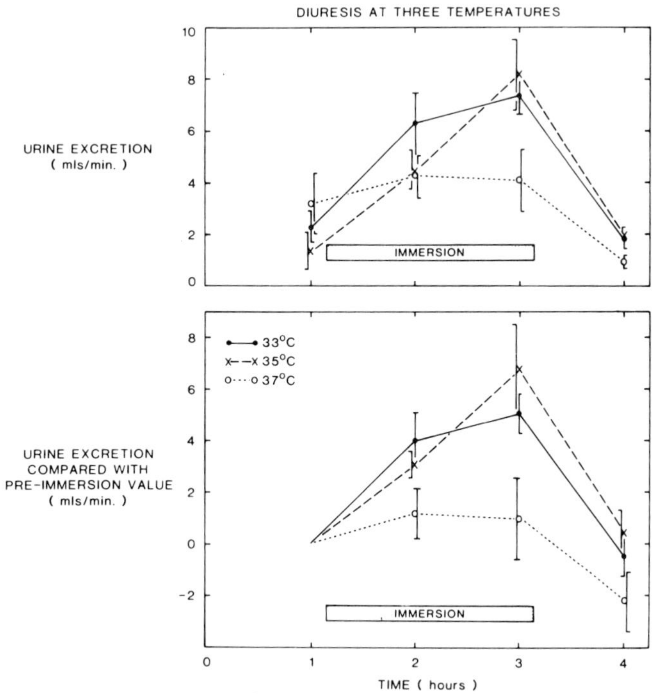
DIURESIS AT THREE TEMPERATURES
URINE EXCRETION (mls/min.)
10
8
6
4
2
0
IMMERSION
33°C
35°C
37°C
URINE EXCRETION COMPARED WITH PRE-IMMERSION VALUE (mls/min.)
8
6
4
2
0
-2
IMMERSION
0
1
2
3
4
TIME (hours)

Fig. 4.7.

subjects immersed in tap water and this suggests that the important factor of any effect of immersion in thermal spas is not due to the minerals *per se* but the shared physical aspects of the water, namely the temperature and depth of immersion. A depth of at least 3 ft 6 ins is required to produce these physiological changes and these effects would not be achieved by immersion in any ordinary domestic bath.

Indeed when a spectrum of temperatures was studied and depth left constant at 3 ft 6 ins it was shown that at higher temperatures much greater increases in cardiac output and haemodilution were apparent. Bath spa water emerges from its source at 40°C (Thomson 1978; O'Hare *et al.* 1985) but temperatures in the bathing pools and in many other spas across the world vary between 33 and 39°C (Thomson 1978). Clearly any difference in the physiological changes across this range may be important in the understanding of a proposed therapeutic effect of immersion in thermal waters of spas in general. It is of interest that immersion in the hotter pools was in fact prohibited by many doctors supervising spa therapy in those with acute heart failure and fever (Hoffman 1731) in whom the profound tachycardias and increases in cardiac output via vasodilation that would have occurred during immersion may well have been harmful and not therapeutic.

We speculate that the profound physiological changes in hormones, the changes in the composition of blood with a haemodilution, the diuresis and natriuresis and the cardiovascular response that occurs when man immerses himself in thermal waters may constitute part of the scientific rationale for the use of spa treatment as it was practised in many states of disease.

References

Anderson, J. V., Millar, N. D., O'Hare, J. P., Mackenzie, J. C., Corrall, R. J. M. & Bloom, S. R. 1986. Atrial natriuretic peptide: physiological release associated with natriuresis during water immersion in man. *Clinical Science,* **71**, 319–322.

Arborelius, M., Balldin, U., Lilja, B. & Lungren, C. 1972. Haemodynamic changes in man during immersion with head above water. *Aerospace Medicine,* **43**, 592–598.

Epstein, M 1978. Renal effects of head-out water immersion in man: implications for an understanding of volume homeostasis. *Physiological Reviews,* **58**, 529–581.

Evans, J., O'Hare, J. P., Weston, C., Davies, J., Lewis, G. & Corrall, R. J. M. 1986. Haemodynamic changes in man at different temperatures of water immersion. *Clinical Science,* **70**, 513.

Hoffman, F. 1731. New experiments and observations upon the mineral waters directing the further use for the preservation of health and diseases. J. Osborne and T. Longman, (transl. P. Shaw), London.

O'Hare, J. P., Dalton, N., Roland, R. J. M., Gooding, J., Payne, B., Walters, G. & Corrall, R. J. M. 1986. Plasma catecholamine levels during water immersion in man. *Hormone and Metabolic Research,* **18**, 713–716.

——, Heywood, A., Summerhayes, C., Lunn, G., Evans, J. M., Walters, G., Corrall, R. J. M. & Dieppe, P. A. 1985. Observations on the effects of immersion in Bath Spa water. *British Medical Journal,* **291**, 1747–1751.

——, Watson, M., Penney, M. D., Hampton, D., Roland, J. M. & Corrall, R. J. M. 1985. Urinary prostaglandin E and antidiuretic hormone during water immersion in man. *Clinical Sience,* **69**, 493–496.

Sutherland, A. 1764. *An Attempt to Ascertain and Extend the Virtues of Bath and Bristol Waters.* Frederick and Leake, London, pp 352–353.

Thomson, W. 1978. *Spas That Heal.* Adam and Charles Black, London, p 4.

From Kellaway, G. A. (ed.) *Hot Springs of Bath*, pp 77–88

5. Lead, gout and Bath Spa therapy

Audrey Heywood

In the middle of the first Century A.D. the Romans converted the old Celtic shrine of Sulis into the splendid sanctuary of Sulis Minerva, part of which remains today. That the shrine was dedicated to the goddess Minerva could be significant; not only was she the goddess of wisdom and war but curative powers were also attributed to her. However, no records of cures from this period exist. After the Romans left the baths fell into disrepair. The baths came under the control of the church until the dissolution of the monasteries by Henry VIII. Although not widely used there is an account of a Leper's bath being built in 1138 fed from the overflow of the Hot bath (Freeman 1888).

The reputation of the healing properties of the waters was enhanced when in 1562 Dr William Turner published his book comparing the properties and virtues of the bath at Bath with those in Germany and Italy. He complained about the appalling condition into which the baths had fallen and suggested practical ways in which they could be reinstated (Turner 1567). Palsy, colic and gout were listed amongst many other conditions that he believed should respond to treatment in the baths of Bath, the water of which he compared with those used in the continental baths. After this book was published others also appeared commending the virtues of the Bath waters. This resulted in an increasing number of people visiting Bath in search of a "cure". So many claims were made that cures had indeed resulted from using the Bath waters that a certain amount of scepticism arose especially amongst physicians who did not practise in Bath. The therapeutic properties of the waters were further extolled in many books and pamphlets published in the 17th Century but gradually it was appreciated that scientific methods of analysis rather than polemic offered the best way of arriving at a useful assessment of the healing properties of the waters. The view of the Royal Society was expressed by Joseph Glanvill, the Dean of Bath. In his book Ultra Plus (Glanvill 1668) he said that the business of the experimental philosopher was to seek, gather, observe and 'lay up in bank for the ages to come'.

Many years later these ideas were expressed when the Bath Hospital was set up in 1741 in an attempt to demonstrate that Bath Spa therapy, as practised in the 18th Century was a very effective treatment for many diseases. In fact the evidence collected from the records kept in the Bath hospital for this paper show that the treatment in Bath was successful for at least one disease: chronic lead poisoning.

The symptoms of lead poisoning are so characteristic that it is possible to use the evidence collected 200 years ago and then by using modern immersion studies suggest a mechanism by which the recorded 45% cure rate was achieved.

Lead poisoning was a widespread problem from Roman times until the end of the 19th Century. A large proportion of the population would have been affected to some degree as lead was used for pipes to carry the water supply; however this was not a problem in a hard water area like Bath where calcium salts soon form a protective layer. Lead glazes were used on most earthenware and large wooden food storage boxes were often lead lined. Copper pans were lined with a mixture of tin and lead and the fact that there were tradesmen whose sole occupation was relining these cooking utensils indicates that some lead must have found its way into the food cooked in them. Cosmetics and even some medicines in common use also contained lead salts. Even in this century children in Queensland became seriously lead poisoned after playing on their lead painted verandahs. For many people this level of poisoning would have caused tiredness, aching muscles, headaches and a vague feeling of malaise but others were more seriously affected. These were the people who added to their base load of lead repeated doses of comparatively small amounts of lead from occupational exposure or lead adulterated alcohol.

Lead is soluble in weak acids such as acetic acid which is formed when alcohol is exposed to the air. Thus alcohol may easily be contaminated with lead and was sometimes adulterated deliberately by vintners following the Roman tradition of using lead acetate as a sweetener, improver and fungicide.

The results of this more serious exposure were colics and peripheral neuropathies. Even higher dosage can result in epilepsy, coma and death. This retrospective study has been possible because this more serious severe lead poisoning presented in such a readily recognizable and characteristic way: a peripheral paralysis of the limbs following attacks of severe colicky abdominal pain associated with constipation.

A concise description was given by Dr Rice Charleton (1770) in his "*Tract on the Bath Waters*".

> "In consequence of a most obstinate costiveness attended with exquisite pain in the bowels, upon the constipation being removed and the pain diminished, the patient loses the use of his limbs. The arms and hands most commonly. Rheumatic pains sometimes attack the limbs before they become paralytic. The colic which gives rise to this species of palsy, was not unknown to the ancients. It was twice mentioned by Paulus Aegina. Lead we know is remarkably productive of this complaint."

Three years earlier, George Baker (1767) had published a paper which established that lead was the cause of this syndrome, then called the colica pictonum, and that the Devonshire colic was the same disease. In this paper he quoted results from the records in the Bath hospital.

The other serious disease which lead poisoning can cause is gout; which reached almost epidemic proportions amongst prosperous men during the 18th Century.

It has been shown that lead poisoning can be implicated as the cause of some cases of gout. Working in Queensland, Emmerson (1963) showed that of patients who developed lead nephropathy following childhood ingestion of lead-based paint, 50% subsequently developed gout in middle age. Lead may damage the renal tubular excretion mechanisms, and thus by impairing the urinary urate excretion can cause gout. But other factors are involved, some of which are still to be identified. Although lead poisoning, by impairing urate excretion, may produce a gouty arthritis, a patient may have gout without any exposure to lead at all or may have a markedly raised serum urate level without developing gout.

In the 16th and 17th Centuries gout was considered to be a disease of the idle well-to-do. Their typically high-protein diet would increase the amount of urate to be excreted by their kidneys. If their renal tubules were already damaged by lead from adulterated alcohol, e.g. port, the urate level in the blood would rise, leading eventually to gout in susceptible people. The poor had a low protein intake; very few cases of gout were seen at the Bath Hospital, or elsewhere amongst Charity patients until the middle of the 19th Century. For reasons that are not yet understood a high protein diet appears to protect the subject from paralyses but allows lead to damage the renal tubules. The greedy got gout and the poor paralysis. A visit to take the waters at Bath was a popular practice for those who had or fancied that they had the gout in the 18th Century. But almost all of these patients were treated privately and no accurate records were kept; nor was the treatment controlled except in the most general way.

Chronic lead poisoning may also result in subfertility or early abortions. By the 17th Century Bath had a well established reputation for the treatment of barrenness. The most famous, if controversial, case was that of Mary of Modena, the wife of James II, whose subsequent pregnancy was attributed to her having had a course of bathing in the Cross Bath. The birth of an heir was celebrated by the erection of an elaborate cross in the Cross Bath.

Fig. 5.1. The Cross Bath in 1739 by John Fayram. Bath Reference Library.
This shows the Cross presented to Bath by the Earl of Melford to commemorate the birth of a male heir to the Catholic Queen of James II, Mary of Modena, in June 1688 which was attributed to her having bathed in this bath. Note that the bathers are standing up to their necks in the (35°C) water.

Barren women were not considered to be suitable "objects for charity" so they were not admitted to the Hospital.

However, as (3377 cases) 8% of the patients admitted to the Bath Hospital from 1761–1879 are recorded as having had a paralysis due to lead it is possible to study the results of Spa therapy in this condition.

As the concept of medical proof shifted from the philosophical and dogmatic Galenic concepts to the new scientific ideas inspired by Francis Bacon, able physicians such as Citious (1639) decribed distinct disease patterns (syndromes). The colica pictonum was one of these.

In Bath Robert Peirce (1713) recorded the cases he had seen during 60 years of practice from the middle of the 17th Century. As a result he was able to differentiate "the palsy after the colic" from the paralyses that were the "consequence of apoplexes, epilepsies or convulsions," and identified it as the Colica Pictonum. He said "I find no one Distemper more frequent amongst my Adversaria, nor in none more eminent recoveries than in this, of persons thus disabled to have been restored by the assistance of the Bath, and the Bath Water."

He presents the most enlightening case history of the Rev. Mr Pilkington, aged 33, who before coming to take the Bath waters could not even dress or feed himself but after treatment was able to walk to Dr Pierce's house and "put off his hat to me; for tho he was a clergyman, his disease had made him so much a Quaker that he could not perform that accustomed civility until after a whiles bathing".

Fig. 5.2. Bathing in the King's Bath *c.* 1800. Detail from a painting by John Nixon. This shows one bather using the 'wet' pump (the number of strokes of the pump was prescribed) and another patient using a flesh brush.

Defoe (1983) also recorded that the waters in Bath had acquired a longstanding reputation as a cure of paralyses. The collection of unwanted crutches left by grateful patients was he said "legendary". John Wynter (1727) recommended the use of the Hot Baths in Bath for the treatment "of the great crop of paralytics that occur amongst tradesmen" (handling lead) and also to treat those returning from the West Indies with the "Dry Gripes", "their paralyses being the successor of the most painful colic imaginable".

Workers using lead, including painters, potters, plumbers and printers as well as smelters, found that they frequently suffered from colic and constipation which they treated themselves with traditional medicines such as the regular use of emetics and purges.

Opiates were also used if the pain became too severe. This regime was also used for the colic and constipation which also often affected the general population (Anon 1778).

Continuing exposure could lead to weakness and later paralysis of the most used muscles, typically a dropped wrist. If treated early enough this paralyis often responded to conventional treatments: a bland diet and removal from the source of lead. But this was difficult for a working man and if the paralysis persisted in spite of treatment it was a disaster as it was impossible for the man to earn his living. Half of these cases, considered to be incurable elsewhere, responded if they were sent for treatment in Bath.

It was in these apparently incurable cases that the cures obtained at Bath must have appeared to be so remarkable. They were acknowledged as such even by the established London physicians such as Willis in the 17th, Hebeden and Baker in the 18th Centuries.

We have an example from one of the letters in the referral Register 1751–1758 of Newton Highborn, a painter from St Georges, Bloomsbury, who was referred from Guy's Hospital on October 1751 and was discharged better after 186 days.

In the second half of the 17th Century, watering places throughout Europe were becoming widely used by the aristocracy and the growing wealthy middle classes for pleasure and therapy. A visit to Bath during the season was recognized to be helpful to those suffering from gout.

Peirce (1713) also recorded cases of childless women who became pregnant after using the Bath waters.

So by the beginning of the 18th Century Bath had acquired a well documented reputation for curing paralyses, especially those which followed severe colic, relieving gout and offering hope to infertile women.

Patients with skin diseases and rheumatic pains also got relief from bathing in Bath water. The citizens of Bath were responding to this situation by developing the city and increasing the bathing facilities. This was also happening in other parts of England and Europe. As the fashion for spas boomed it became necessary for the citizens of Bath to establish without doubt that the healing properties of the waters of their city were superior to those of their rivals. It was essential to be able to produce "irrefutable proof" of this in a modern and scientific way.

By 1730 the concept of the provincial charity hospital was being developed, in this the newly rich and the aristocracy with commercial interests were able to achieve virtue and social status as well as looking after the welfare of their labour force, by providing local hospitals for the treatment of acute illnesses or the injuries of their workers. This group of workers was not catered for within the parish poor law system (Porter 1988). A group of influential people in Bath, aware of the danger of competition to their prosperity by the development of other spas, realized that they could adapt the concept of the provincial charity hospital into one that would serve their commercial interests. If the charity hospital in Bath would treat only those "objects of charity" who were suffering from the chronic diseases that were known to respond favourably from treatment with the Bath waters those patients would benefit greatly. Then by allowing them to continue their treatment for as long as was necessary to obtain a cure or maximum improvement it should be possible to demonstrate, scientifically, the efficacy of the treatment available only in Bath. This combination of virtuous charity and self-interest enabled enough money to be collected to found the Bath General Hospital, which opened its doors in 1741 (Heywood 1990).

The founders of the Bath hospital knew that they would have no problems persuading the tradesmen with peripheral neuropathies to stay in the hospital for the 5–8 months that might be needed to obtain a cure. Paralysis meant that the man was unable to earn a living. People with severe rheumatism, hemiplegias and spinal injuries were also unemployable. Unfortunately the poor did not present with gout in the 18th Century so it was not possible to provide scientific evidence to support the anecdotal evidence of the curative powers of the Bath waters to this potentially large market of well-to-do patients with gout.

By the time the hospital was opened to admit patients in 1741 the foundations for a scientific trial of the Bath waters had been laid down (Falconer 1988). The hospital environment meant that the treatment could be controlled and the patients were carefully selected on admission. By clearly stating that the final outcome was to be assessed by a committee of doctors, not only the one in charge of the case, no change of bias could be brought. These results were recorded immediately in the minutes of the committee meeting and then transcribed into the referral register. Because the hospital was a charity an analysis of these reports was published annually.

Ten years later Dr Summers (1751) was quoting results from the hospital records to refute a misleading statement published in the English translation of "Medical precepts and cautions" by Dr Richard Mead (1751) which alleged that it was dangerous, and probably fatal, for any one who was paralysed to bathe in the hot waters of Bath. Dr Summers wrote "when patients are discharged it is what the whole committee sees and examines which is recorded. It is therefore a Testimony to be relied on. Proof as none can dispute". He was then able to quote the results of the first nine years of treatment at the Hospital of a great variety of paralytic patients. Of 310 admitted only 12 had died and 208 were improved.

By 1760 Dr Rice Charleton had classified the results. Outcome (of treatment) had fallen into six groups:

(i) cured, when no sign of the original complaint remained;
(ii) much improved;
(iii) no better;
(iv) improper, patients with ulcers or fevers or who were otherwise unsuitable for treatment in the hospital, these were often admitted as a result of misleading or inadequate referral letters, a fact of which Charleton complained;
(v) irregular patients, those discharged before treatment had been completed, either at their own request or for misbehaviour;
(vi) dead.

At the same time the cases were put into seven diagnostic groups: rheumatism, paralysis, leprosy or skin diseases, lameness, hip cases and jaundice. Paralysis due to *Colica pictonum* was added the following year. Gout was not included until 1865. These classifications were not radically changed until after 1880.

Only one Referral Register still exists. This contains the handwritten copies of the referral letters of the 1590 patients admitted to the hospital between May 1751 and September 1758 together with the dates of admission and discharge also the outcome of the treatment. Frequently these entries were initialled by the the physician and surgeon who had discharged the patient. By including only those patients with paralyses who were stated to have had occupational exposure to lead or who gave a clear history of attacks of severe colic preceding the peripheral paralysis it has been possible to extract a series of 108 cases with definite symptoms of chronic lead poisoning.

These 108 cases fell into two distinct groups: firstly, those in which there was evidence of occupational exposure and secondly those in which there was a clear history of a peripheral paralysis following an attack or several attacks of severe colic but no definite history of lead exposure.

Occupational exposure to lead

37 admissions (31 patients, 6 re-admissions)

Occupation

11 Painters, 7 Plumbers, 4 Glaziers, 5 Lead manufacturers, 2 Glass grinders, 1 Potter, 1 Gilder.

Place of origin

South West 5, South East 5, Referred from London Hospitals 15, Others 6

Average stay 149 days

Results of treatment

cured 22 (59.5%), much better 5, better 7, no better 2, dead 1

total improved 34 (92% of cases)

N.B. That 15 of the 31 patients had been referred from London hospitals as incurable and that 8 of these were later discharged as cured, 7 were improved. It is noteworthy that 20 patients came from London or the South East and that more than half (59.5%) were cured. 92% of the cases admitted after occupational exposure to lead were improved.

This evidence supports the claim that the treatment in Bath had something special to offer as in London they would have been removed from exposure, given purges, emetics and a bland diet to no avail.

In the other group no definite exposure to lead could be demonstrated. They presented with the classical symptoms of the Devonshire colic; colic followed by paralysis.

More patients came from the South West than in the previous group and were found to have a longer stay in the Hospital. Compared to the cases due to occupational exposure, a smaller proportion, 42% of cases, were cured although a total of 93% were improved.

Paralysis after an attack of colic

71 admissions (63 patients, re-admissions once 6, twice 1)

Place of origin

South West 43, London 5, Other 15

Average stay 168 days

Results of treatment

cured 30, much better 36, no better 2, dead 1, misbehaviour 1, afraid of smallpox 1

total improved 66 (93% of cases)

It was well known outside Bath that these records were being kept. Sir George Baker giving his famous paper on the Devonshire Colic showing that it was caused by drinking lead-adulterated cider, quoted from the records of the Bath and Exeter Hospitals.

Between 1762 and 1767, 285 patients with Devonshire colic (*Colica pictonum*) were admitted to the Exeter Hospital; of those 209 (73%) were cured or improved.

During the same period, 281 patients with Colica Pictonum were admitted to the Bath Hospital of whom 259 (92.5%) were cured or improved. It is worth noting that during this five-year period 20% more patients were cured in Bath than in Exeter. This included some patients referred from Exeter as incurable.

Dr Andrew, the Physician in Exeter, acknowledged this in a communication to Sir

George Baker (1767). "When the disease proves obstinate we always endeavour to get our patients to the Hospital in Bath. The Bath water though not a specific, being esteemed by us the most effectual remedy both internally and externally used."

In 1770 Dr Rice Charleton published three tracts on the Bath waters quoting cases seen by himself and Dr William Oliver. Included was a table "A State of the paralytic Patients admitted into the Bath Hospital, from May 1751, to May 1764."

The total number of cases admitted was 1053 of which 813 were cured or benefited. Of 237 cases admitted with "Palsies from Cyder and Bilious Cholics" 218 (92%) were said to have benefited. Of 40 with palsies due to metallic effluvia, 38 (95%) benefited.

That this remarkable Table was published in 1770 indicates very clearly that the Hospital Physicians considered that the treatment of paralysis was a very important part of their practice and that they were aware of the value of the records that they were keeping. Rice Charleton had been responsible for the Disease classification presented in the Annual Reports from 1760 onwards.

The earlier statistics could have been gathered from the Referral Registers and from his own records.

Other results published included those of Dr William Falconer (1790) who published the results from 1776 to 1785:

264 cases of lead poisoning;
117 (44.3%) were cured and 143 (54.2%) were much better;
98% had therefore benefited from the treatment.

The Annual Reports 1760–1879 show similar remarkable results:

3377 cases of paralysis due to lead;
1533 (45.4%) cured and 1629 (48.6%) better;
3160 (93.5%) improved.

These consistent results were found throughout the series:

1771–1778	244 cases	93.4% cured or improved.

A hundred years later:

1871–1878	228 cases	93.4% cured or improved.

It might appear from this high success rate that this was because the patients were highly selected or the results of outcome too optimistic but if one compares the cure rate of the paralyses due to lead with those of paralyses from other causes a marked difference is seen.

This was illustrated clearly in a Broadsheet printed in 1829 (Hunt Collection, Bath Central Reference Library, vol. 1, p 232) giving the aggregate of the Patients of the Bath General Hospital between 1799 and 1828. This showed that the percentage of cures for all cases of paralyses was 20%, though when this was broken down it showed that only 6% of the paralyses due to deformities of the spine and 11% of other paralyses not due to lead (often due to strokes) were cured. However, 49% of the cases due to lead poisoning were discharged as cured.

Aggregate of Cases of Paralysis admitted to the Bath General Hospital (1799–1828)

3150 cases of paralysis admitted	627 cases (20%) cured

182 cases due to deformity of spine 11 cases (6%) cured

2204 cases of paralysis CVA, and unknown causes 242 cases (11%) cured

764 cases due to lead 374 cases (49%) cured

The variable cure rates indicate quite clearly that the Doctors showed clinical judgement when assessing the outcome of the treatment of paralysis from different causes.

The historical evidence shows that Bath Spa therapy was a very effective treatment of paralysis due to lead poisoning.

Bath Spa therapy was also reputed to be useful in the treatment of gout. Few of the charity patients admitted to the Bath Hospital in the 18th and early 19th Centuries had a sufficiently high protein intake to develop gout. It was not until 1865 that gout was introduced as a diagnostic category in the Annual Report.

However, it had been established in the 17th Century that a visit to "The Bath" was helpful to people with gout. So it was not surprising that it entertained many who succumbed to the epidemic in the 18th Century. In fact it was the massive influx of the well-to-do and influential gouty in the 17th and 18th Centuries that resulted in the beautiful city we have today. As it is probable that many of the cases were of Saturnine gout from drinking lead adulterated Port, in vast quantities, the Bath waters would have been doubly useful.

Fig. 5.3. One of Thomas Rowlandson's Comforts of Bath (Courtesy Bath Museum Services). Gout sufferers taking exercise in Bath. Bathing was not carried out during an acute attack (fit) of gout.

As almost all of these patients were treated privately no records have been kept, nor was the treatment controlled except in the most general way. Bathing increased the agony during an acute attack, a fit, of gout so patients only took the waters when this had subsided. They often reported that they had a further acute but mild fit of the gout after a few weeks treatment but later on returning home felt that they had benefited from the treatment and returned next year. Immersion has been shown to increase the urinary urate excretion in gouty patients. Modern treatments which lower the serum urate level sometimes provoke a similar acute attack.

From the Hospital records it appears that about 90% of the patients admitted with gout, after 1865, were considered to have been improved on discharge though it is difficult to know on what criteria this was based.

By using modern techniques it is possible to show how these high cure rates could have been achieved. Sitting in warm water up to the neck formed a vital part of the treatment in Bath. Since 1970 extensive studies of the effects of head-out water immersion (HOWI) have been carried out.

Murray Epstein (1978) working for the American space programme was able to show that by using HOWI at 35°C it was possible to simulate the physiological effects of weightlessness. It was found that HOWI resulted in a profound increase in the rate of urinary excretion of sodium, a significant increase of water excretion in the urine and an increase in the rate of urinary calcium excretion. As the human body handles calcium and lead in a similar way it was decided to set up a small trial to study the effect of HOWI on the urinary excretion of lead (Heywood *et al.* 1986).

Lead poisoning is very rare these days but it was possible to find three men who were working in a factory making lead shot in the traditional manner. They volunteered to take part in the immersion experiment which entailed sitting for three hours in a tank of warm water up to the base of their necks. They sat for a further hour outside the tank before and after the immersion. Samples of blood was taken and urine was collected every hour for analysis. A control experiment was set up when they sat for four hours outside the tank when blood and urine samples were taken as before. Although they were all symptom-free with blood lead levels well below the legal safety level their lead levels were higher than those in the general population.

Measurement of the urinary lead levels showed that although very small amounts of lead were in fact excreted, during immersion the rate of urinary lead excretion was significantly increased (mean total amount of lead excreted in the urine for each hour during immersion (μ g): pre-immersion, 5.85; hour 1, 9.79; hour 2, 24.25; hour 3, 15.0).

As the three lead workers taking part in the trial were symptom-free it is reasonable to assume that their blood levels would have been lower than those of the patients with paralyses admitted during the 18th Century. It is possible therefore that the effects of immersion could have been even more marked in these patients. The hospital records show that the patients bathed three times a week for an average of 24 weeks, during this prolonged period of repeated immersions a significant amount of the body lead load could have been removed, resulting in recovery from the paralysis.

It is necessary to appreciate that spa therapy is a multi-factorial process, so it is essential that we do not disregard the other elements involved in the Bath Spa treatment plan which included removal from the source of exposure, good food, the beneficial effects of exercising in warm water and the therapeutic effects of drinking the mineral water. An intake of one to one and a half pints was usual. The mineral waters arising from the sources in Bath have a high calcium content (300 mg per litre) and iron salts as the readily absorbable soluble ferrous salts, although this becomes converted to the insoluble ferric form when exposed to the oxygen in the air. This explains why it was considered important to drink the water as soon as possible after it had arisen from the spring and also why patients with 'delicate stomachs' were advised to allow the water to stand in an open container overnight and then to decant the clear fluid from the deposit of orange ferric salts. Work done in North America (Mahaffey *et al.* 1973) during the 1970s has shown that calcium and iron deficiency increases the amount of lead absorbed and retained by the body and increases the toxicity of that already in the body (Barton *et al.* 1978). A recent U.S. report suggested that giving calcium and iron supplements to deprived infants with raised blood lead levels was helpful in reducing the morbidity in these cases. Drinking the

Bath water would also counteract any calcium loss which might arise from the increase in the rate in urinary calcium excretion during prolonged immersion.

So drinking the water could be helpful too. The 18th Century doctors certainly thought was an important part of the therapy. The effects of drinking the water have yet to be studied.

Analysis of the results of the treatment of paralysis due to lead poisoning in the Bath Hospital from 1760 to 1879 shows that of the 3377 cases of paralysis due to lead admitted during this period, 1533 patients had no residual paralysis on discharge, therefore 45.4% were said to be cured. A further 1629 patients were said to be greatly improved. Thus 93% of the patients suffering from chronic lead poisoning were improved by treatment using the Bath waters. Even by modern standards this would be considered to be a successful therapeutic trial. The immersion studies indicate one of the factors that could have contributed to this high cure rate.

The founders of the Bath Hospital believed that by building up a bank of "indisputable evidence" it would eventually be possible to show that the Bath Spa therapy could be recognized as an effective cure, not merely a pleasant experience. This paper is an attempt to justify their endeavours in the case of chronic lead poisoning.

Acknowledgements. I would like to thank Bath City Council, The Swainn Foundation and the Bristol Kidney Fund for their financial support. I acknowledge the encouragement and the assistance that I received from Dr A. H. Waldron with the urinary lead analysis and from Mr C. Quinnell for access to the Bath Hospital records. I also thank Dr P. O'Hare, Dr N. Millar, Professor P. A. Dieppe and Dr J. Campbell McKenzie for their help and for permission to use the immersion tank in the Bristol Royal Infirmary.

This work was previously published in *Medical History Supplement* 10 (1990, pp 82–101) and this revised presentation appears by permission of the Trustees of the Wellcome Trust.

References

Baker, Sir Geo. 1767. An inquiry concerning the cause of the endemic colic of Devonshire. *Trans. Coll. Phys.* **29**.

Barton, J. C. *et al.* 1978. Effects of iron on the absorption and retention of lead. *J. Lab. Clin. Med.* **92**(4), 536–547.

Charleton, Dr Rice 1774. *Three tracts on Bath Water.* Wm Taylor. 2nd Edn, Tract 2(2), 77–79; 12–13 (table).

Citois, Francis 1639. De novo et Populari apud Pictones dolore colico bilioso diatriba, Paris. Quoted by Eisinger Jos, *Lead and Wine.* Erbhard Glockel and the Colica Pictonum. *Medical History* 1982, **26**, 279–302.

Cunliffe, Barry 1986. *The City of Bath.* Alan Sutton, Gloucester, 16–39.

Defoe, D. A. 1983. *Tour through Great Britain.* London Folio Soc, Vol. 2, p 170.

Emmerson, B. T. 1973. Chronic lead neuropathy. *Kidney International,* **14**, 1–5.

Epstein, M. 1978. Renal effects of head-out immersion in man. *Physiol. Rev.,* **58** (3), 529–581.

Falconer, W. 1790. *A Practical Dissertation on the Medicinal Effects of the Bath Waters.* Cruttwell, Bath, 69–70.

Falconer, W. R. 1888. *History of the Royal Mineral Water Hospital,* 20–21.

Freeman, H. 1888. *The Thermal Waters of Bath.* Hamilton, Adams, London, 53.

Glanvill Rev Jos 1668. *The Progress and Advancement of Knowledge since the Days of Aristotle.* Collins, London.

Heywood, A. 1990. A trial of the Bath Waters. *Medical History,* Suppl 10, 82–101.

——, Waldron, H. A., O'Hare, P. & Dieppe, P. A. 1986. Effect of immersion on urinary lead excretion. *British Journal of Industrial Medicine,* **43**, 713–715.

Mahaffey, K., Goyer, R. & Haseman, J. K. 1973. Dose response to lead ingestion in rats fed low dietary calcium. *J. Lab. Clin. Med.* **82**, 92–100.

Mead, R. 1751. *Medical Precepts and Cautions.* Brinsley, London, 69.

Peirce, R. 1713. *The History and Memoirs of the Bath.* Hammond, London, 189–210.

Porter, R. 1988. The gift relationship. *In*: Granshaw, L. & Porter, R. (eds) *The Hospital in History*. Wellcome Institute Series in the History of Medicine, Routledge, London, 150–163.
Summers, J. 1751. *A Short Account of the Success of Warm Bathing in Paralytic Disorders*. Hitch & Hawes, London, 4–7.
Wynter, J. 1728. *Of Bathing in the Hot Baths at Bathe*. Leake, London, 53.

From Kellaway, G. A. (ed.) *Hot Springs of Bath*, pp 89–96

6. Pathogenic *Naegleria* amoebae in the waters of Bath: a fatality and its consequences

S. Kilvington, P. G. Mann & D. C. Warhurst

In July 1978 a child was admitted to the Royal United Hospital, Bath, apparently suffering from bacterial meningitis. Standard antibiotic therapy was instituted but produced no improvement. During a routine cell count on the cerebro-spinal fluid, amoebae were observed by an alert MLSO, Alan Yates, moving actively in the counting-chamber.

It was realized that the patient had participated a few days earlier in a swimming gala at the Beau Street baths. The organism was identified presumptively as *Naegleria fowleri*, a small free-living amoeba generally acquired from thermal water sources. Appropriate therapy with amphotericin was begun, but although amoebae were eliminated from the CSF the patient died on the ninth day of illness from damage already inflicted on the brain (Cain *et al.* 1981). This has been the unfortunate outcome of nearly all the cases of *Naegleria* infection reported throughout the world.

A review of the world literature on pathogenic small free-living amoebae has been published (Warhurst 1985) and can be consulted for references to points made in what follows. The pathogenic potential of free-living amoebae was discovered by Clyde Culbertson in 1958, when working on Salk poliomyelitis vaccine. Culbertson and his colleagues found that a free-living soil amoeba (*Acanthamoeba*) contaminating kidney tissue-culture cells produced fatal meningoencephalitis in mice and monkeys when instilled intranasally. In the 1960s reports of acute, fatal human brain infections due to free-living amoebae came from Australia, Czechoslovakia and the United States. The organism responsible was eventually identified as a species of *Naegleria*. *Acanthamoeba* was also recognized as the cause of a more chronic, fatal brain disease in man, mainly in the immuno-suppressed, as in Hodgkin's Disease and, latterly, in AIDS. The same amoeba also causes intractable corneal ulcers in immuno-competent individuals.

Both *Naegleria* and *Acanthamoeba* are found living normally in soil and water in the absence of a human or animal host, feeding on bacteria, yeasts and algae in the environment. They are facultative parasites, representing a transitional stage between free life and parasitism.

Naegleria fowleri infects healthy individuals, causing a swiftly fatal meningoencephalitis. Amoebae in their active stages enter the nose by inhalation of contaminated, warm, fresh water. The amoebae penetrate the epithelium at the back of the nose and travel along the olfactory nerves into the brain. Infection spreads over the meninges and along the perivascular channels within the substance of the brain.

Naeglerias are amoebo-flagellates, with a transient, non-feeding biflagellate stage in their life-cycle. This is thought to provide in nature for the rapid spread of the organism to fresh pools after rain. This flagellate stage can be induced to form in cultures by diluting the medium with a small amount of distilled water. In common with many other amoebae there is also a resistant cystic stage. Cysts will survive a cold winter if kept moist, but are killed rapidly by desiccation. The feeding stage (trophozoite) has a lobose pseudopodium, used for locomotion. This is generally rapid, in the region of two body lengths per minute. The cell nucleus has a large central body (nucleolus) composed of ribonucleoprotein, and scanty peripheral DNA. The three morphological forms of *N. fowleri* are shown in Fig. 6.1. Multiplication is by binary fission, with a generation time of the order 6–8 hours. Populations grow rapidly by comparison with other protozoa, but more slowly than bacteria.

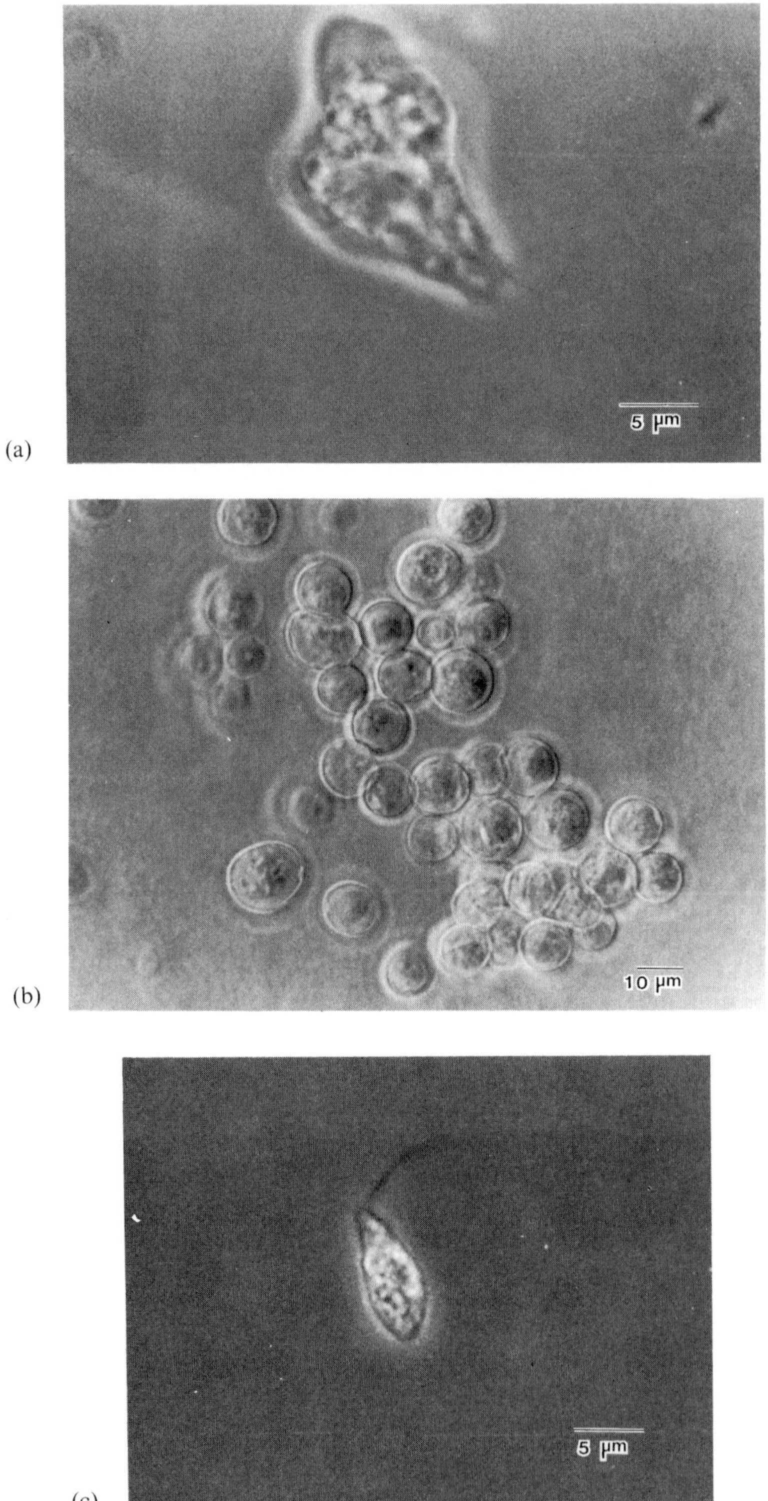

Fig. 6.1 The three morphological forms of *Naegleria fowleri*: (a) trophozoite (motile vegetative amoeba); (b) cyst; (c) flagellate.

The role of free-living amoebae in primary amoebic meningoencephalitis (PAM) was first recognized in 1965 by Fowler & Carter. At first the pathogen was confused with the ubiquitous *N. gruberi*, but distinguishing features were soon recognized, including preference for warm waters, ability to grow at 45°C (113°F), cyst morphology (size, number of pores) and pathogenicity for mice (*N. gruberi* is non-pathogenic). In 1970 the pathogen was given the trivial names *fowleri* and *aerobia*, the former achieving publication priority by a short head. Differentiation between the two species is now much strengthened by biochemical studies. Indeed *N. gruberi* is now recognized to be a species complex. More than 150 cases of PAM have been reported in the world literature. Only two patients with fully authenticated infections have survived. Although clinical isolates from all over the world are apparently identical serologically, restriction endonuclease digests of nuclear DNA indicate the existence of several genetically distinct lineages from different geographical locations (De Jonckheere 1987). The isolation of *N. fowleri* from the environment has proved difficult. Incubation of samples at elevated temperatures was tried early, with some success. A major source of confusion was the unsuspected existence of a second thermophilic species of *Naegleria*, morphologically and serologically very similar to *N. fowleri*, sharing the same econiche, but non-pathogenic for mice. Not until 1980 was this recognized to be a new species and given the specific name *lovaniensis* by Stevens and her colleagues. The fact that cases of PAM the world over occur in clusters, and that in several instances a succession of fatalities has persisted until measures were taken to eliminate the common source of infection, makes it mandatory to search for the source of infection whenever a case is encountered. Particularly striking was the series of 16 deaths recorded at Usti in Czechoslovakia over a period of four years.

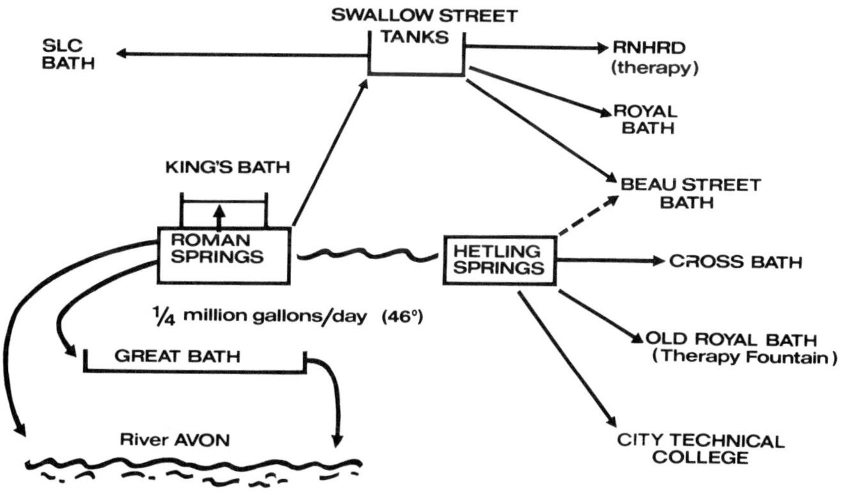

Fig. 6.2 Distribution of spa water from thermal springs to recreational baths and other sites in Bath.

In 1978 in Bath we had first to consider as possible sources a small swimming pool and a fish-tank at the child's school. Both proved negative. Proceeding then to the Beau Street baths, where the school held a swimming gala shortly before the child fell ill, samples from a holding tank above the former spa laundry in Swallow Street yielded a *Naegleria* amoeba which grew at the same high temperature as the CSF isolate, readily produced a flagellate phase, and reacted in the indirect fluorescent antibody test with anti-*N. fowleri* serum at the same titre as the CSF isolate. It now became imperative to examine the Roman springs beneath the King's Bath. This necessitated draining the bath and providing lighting and ventilation for the enclosed space where the springs rise. Samples yielded several more apparently identical *Naegleria* isolates. This evidence was sufficient at that time to identify the organism as *Naegleria fowleri*. In view of the association with a fatal case of PAM it was necessary to recommend the closure of the thermal springs and of the baths they served (Fig. 6.2).

Access to the Roman springs for our amoebal investigation also revealed to the City Engineers that the foundations of the Pump Room were being severely eroded. Substantial underpinning of the buildings and measures to avoid further erosion were seen to be necessary. To facilitate this work it was decided to cut away the concrete floor placed under the King's Bath by Major Davies at the end of the 19th Century. This in turn provided an opportunity, magnificently seized, to excavate the Roman temple remains known to be lying beneath the Pump Room and Abbey Churchyard (Cunliffe 1971).

The description in 1980 of *N. lovaniensis* as a new thermophilic, non-pathogenic *Naegleria*, distinct from *N. fowleri*, necessitated a reappraisal of our 1978 environmental isolates. Development of a cellulose acetate isoenzyme electrophoresis system for naeglerias enabled us to show that glucose phosphate isomerase and phosphoglucomutase provided clear distinctions between 'non-pathogenic' and 'pathogenic' isolates. Moreover Ann Stevens and her colleagues had shown that electron microscopy reveals important structural differences between *N. fowleri* and *N. lovaniensis*. The nucleus of *N. lovaniensis* has around it a wrapping of rough endoplasmic reticulum, which is not present in *N. fowleri*. Use of the new isoenzyme technique and electron microscopy showed that all our 1978 environmental isolates were *N. lovaniensis* and not *N. fowleri*. A fresh hunt for the pathogen was therefore begun in the summer of 1981. By December of the same year *N. fowleri* had been identified in samples from the NE corner of the Great Bath (Fig. 6.3).

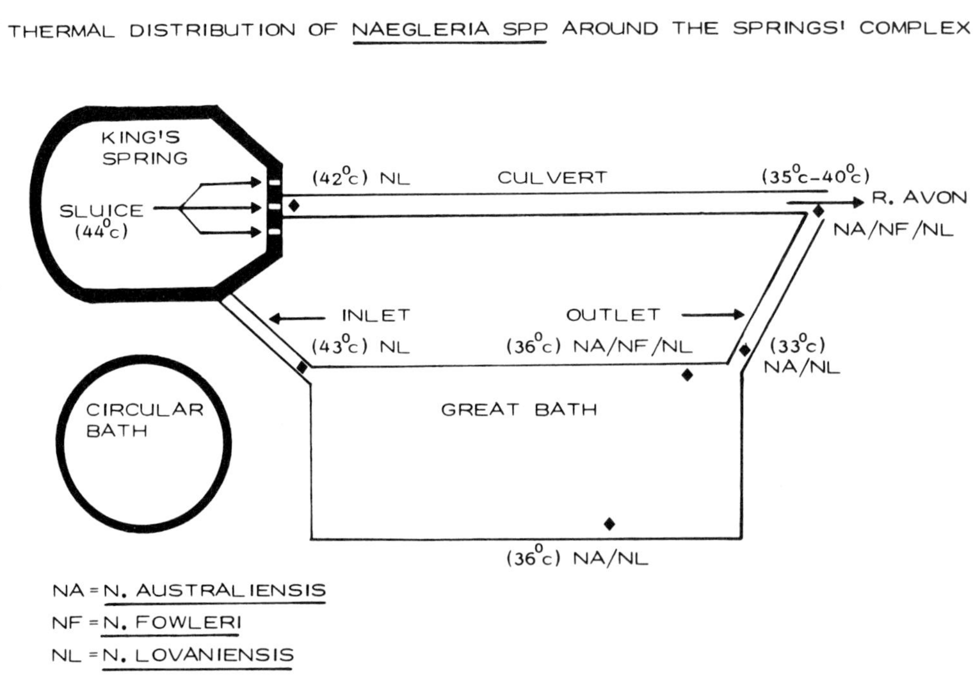

Fig. 6.3 Diagram of Roman baths indicating the location of sites sampled for thermophilic *Naegleriae* 1981–1984.

Identification was carefully validated by electron microscopy, isoenzyme electrophoresis and mouse pathogenicity. Additional enzyme tests since developed in Bath (Kilvington & White 1985) have provided still further confirmation. Further systematic attempts to recover *N. fowleri* from the thermal springs continued, with negative results, throughout 1982 and the first half of 1983. In May of that year sampling was concentrated solely on the point at which the overflow from the Great Bath reaches the main drain into the River Avon. In June 1983 several further fully validated isolations were made.

While the second search for *N. fowleri* was in hand, public demand for a supply of safe spa water for Bath led to a detailed examination of the fault under Bath, up which the Roman thermal springs rise. Dr Geoffrey Kellaway skilfully drove a new oblique borehole to meet the relevant aquifer at a depth of some 60 metres (197 ft), so bringing the water to the surface without contamination by the superficial soil which is the primary home of *Naegleria*. It is unlikely that water derived from such a depth will harbour aerobic amoebae, and prolonged monitoring of the water from its new source has failed to detect any naeglerias to date.

In 1984 a search for a third thermophilic *Naegleria*, *N. australiensis*, was initiated. This species was first definitively described in 1981 by De Jonckheere. Although not yet associated with human disease, it is pathogenic for mice.

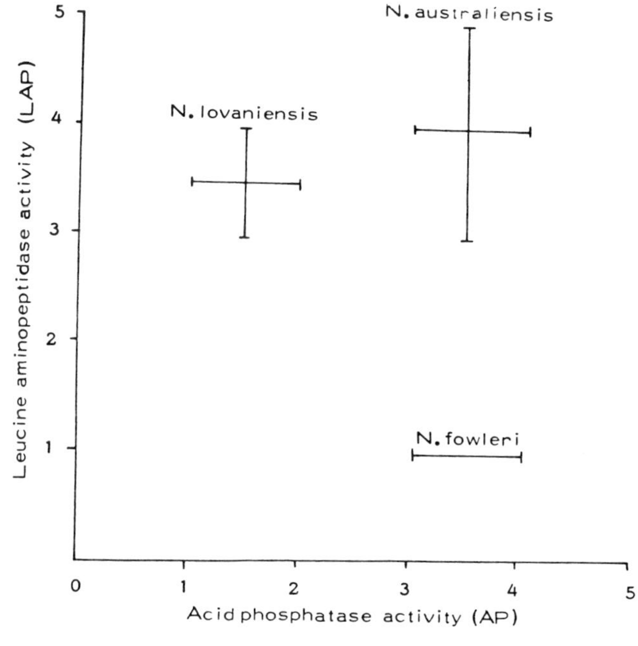

Fig. 6.4 Differentiation between species of thermophilic *Naegleria* by activity patterns for the enzymes Leucine aminopeptidase (LAP) and Acid phosphatase (AP).

The methods previously used to isolate and identify *N. lovaniensis* and *N. fowleri* in the hot springs used incubation at 43/44°C to favour the growth of *N. fowleri*. Since *N. australiensis* will not grow at temperatures above 42°C, cultures were now incubated at 37°C. Primary cultures were made as before on non-nutrient agar supplemented with *Esch. coli*. Plaques of amoebae which resembled *Naegleria* morphologically and produced amoebo-flagellates in distilled water were subcultured in liquid TRIS-buffer/*Esch. coli* medium. After incubation for 48–72 hours the medium was tested for *N. australiensis* antigen by staphylococcal coagglutination (Kilvington & White 1986). Antigen positive cultures were transferred to NNA/*Esch. coli* medium. Amoebal growth on this medium was tested by API–ZYM for production of acid phosphatase and leucine aminopeptidase. Distinctively different patterns of activity for these two enzymes are produced by *N. fowleri*, *N. australiensis* and *N. lovaniensis* (Fig. 6.4, Kilvington & White 1985). Candidate strains of *N. australiensis* selected in this way were then tested by cellulose acetate isoenzyme electrophoresis for glucose phosphate isomerase. Strains identified by this test as *N. australiensis* were tested for mouse-pathogenicity by intracerebral inoculation. Brains taken from dead or moribund mice were examined microscopically for *N. australiensis* after specific immuno-peroxidase staining. Amoebae were also re-isolated from brain-tissue by culture and confirmed as *N. australiensis* by immuno-fluorescence and GPI isoenzyme production. By these means *N. australiensis* was isolated on a number of occasions and from several sites during 1985. These were the first isolations of this *Naegleria* from environmental material reported from the United Kingdom.

In 1986 further inquiry was made into the distribution and relative numbers of the three thermophilic naeglerias in the hot springs system, with particular regard to the temperatures prevailing at the sampling sites. Three sites were examined: the waterfall pool at the foot of the overflow from the King's Spring, the sluice-gate at the NE corner of the Great Bath and the overflow from the Great Bath at its junction with the main drain to the River Avon (Fig. 6.3). Samples of mud were inoculated onto NNA/*Esch. coli* medium and incubated at 37°C. Presumptive *Naegleria* isolates, selected by morphology and flagellate formation, were then tested for specific soluble naegleria antigen by staphylococcal coagglutination. The results of this survey are shown in Figs 6.3 & 6.5.

The waterfall pool yielded *N. lovaniensis* only, in numbers that were constant through the year. The water temperature at this site is always between 42° and 43°C. This may account for the absence of *N. australiensis*, which will not grow at temperatures above 42°C. *N. lovaniensis* and *N. australiensis* were both found at the NE corner of the Great Bath, where temperatures ranged from 29–34°C. The latter species was present only from April onwards, but was dominant during the summer months. Water temperatures at the outlet from the Great Bath to the main drain ranged from 30 to 41°C. Only *N. lovaniensis* was found here, the numbers being highest from July to September.

N. fowleri was not detected on any occasion at any of the three sites throughout the whole 1986 survey. The samples taken during the initial search for *N. australiensis* in 1984/85 had also been negative for *N. fowleri*, as were all the samples taken between January 1982 and June 1983. Impressed by the infrequency of isolations of *N. fowleri*, efforts were directed to devising a selective medium that would permit the growth of this species, while inhibiting *N. lovaniensis* and *N. australiensis*. After trial of a series of diamidines, a suitably selective medium was achieved by incorporating M&B 800A (pentamidine isethionate). In February 1987 this medium was inoculated with samples from the hot springs system. Two strains of *N. fowleri* were recovered from the NE corner of the Great Bath. Duplicate samples examined with the non-selective NNA/*Esch. coli* medium yielded only numerous isolates of *N. lovaniensis* and *N. australiensis*. It seems that the apparent infrequency of *N. fowleri* may have been due to the failure of non-selective media to detect small numbers of *N. fowleri* in the presence of much larger numbers of other thermophilic naeglerias.

In summary, investigations into the tragic death of a child established that the cause was a small free-living amoeba, *N. fowleri*. Examination of the hot springs revealed the presence of three species of thermophilic *Naegleria*, not previously reported from environmental sources in the United Kingdom, viz. *N. fowleri*, *N. lovaniensis* and *N. australiensis*. Local variations in temperature affect the distribution of these amoebae. The hitherto apparent infrequency of *N. fowleri* relates, at least in part, to the difficulty of finding this

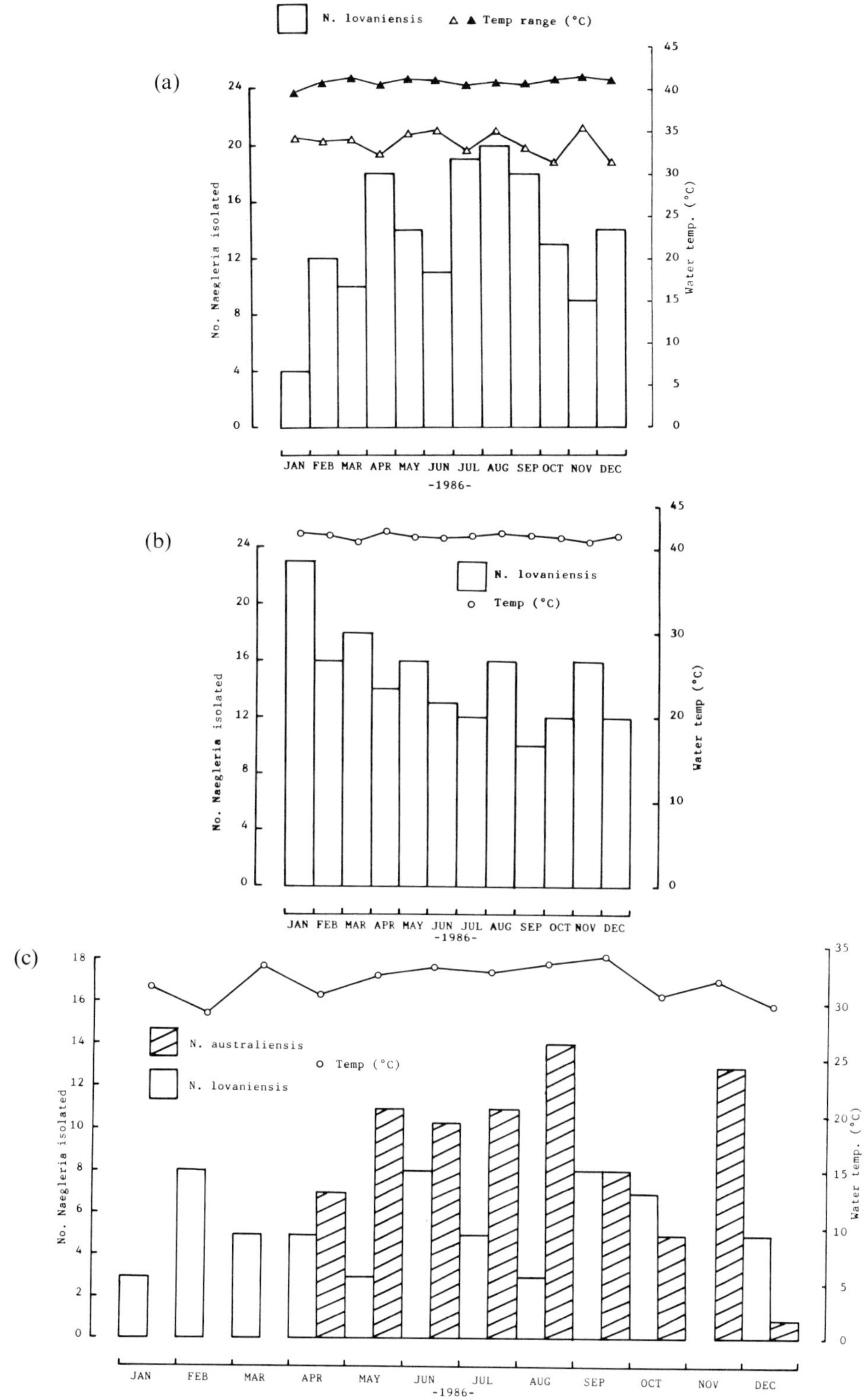

Fig. 6.5 Seasonal variation in relative numbers of thermophilic *Naegleria* by site and temperature (1986) survey) (see also Fig. 3): (a) outlet from Great Bath to Roman drain; (b) pool below waterfall-outlet from King's Spring; (c) outlet from Great Bath to sluice leading to Roman drain.

amoeba when in a minority among larger numbers of the other two thermophils. This may also contribute to the rarity of infection from a source which is probably constantly positive for *N. fowleri*, but generally in small numbers only.

The need to search for a source of infection following a case of PAM had some surprisingly far-ranging consequences for Bath. In searching for amoebae in the Roman springs, erosion of the Pump Room foundations was revealed. Repair work gave an opportunity to uncover the remains of an important Roman temple-complex. Public concern that safe spa water be once more available led to hydrogeological investigation of the system of faults beneath Bath. This provided the basis for sinking a new borehole to the deep aquifer which supplies the healing waters of Bath, thus avoiding the potentially infected spring-pipe which was their original source. The importance of protecting this new source from amoebal re-contamination called for a degree of hygienic care unusual in most water-extraction operations. It is to be hoped that a new range of safe, contemporary spa facilities may yet arise on the site of our earlier establishments.

Acknowledgements. All three authors worked within the uniquely supportive framework of the Public Health Laboratory Service. Simon Kilvington also received support from Bath City Council. Our thanks for much encouragement and working facilities generously given are due to Dr D. G. White, Director of the Public Health Laboratory, Bath. Thanks are also due to the editors of *J. Clin. Path.* for permission to reproduce Fig. 6.4.

References

Cain, A. R. R., Wiley, P. F., Brownell, B. & Warhurst, D. C. 1981. Primary amoebic meningo-encephalitis. *Arch. Dis. Childhood,* **56**, 140–143.

Cunliffe, B. 1971. *Roman Bath Discovered.* Routledge & Kegan Paul, London.

De Jonckheere, J. S. 1987. Characterization of *Naegleria* spp. by restriction endonuclease digestion of whole-cell DNA. *Molecular and Biochemical Parasitol.,* **24**, 55–66.

Kilvington, S. & White, D. G. 1985. Rapid identification of the thermophilic *Naegleria*, including *N. fowleri*, using the API-ZYM system. *J. Clin. Path.,* **38**, 1289–1292.

—— & —— 1986. Identification of *N. fowleri* in fresh isolates of environmental amoebae using a staphylococcal coagglutination technique. *Trans. Roy. Soc. Trop. Med. Hyg.,* **80**, 564–569.

Warhurst, D. C. 1985. Pathogenic free-living amoebae. *Parasitology Today,* **1**, 24–28.

From Kellaway, G. A. (ed.) *Hot Springs of Bath*, pp 97–126

7. Investigation of the Bath hot springs (1977–1987)

G. A. Kellaway

Geological investigations of Bath Hot Springs commenced in 1978 following the discovery of *Naegleria fowleri* in the thermal water installations of the Spa and at a time when structural damage to the south wall of the Grand Pump Room was undergoing repair. The presence of *N. fowleri* resulted in the closure of the Hot Bath, the Cross Bath and the drinking fountain in the Pump Room. The King's Bath had already been closed in order to deal with structural problems, removal of the mass concrete floor of the bath and the damaged arches which supported it having revealed the remains of the Roman Reservoir beneath. At this stage the cause of the damage to the Pump Room was unknown, but it was clear that movement had been in progress for some time. On the other hand the collapse of the supporting arches to the floor of the King's Bath can only have post-dated their construction at the end of the 19th Century. The most puzzling feature of the damage was the lack of any obvious deterioration of the Roman or Mediaeval masonry surrounding the spring.

At the time that these untoward events took place, there was very little information as to the nature of the ground beneath the bed of the King's Spring. This spring and, to a lesser extent, the Cross Bath Spring, have always deposited some sand and silt and specimens of this material were investigated by Rastall (1925, 1926). The results indicated the presence of alluvial deposits, including sand and silt, through which the waters emerge, the bedrock being Lower Lias Clay which forms the floor of the Avon valley at Bath (Fig. 7.1). Beyond these facts, however, little was known of the character and thickness of the Drift (Quaternary) deposits, or of the detailed structure of the underlying rocks.

The Geological Survey's one inch to one mile map of Bath (Sheet 265) published in 1947 shows that a belt of E–W to ESE–WNW trending post-Jurassic faults crosses the Chew Valley at Burnett and continues eastwards by way of Newton St. Loe and Twerton. The Pennyquick Fault, which forms part of this group of structures extends towards the centre of Bath and may possibly continue along the northern margin of Bathampton Down (see Plate B, p 276). The structural belt cannot be traced continuously through the Drift-covered tract of Lower Lias Clay in the valley floor which is heavily built up between Twerton and Bathwick and in 1977 the lack of geological evidence made it impossible to explain the reason for the damage to the Pump Room. The only practical solution open to the Consulting Engineers, Messrs Mander, Raikes and Marshall, was to utilize the support given to the walls and floor of the Pump Room by sound Roman masonry, stiffening the basement of the building and thus bridging the area where subsidence had taken place.

In the Summer of 1978, the discovery of *N. fowleri* in the thermal water led to complete closure of the Spa, though the contaminated water of the King's Spring continued to flow into the Roman Baths. For the first time since Mesolithic man visited the hot springs, the entire output of the thermal springs of Bath was now running to waste. Records of earlier restorative activities were examined in the hope that they might provide some help in dealing with the double problem of structural instability and microbiological contamination, but nothing of practical value was found. Like his predecessor, Dr R. H. Rastall, the writer had consistently advised the City Council not to interfere with the flow of the thermal springs by drilling or excavating except in case of extreme need. In the face of this unprecedented disaster, however, action was called for and a plan was drawn up for drilling the bed of the King's Spring in order to ascertain its geological structure and to test the quality of the rising thermal water at depth before it made contact with any contaminated material within or around the spring basin.

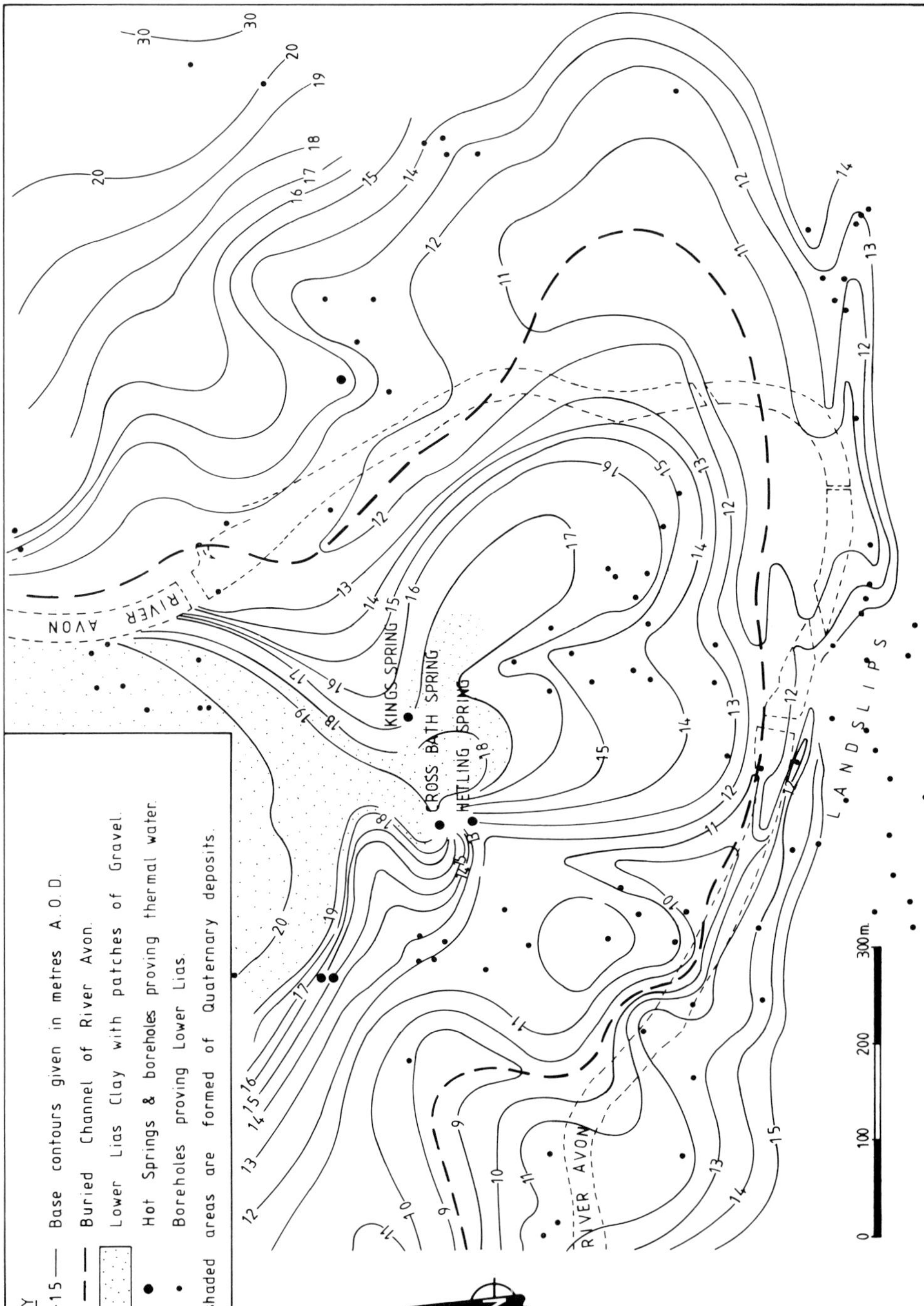

Fig. 7.1. Late Quaternary relief of Bath.

In the decades preceding the closure of Bath Spa a number of investigations of the geothermal resources of the British Isles were carried out by officers of the British Geological Survey in collaboration with other scientists from U.K.A.E.A. (Harwell) and Bath University. The resources include the thermal waters of Bristol and Bath which form the principal hot springs of the British Isles. This research provided much new evidence relating to the chemistry and radioactivity of the thermal water and its exsolved gases and the possible origin of the thermal water issuing in the Avon Valley at Bristol and Bath (Andrews *et al.* 1982). Both groups of hot springs issue at or near the margin of the Upper Carboniferous rocks of the concealed Bristol and Somerset Coalfield (Fig. 7.2). They have attracted much attention in the past and have a voluminous literature (Richardson 1928).

Fig. 7.2. Distribution of the Upper Carboniferous rocks of the Bristol and Somerset Coalfield and position of the hot springs. The outline of the coal basin has been generalized and Variscan faults omitted. The map also serves to identify many locations mentioned in the text.

Exploration of the basin of the King's Spring (1978–1980)

Following the removal of the floor of the King's Bath (Plate 12), archaeological investigations revealed the mode of construction of the Roman Reservoir (Cunliffe & Davenport 1985, pp 39–53) and gave access to the bed of the spring which was found to consist primarily of gravel. This deposit is virtually indistinguishable from the material found in the flood plain terrace and low-level river gravels of the River Avon between Bath and Keynsham. After completion of the archaeological excavations, staging was erected above the spring bed and a series of trial boreholes was made by Geotechnical Services of Gloucester (February–March 1980). The sites of the trial borings and levels of the interface between the Drift deposits and the Lower Lias Clay are shown in Fig. 7.1. From this it will be seen that the thermal water issues from a deep conical depression extending to a depth which is now known to be more than 6 m below sea level, the natural ground surface being about 17 m above O.D.

It appears that the Romans constructed their lead-lined masonry reservoir around the margin of an oval or elliptical depression floored by gravel partly covered by alluvial deposits. For the most part the Roman walls are founded on the Lower Lias Clay but where this was not possible, as for example in the channel where the thermal water drained away to the river, the foundations were supported on wooden piles caulked with puddled clay. Beneath the spring bed, however, a deep conical depression or spring pipe is filled with fluviatile silt, sand and gravel interbedded with layers or lenses of recycled or slumped Lower Lias Clay. Owing to the strong artesian flow of the thermal water it is impossible to obtain undisturbed cores of the gravelly and sandy material, though some of the recycled Lower Lias Clay was cored. The presence of scattered pebbles of Middle Jurassic Limestone in this material proved that it had been redeposited and was not Lower Lias Clay in situ. The coarse gravelly material in the spring pipe includes Old Red Sandstone, Carboniferous Limestone and Triassic conglomerate from the eastern Mendips, Jurassic, Cretaceous and Tertiary rocks from West Wiltshire, with Lower and Middle Jurassic rocks from the South Cotswolds and Bath.

The water issuing from the trial boreholes in the King's Spring gave temperatures up to 46.5°C (115.7°F). Tests carried out over a period of more than one year at the Pathology Laboratory of the Royal United Hospital at Bath showed no *N. fowleri* in the water samples. Specimens of stony clay recovered from the spring pipe at various depths were also free of *N. fowleri*, thus confirming the view on which the trial boreholes were predicated, namely that pathogenic amoebae are unable to survive under anaerobic conditions in water with redox potentials and extremely low concentrations of dissolved oxygen such as those recently recorded in the thermal water of the King's Spring and Stall Street Inclined Borehole (Chapter 9). Oxidation of the thermal water produces the characteristic orange–brown ferric hydroxide stain which can be seen on the stone basin of the new Pump Room fountain. Once oxidation has been completed, the thermal water becomes a potential medium for the growth of *N. fowleri*. Therefore the new Pump Room fountain has been designed for continuous flow, unused thermal water being immediately drained away to waste (Plate 5).

During the drilling of the basin of the King's Spring, direct measurement of the hydrostatic head of the thermal water in trial borehole No. 4 (Fig. 7.7) was carried out by connecting a flexible pipe to the drill head. This gave a maximum standing water level of about 30 m A.O.D. Later measurements gave slightly lower readings, probably because water was issuing from a number of holes drilled in the spring pipe as well as from the central area of the spring basin. As will be seen later, the standing water level in the Stall Street Borehole is 27 m A.O.D. It is possible that some higher values may have been obtained at Pinch's Well near the Kingsmead Borehole (p 106) but no readings in excess of 30 m A.O.D. have been observed in the centre of Bath since the original observations were made in 1977.

Sinking of the shallow trial boreholes in the basin of the King's Spring provided evidence of the temperature (46.5°C) and pressure of the thermal water and also demonstrated in no uncertain fashion the ease with which silt, sand and fine gravel could be washed out of the spring pipe. The need to remove sand from the floor of the King's Bath has already been mentioned and this process of washing out of sand and silt must

Plate 7. The Stall Street inclined borehole: drilling in progress, 1983. Photo – Stephen Bird.

Plate 8. Stall Street inclined borehole; thermal water issuing from the blow-out preventer in the underground chamber beneath Stall Street. Photo – Stephen Bird.

Plate 9. Thermal water overflowing from the Roman Reservoir stains the stonework as it deposits hydrated iron oxides. Photo – Bath City Council.

Plate 10. Thermal water and vapour issuing from the borehole during drilling of the Stall Street inclined borehole (1983).

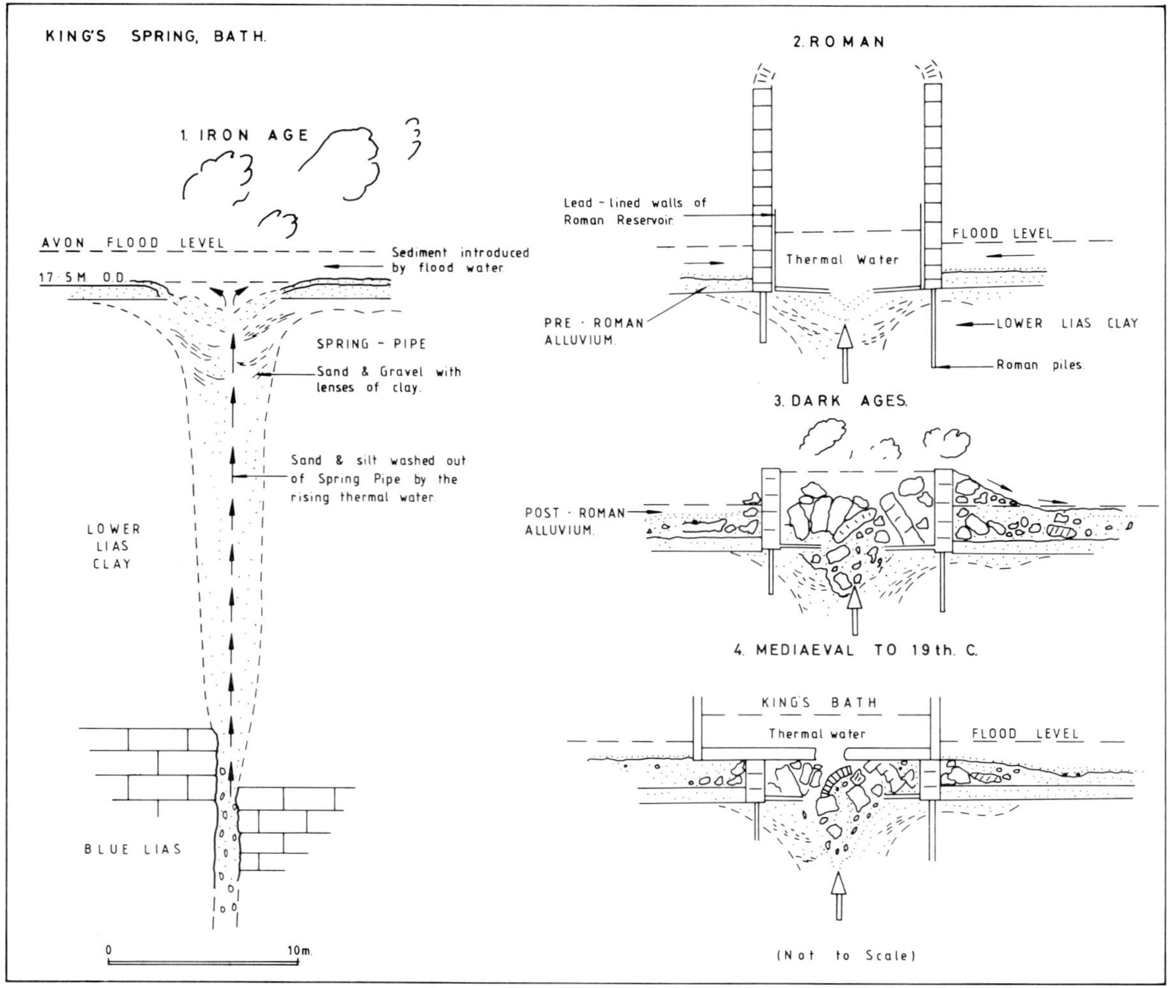

Fig. 7.3. The King's Spring. Diagrammatic representation of its history from the Iron Age to the 19th Century.

have been operating over a long period of time. The natural ground level of the King's Spring (about 17 m A.O.D.) shows that the basin is situated on the alluvial plain of the River Avon and must, therefore, have been liable to inundation by the turbid waters of the Avon at times of flood (Fig. 7.3). The effect of these inundations would have been to deposit silt, sand and gravel on the floodplain. Some of this would have lodged in the spring basin, though most of the silty material would have been quickly removed by the flowing thermal water. Nevertheless, there would have been a residual deposit retained within the basin and this would have served to replace losses due to washing out by rising thermal water.

By building a massive lead-lined masonry enclosure with piled and puddled foundations, the Romans effectively isolated the top of the spring pipe from the flood waters of the river (Fig. 7.3). Thenceforward replacement of silty or sandy material lost from the spring pipe was prevented, though the powerful King's Spring continued to convey sand and silt to the surface with the result that the volume of material filling the spring pipe suffered progressive loss. The process briefly outlined above accounts not only for the localized subsidence of the spring basin but also for the difference in the grade of clastic material found in the filling of the King's Spring and the Cross Bath Spring basins. The flow of the thermal water in the Cross Bath Spring is less than 10% of that of the King's Spring. Both springs have tapering or conical spring pipes with infillings of Quaternary sediments, but the material beneath the King's Spring is mainly coarse gravel while that of the Cross Bath spring basin is dominantly composed of fine-grained silty, sandy or carbonaceous deposits.

Following the close of the Roman occupation, the buildings of the thermae at *Aquae Sulis* collapsed. The destruction of the strong well-founded structure enclosing the King's Spring is something of a mystery as there is nothing to suggest that subsidence of the foundations of the Roman Reservoir took place, though the foundations of the adjacent Roman Temple do show some evidence of movement, including lateral shift. Flooding or seismic shock may have played a part but whatever the cause of its collapse, the roof and upper walls of the Roman building were broken down and buried the floor and lower walls of the reservoir. Many of the masses of brickwork were very large and the debris, partly buried by alluvial deposits, partly protruding above the surface of the flood plain, was invaded by the rising thermal water. Some of the thermal water may have continued, at least for a time, to find its way into the Great Roman Bath and the Roman Drain. A substantial part eventually adopted a southerly route roughly coinciding with the original gully leading to the Avon in Iron Age times (Fig. 7.4). The debris deposited in and around the spring basins must have had an important damping-down effect on the springs, causing the formation of many small distributaries and expansion of the area invaded by the thermal water. With the exception of Wood's Hot Bath the baths erected in the 16th–18th Centuries were mainly situated above the hot springs, but some, like the Kingston Baths in the 19th Century made use of thermal water which had leaked away through the layer of broken and collapsed Roman masonry. The Roman debris also trapped quantities of mud, silt and fine granular material and must eventually have exerted a considerable influence in checking erosion caused by the emerging thermal waters (Fig. 7.3).

These conditions prevailed almost unchanged until the end of the 19th Century when Major C. E. Davis undertook the task of excavating the Roman thermae. In carrying out this work, Davis was obliged to remove a large part of the broken masonry and debris which had served to protect the foundations of the post-Roman buildings. Thus the damage to the 18th Century Pump Room probably took place from about 1880 onwards and, like the damage to the floor of the King's Bath (late 19th Century), was restricted to the area immediately above the spring pipe. The remainder of the Pump Room and the walls of the Roman Reservoir and the mediaeval walls erected on the Roman foundation, are built on or above Lower Lias Clay or on the alluvial deposits surrounding the spring pipe. These parts of the Pump Room were therefore unaffected by the increased subsidence of the floor of the spring basin after 1880.

During the drilling of the deeper boreholes in the basin of the King's Spring there was some evidence for the presence of water-filled cavities in the clay-bound sediments filling the spring pipe and considerable anxiety was felt as to the possible collapse of the material while drilling was in progress. As a precautionary measure, all the trial boreholes were, therefore, backfilled with pea gravel or, if lined with steel tubes, were immediately cemented up when drilling was completed.

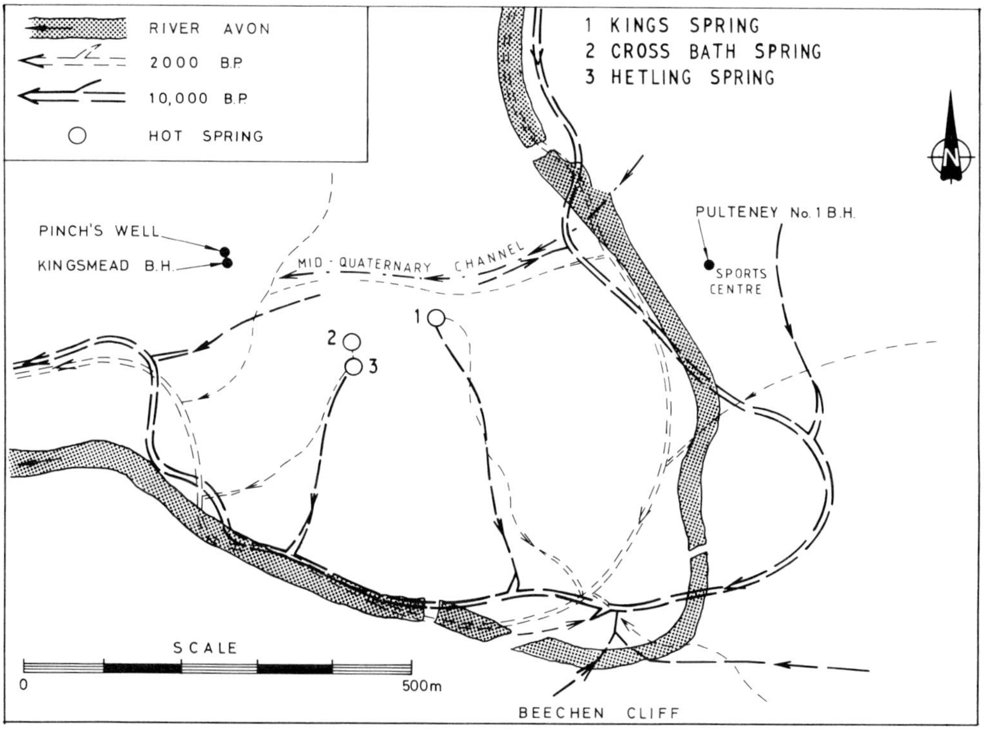

Fig. 7.4. Changes in the course of the Avon at Bath.

The results of the trial drilling were of considerable significance and formed a basis for all subsequent work. The presence of the deep drift-filled spring pipe, and absence of *N. fowleri* in the unoxidized thermal water were two most important discoveries. On the other hand, it became evident that there was no way in which an acceptable method of extracting sediment-free thermal water could be devised using vertical boreholes sunk in the floor of the King's Spring. Any attempt at extracting thermal water by means of boreholes would, therefore, have to prove the water at depth in some structurally stable medium, and recover it without the aid of a filtration plant or special linings requiring frequent cleaning.

Moreover, the temporary linings used in the King's Bath trial boreholes were of normal mild steel type employed primarily because of their strength, and it is known that these are liable to chemical attack by the thermal water. Non-corrodible plastic linings would therefore be required in order to construct permanent tube wells.

The Kingsmead and Sports Centre (Pulteney No. 1) boreholes (1981–1982)

In the next phase of the work, the Wessex Water Authority was able to provide welcome assistance by drilling, under contract, two trial boreholes to prove the thermal water at depth. One of these was sunk west of the King's Spring in 1981 in the public car park near Kingsmead Square, the second (Pulteney No. 1 Borehole) sited east of the King's Spring, was sunk at the Sports Centre in 1982.

Andrews *et al.* (1982), concluded that the red Triassic rocks beneath the Avon valley are an important aquifer in which the thermal water is circulating in significant quantities. Thermal water is also circulating in the underlying Carboniferous Limestone, pointing to the contact of the Triassic and Carboniferous rocks as being a critical zone for testing purposes. It was also required to determine whether the presence of an open fissure system involves any danger of the introduction of oxygenated water, carrying *Naegleria*, as a result of changes in pressure induced by groundwater movements associated with drilling.

To obtain biologically "clean" sediment-free thermal water by means of a vertical borehole penetrating a structurally stable fissured aquifer, a large number of criteria have to be satisfied. Kingsmead Street was chosen as being an open site some 300 m WNW of the King's Spring, where thermal water was encountered in Pinch's Well in 1835. Armstrong's account (1838) of the deepening and closing of Pinch's Well is the best documented and most interestingly written account of its kind, although there have been a number of other instances of thermal water having been encountered in wells, shafts and borings in and around Bath, dating from the sinking of Batheaston Shaft in 1804–1805 up to recent times. The last influx of this kind took place in Lower Lias Clay in Shaft 12A [74496444] south of the Lower Bristol Road during the construction of the southern Sewer Tunnel in 1979.

The house occupied by Pinch's Brewery (29 Kingsmead Street) was destroyed during the war of 1939–1945 and its site now lies in the northern part of Kingsmead Car Park (Fig. 7.5). The geological section as published by Armstrong (1838, p 3) has been reproduced in a number of later publications (Moore 1867; Woodward 1876; Richardson 1928) usually without any explanation of the anomalous thicknesses which it records. Either the driller's memory was at fault, or there were unsuspected structural complications which were not recognized at the time. Armstrong found Mr Stone, the driller, to be a very reliable person and in the light of the evidence now available from the Kingsmead Borehole, it is clear that the reason for the apparent discrepancy in the log of the borehole sunk in the bottom of Pinch's Well is that the boring penetrated one or more faults (Fig. 7.5).

There have been many changes in level and road layout since Kingsmead Street was partly destroyed by bombing in 1939–1945. Pinch's Well was sunk at the Brewery floor level which is thought to be approximately the same as that of Kingsmead Car Park. This level as determined by Bath City Engineer's Department is about 19.36 m A.O.D., almost identical with that of the top of the Kingsmead Borehole (19.5 m) situated about 10 m SE of the presumed site of Pinch's Well. The Kingsmead Borehole was cored throughout and no evidence of faulting was found, either in the Lower Lias or in the underlying Cotham Beds and Westbury Beds. It seems likely that a normal fault reaches the surface roughly midway between Pinch's Well and the Kingsmead Borehole and that other faults are present in this area. The presence of these structures may explain the original siting of the old dug well at Pinch's Brewery, the well having been situated where there was a small seepage or boggy area where water was issuing from faulted Lower Lias Clay.

It has been suggested that the water which flooded Pinch's Well was derived from a Triassic aquifer, but it appears that the red material in which the thermal water was struck, though possibly of Triassic origin, was situated in a fault. It is true that a substantial flow of thermal water was encountered in the red and green Triassic rocks beneath the black shales of the Westbury Beds in the Kingsmead Borehole but these rocks were themselves strongly fissured, the boring having penetrated the Triassic mudstones and sandstones at a point where a belt of unmineralized fissures is developed. Elsewhere, as in the Cross Bath Inclined Borehole, petrologically similar Triassic rocks have been encountered in a massive unfissured condition and have yielded not a drop of thermal water. In general the Cotham Beds and Westbury Beds seldom yield water and are effective aquicludes, due to their fine grain and the ease with which they break down to give clay or silt.

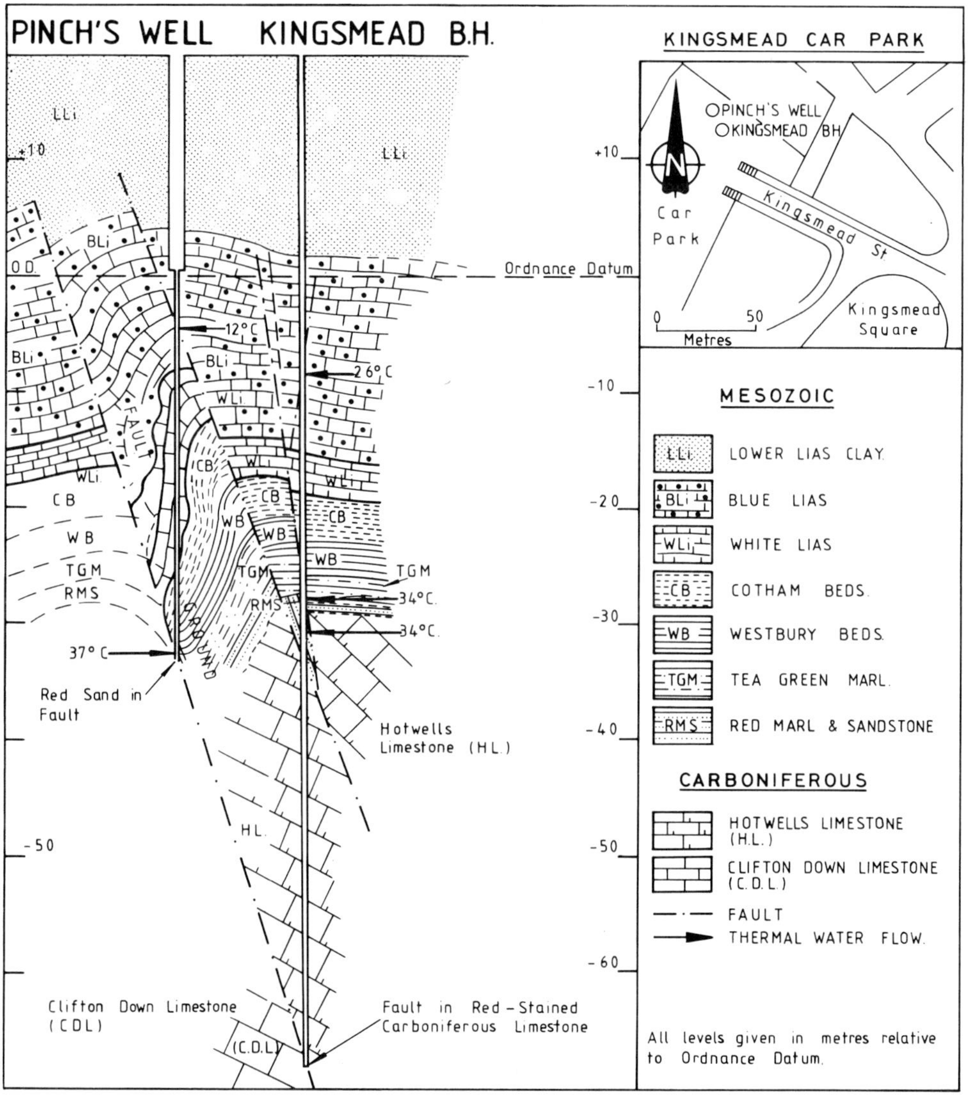

Fig. 7.5. Section through Pinch's Well and the Kingsmead Borehole. The structure may represent the northern side of a trough fault.

Prior to the sinking of the Kingsmead Borehole, it had been supposed that the pre-Triassic rocks beneath the centre of Bath are of Lower Carboniferous age, though there was little positive evidence for this. It has long been known that Coal Measures are present beneath the western part of the City at Twerton (Anstie 1873), and productive Middle Coal Measures were formerly worked in a Variscan tectonic belt at Pennyquick Bottom [ST 71476453]. Other Palaeozoic rocks, probably of Lower Carboniferous age were proved in Batheaston Shaft [ST 78186773] and rocks of Lower Old Red Sandstone age were encountered in the British Geological Survey's Hamswell Borehole [734708] about 6.4 km north of Bath (Cave 1963).

At Kingsmead the borehole proved the base of the White Lias at 18.7 m below O.D. and penetrated the base of the Triassic rocks resting on Hotwells Limestone at a depth of 30.7 m below O.D. Contact of the Hotwells Limestone and Clifton Down Limestone was proved at 65.2 m below O.D., showing that the two formations which have a dip of about 45° are in normal (i.e. uninverted) sequence. The direction of dip could not be ascertained but from such geological evidence as is available, the dip of the Carboniferous Limestone is likely to have an appreciable southerly component.

Hydrogeological evidence obtained by monitoring the flow from the Kingsmead Borehole is discussed by Dr Stanton (Chapter 8). The thermal water flow, chemistry, radioactivity and other properties were duly recorded. The thermal water was tested regularly over a period of more than one year but no evidence of the presence of *N. fowleri* was found. In view of the very large flow which took place in the first months of the test period and the lack of any sign of contamination there seems to be little danger of *N. fowleri* finding its way into the thermal water via the higher parts of the recharge mound. The Kingsmead Borehole, therefore, demonstrated that biologically "clean", sediment-free thermal water can be obtained by drilling fissured Triassic rocks underlying the black shales of the Westbury Beds. At higher stratigraphical levels thermal water is circulating with less freedom in the faulted and fissured limestones of the White and Blue Lias. Both these formations include beds of marl, clay or shale which tend to break down and produce turbid water. These weaknesses may have been engendered to some extent by long contact with thermal water, and, as was later proved at the Cross Bath, the stability of the basal beds of the Lower Lias Clay has been seriously affected by a combination of Quaternary structural deformation and the penetration of thermal water.

Although the Kingsmead Borehole continued in Carboniferous Limestone to a depth of 112.2 m and some fissuring was proved, the presence of any substantial quantity of thermal water in the limestone could not be established. A small flow was detected by means of a closed circuit television camera at a depth of about 50.2 m but this can hardly be regarded as more than a minor addition to the main flow. In view of the potential difficulty and cost of searching for water-bearing fissures in the Carboniferous Limestone beneath a thick Mesozoic cover, the borehole was terminated at a depth of 112.2 m. The Kingsmead Borehole is lined with 113 mm steel lining tubes to a depth of 17.7 m and 98 mm casing to 47 m. It was not designed as a permanent source of thermal water for Spa use but is important for monitoring purposes and is fitted with a pressure gauge and flowmeter.

The excessively large flow of water recovered from the Kingsmead Borehole in the early stages of testing was not maintained and the relationship between the yield of the borehole and the rates of flow recorded in the other boreholes and springs shows that there is a limit to the thermal water resources available in the central area of Bath. As Dr Stanton demonstrates, however (Chapter 8), the volume of water is not necessarily the best indication of productivity, thermal yield being a much better guide to the magnitude of thermal water resources.

The Sports Centre (Pulteney No 1) Borehole (1982)

Soon after completion of the Kingsmead Borehole the possibility of using thermal water to supply the swimming bath at the Sports Centre was considered by Bath City Council. The proposed borehole was sited at approximately the same distance east of the King's Spring as the Kingsmead Borehole lies to the west of it. It also appears likely that the Sports Centre Borehole is situated on the same fault system. The temperature of the thermal water at Kingsmead is about 97°F (36°C), i.e. about 17°F (9.4°C) lower than the water of the King's Spring. It therefore seemed reasonable to hope that the water at the Sports Centre would have a temperature in excess of 80°F (26.6°C), this being the temperature required for the baths at the Sports Centre.

The borehole was made under contract by Wessex Water Authority and was eventually lined with 113 mm diameter steel lining tubes to a depth of 36.5 m. At 50.60 m a very strong flow of thermal water was encountered in a fault cutting the Triassic rocks. A section showing the structure of the rocks between Kingsmead and Pulteney No 1 Borehole is seen in Fig. 7.6. The maximum temperature recorded in this hole was 70°F (21°C), a disappointing figure, being about 10°F (5.5°C) lower than was required for supplying the swimming baths. An unexpected result was the thickness of the coarse cavernous Dolomitic Conglomerate which was proved at depths of 62–75.5 m. The top of the red and green Triassic marl which overlies it was not proved in the hole, but at a rough estimate, the thickness of the Dolomitic Conglomerate and red sandy mudstone is likely to be at least 22 m. This result may, at first sight be surprising, as there appears to be only a very thin deposit of red Triassic sandstone at Kingsmead and though this includes a very small amount of conglomeratic material, most of the basal Triassic deposits west of Kingsmead are of a sandy or muddy character. The Dolomitic Conglomerate makes its appearance east of the centre of Bath and it thickens towards Bathwick and Batheaston where it has a total thickness of about 35 m in the Batheaston Coal Shaft.

The chemistry of the water recovered from the Sports Centre Borehole is similar in many respects to that of Batheaston Shaft (Chapter 8). Since the abstraction of water at the Sports Centre and at the Batheaston Shaft affects the flow at the hot springs we may assume that fissured cavernous Dolomitic Conglomerate is acting as a balancing reservoir beneath the Avon valley east of Bath. In the Box ("Lucknam") Borehole, 6 km northeast of Batheaston, black shales of the Westbury Beds rest unconformably on Carboniferous Limestone. The concealed eastern margin of the Dolomitic Conglomerate must, therefore, lie between Batheaston and Box. The discovery of this reserve of low-temperature thermal water in the Dolomitic Conglomerate has a bearing on the maintenance of the hydrostatic pressure of the hot springs as discussed below.

Completion of the Sports Centre Borehole in August 1982 marked the end of the initial investigation preceding the construction of a specially designed borehole or tube well to recover a continuous supply of clean thermal water for use in the Spa. The slightly brackish or saline taste of the water from the Sports Centre Borehole renders it useless for drinking even though it is free of *N. fowleri*. For the same reason it is doubtful whether water of potable composition can be recovered from the Dolomitic Conglomerate beneath the Avon valley between Bath and Box.

The Stall Street Inclined Borehole (1983)

At the time of the closure of the Spa facilities in 1978, it had already been recognized by the City Engineer that the existing arrangements for pumping and distributing the thermal

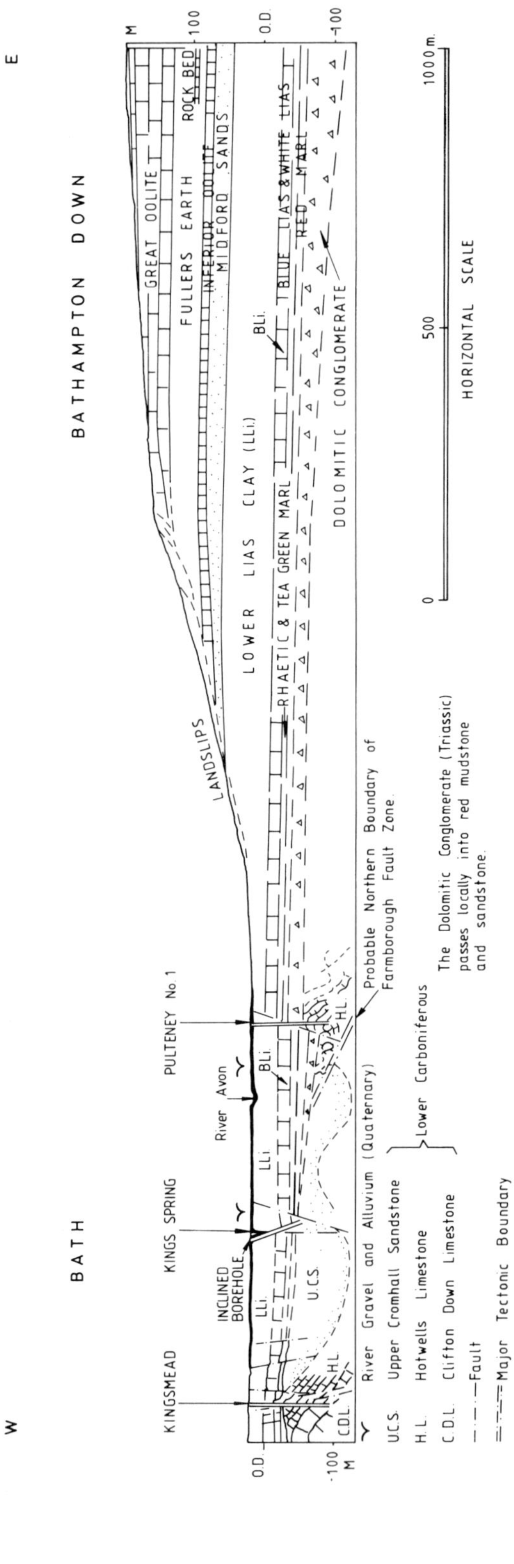

Fig. 7.6. Section showing the geological structure of the Avon valley between Bath and Bathampton Down.

water from the King's Spring were unsatisfactory. The occurrence of *N. fowleri* in the spring basin and thermal installations, notably in the high level storage tank in Swallow Street, now showed that there was also a danger to public health. Under the system then existing the thermal water dissipated its kinetic and potential energy by flowing into the reservoir above the King's Spring where the temperature fell slightly and oxidation took place thus creating a favourable environment for infestation by *Naegleria*. The water was then pumped to the Swallow Street tank where further oxidation took place, followed by distribution of the oxidized and (as we now know) contaminated water to various installations in the City. Whatever method of dealing with the thermal water was to be used in the future, it was clearly imperative to convey it directly from its source to the places where it was to be used without permitting oxidation to take place. Preferably this should be done using the natural artesian pressure, but if boosting should be needed, then this must be by submersible pump or similar method not involving exposure of the thermal water to air prior to its use.

Recovery of the thermal water by tapping it at depth beneath the King's Spring raised a number of geological and practical problems and several alternative methods were considered in the light of the available information. An inclined borehole sited near the Grand Pump Room offered the best hope of success, the balance being in favour of a site near the west side of the Pump Room where there was adequate space for setting up a drilling rig in Stall Street. By good fortune it was found that a tunnel linking the Pump Room basement with that of the former Treatment Centre in Bath Street passed close to the preferred site. This subterranean passage had, on its north side, a disused cellar which was adapted to house the well-head. By this means it was possible to accommodate the controls and monitoring equipment and the terminal pipework in a ready-made underground chamber connected to a tunnel leading directly to the Spa development area with considerable easement of the cost of controlling and transmitting the thermal water.

Seen in plan the spring pipe of the King's Spring is oval or elliptical in shape (Fig. 7.7). This may be taken as indicating that the trend of the main fissure in the Palaeozoic rocks and the overlying formations (including the Blue Lias) coincide with that of the long axis of the ellipse. The depth to the Blue Lias could be calculated with some accuracy on the north side of the fissure and the borehole was therefore oriented to intersect the assumed position of the north side of the fissure at a relatively low angle. This increased the length of the transection, giving the best chance of encountering suitable rocks in which to terminate the borehole.

An inclination of 32° relative to the vertical was adopted in order to pass beneath the estimated position of the bottom of the gravel-filled spring pipe and to ensure penetration of the hard Palaeozoic rocks within the limits of the fissure belt. Although the average level of the Triassic–Palaeozoic contact could be estimated, there were other elements of uncertainty, notably the width of the fissured belt and inclination of the main fissure walls within the Palaeozoic basement. Other factors influencing the design of the well included the need to use linings which would not be affected by the aggressive thermal water and careful grouting of the linings to ensure that the thermal water did not come into contact with air or with cold groundwater carrying a large amount of dissolved oxygen. It was, therefore, decided to use Aquiline linings and to grout all linings down to the base of the Blue Lias and, if necessary, to the depth at which the fissure wall was encountered.

Drilling of the Stall Street Inclined Borehole (Ground level 24.01 m A.O.D.) was carried out by Drillsure Ltd of Warwick and commenced in April 1983 (Plate 7). Slight deviation of the inclination of the hole was recorded, but the average inclination from the vertical is 32°. The top 8.15 m (measured downhole) was drilled through made ground, including a small section of Roman walling. This part of the hole was lined with 400 mm diameter mild

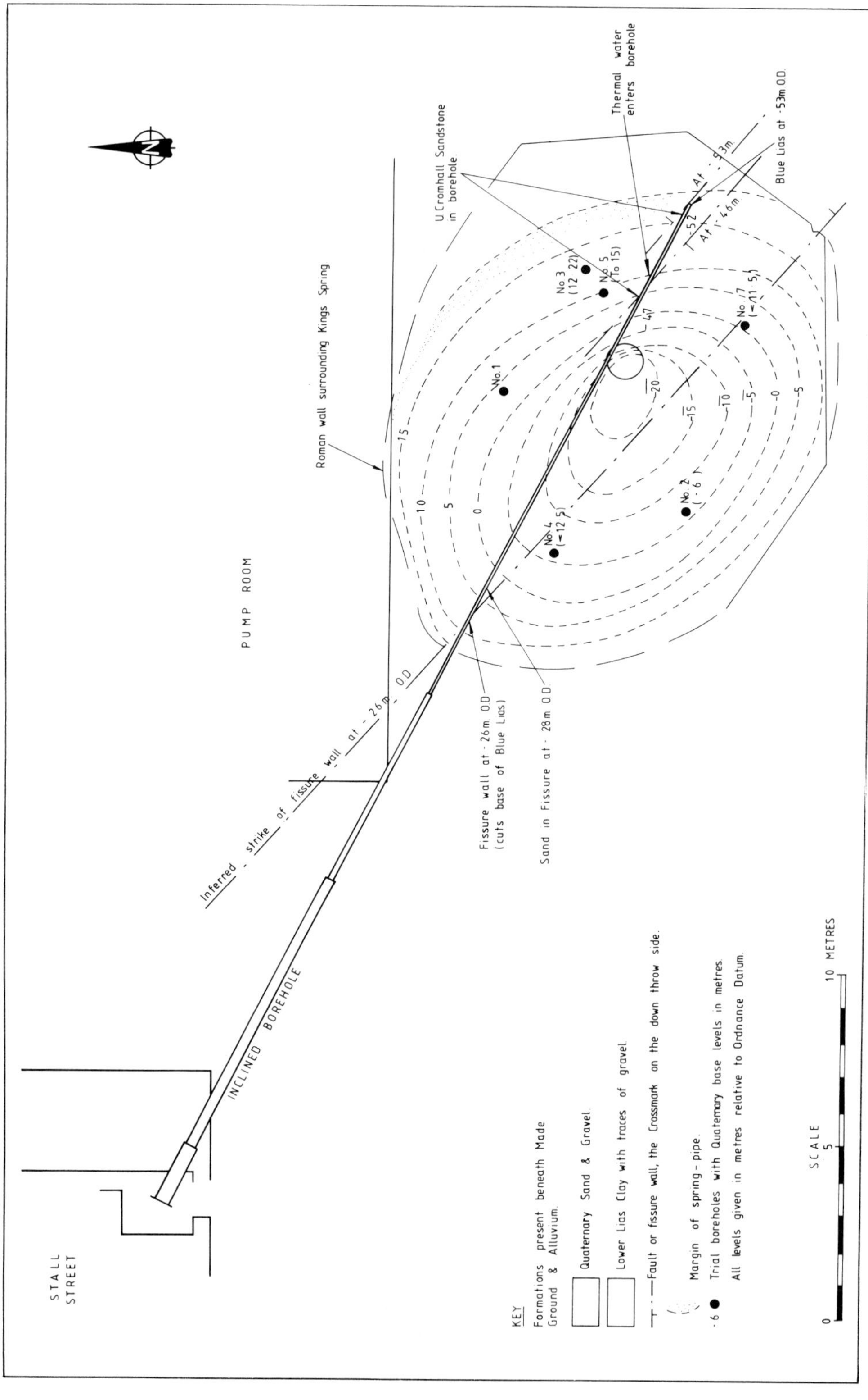

Fig. 7.7. Stall Street Inclined Borehole and plan of the spring pipe of the King's Spring.

steel tubing which was grouted in to facilitate the use of a blow-out preventer and to ensure the stability of the top of the hole and the foundations of the Pump Room (Plate 8).

No evidence of the presence of Alluvium or of a natural soil profile was observed beneath the Roman occupation layer. The top of the Lower Lias Clay was represented by remoulded material so the ground may have been excavated before the Roman foundations were laid. At a depth (downhole) of 34.2 m the borehole penetrated the base of the Lower Lias Clay and entered the Blue Lias (Fig. 7.8). The upper part of the hole was then lined out with 250 mm diameter Aquiline lining tubes to 31.53 m and these were fully grouted in before drilling of the Blue Lias was resumed using a blow-out preventer. At a downhole depth of 55.54 m, the drill penetrated the north face of the main fissure cutting the contact of the Blue and White Lias. Here the temperature and flow of the thermal water increased markedly. From the first appearance of thermal water with a temperature of 63°F (17°C) at a downhole depth of 31.65 m public interest in the work was considerable as jets of hot water and vapour issuing from the opening in the Stall Street pavement indicated that progress was being made (Plate 10). Houses surrounding the drilling compound had viewing spaces provided for public use and a placard was attached to the hoarding giving the temperature of the thermal water as the boring proceeded. The temperatures recorded at depth are given in Table 7.1.

Table 7.1.

Date (1983)	Temperature (°C)	(°F)	Downhole depth (m)	Vertical depth (m)	Formation
May 5	17	63	31.65	29.0	Lower Lias Clay
May 12	32.2	90	36.19	33.21	Blue Lias
May 13	44.4	112	56.00	51.31	Blue Lias
May 14	45	113	56.00	51.31	Blue Lias
May 18	45	113	77.46	70.51	Upper Cromhall Sandstone (in fissured belt)

Drilling established that the north wall of the main fissure extends through the Blue Lias into the White Lias, and, presumably, continues into the Triassic and Carboniferous rocks below. The south wall of the fissure was not proved and the width of this structure is unknown. For reasons discussed below it is thought that the King's Spring may be rising from a nodal position where a post-Jurassic fault belt intersecting a Variscan structure is cut by one or more open Quaternary fissures. From about 55 m (downhole) to about 75 m (downhole) the drill passed through masses of broken White Lias and Blue Lias limestone with a very thin vertical layer of calcreted sandy material at about 56 m. This may mark the downward end of the sandy deposits of the spring pipe but no gravel was encountered. From this point downwards the borehole proved slumped greenish-grey silt and clay (Cotham Beds) and black shaly mudstone (Westbury Beds) all greatly distorted and broken. At greater depths loose debris, mainly derived from Blue Lias and White Lias but with some silt, shale and greenish grey mudstone, extended to 77.46 m (downhole). Very large voids containing nothing but thermal water and gas were encountered; possibly these voids were roofed or enclosed by collapsed Lower Lias Clay as no Middle Jurassic limestone fragments or gravel were seen. The absence of red Triassic rocks is also noteworthy as it is possible that in central Bath the transgressive black shales of the Westbury Beds may have overstepped the older green and red Triassic rocks, just as they have at the Grandmother's Rock [70887120] and Wick inliers northwest of Bath and at the

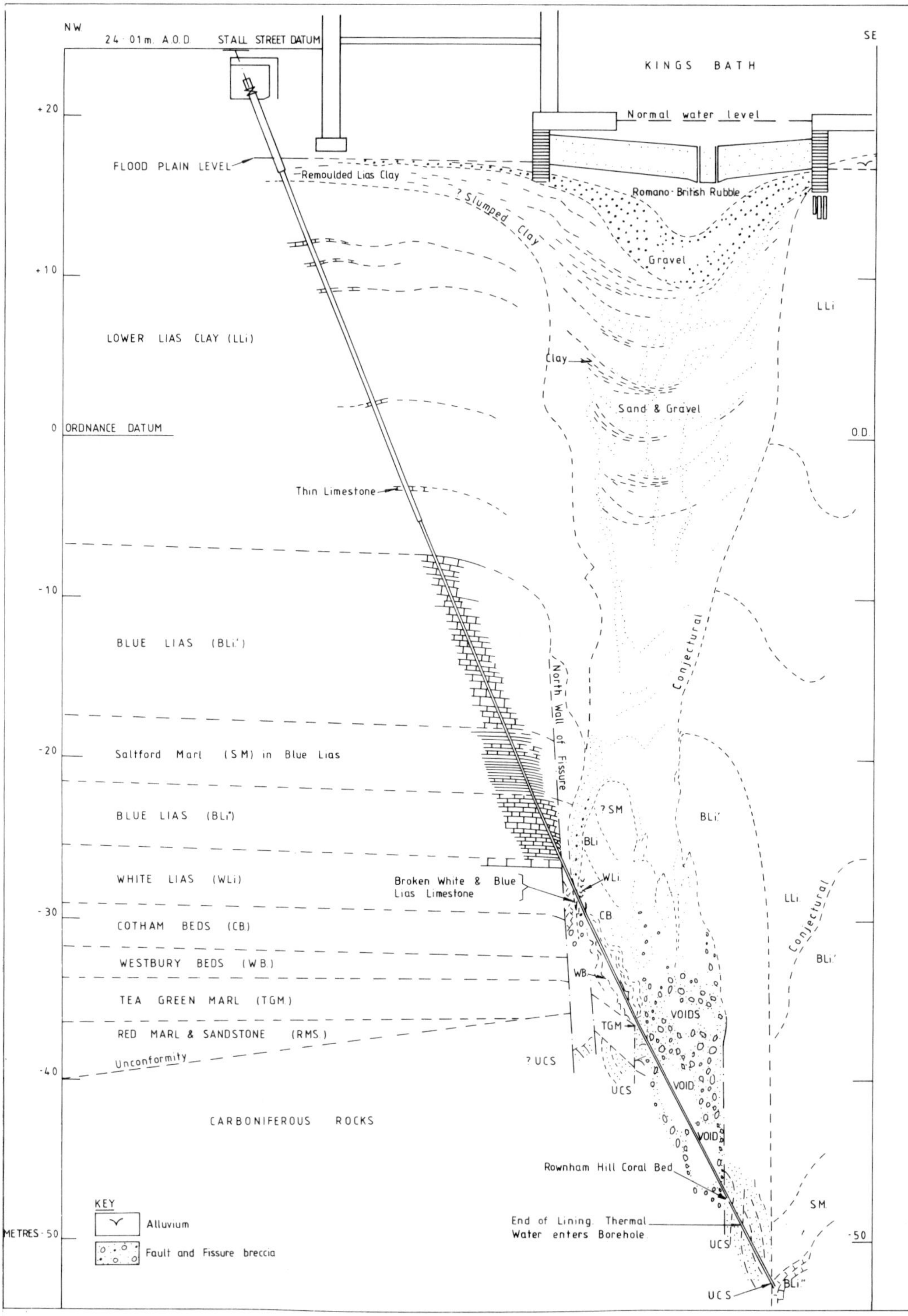

Fig. 7.8. Section through the King's Spring and Stall Street Inclined Borehole.

"Lucknam" Borehole in the valley of the Box Brook, 9.6 km northeast of the City (Fig. 13). At a downhole depth of 77.46 m the drill entered a mass of intensely hard reddish quartzitic and feebly calcareous sandstone. This appears to be the upper part of a faulted and fractured mass of steeply dipping rock with debris of White and Blue Lias limestones mantling its crest and extending down to unknown depths on its two sides. Being intensely hard and siliceous, the sandstone was difficult to drill. Had this very hard material been a loose mass or boulder it would certainly have rotated or moved while drilling was in progress. Stratal dips are very steep, almost vertical in places, but the relationship between the dip and the fracture planes cutting the rock suggests that the quartzitic sandstone is part of a steeply dipping folded and faulted mass cut by geologically recent unmineralized fracture planes. The lithology is identical with that of the Upper Cromhall Sandstone, an arenaceous and muddy deltaic facies of the Carboniferous Limestone Series which is strongly developed in the Bristol Coalfield and at Chepstow and in the Forest of Dean (Kellaway & Welch 1955).

Preserved in a feebly calcareous band in the sandstone, were the remains of a coral which Mr M. Mitchell has identified as *Orionastraea*. This genus is characteristic of the Rownham Hill Coral Bed of the Avon Gorge and the Castle Wood Limestone of Wick (Kellaway 1967). Its other known occurrences all lie in comparable situations where the upper part of the Carboniferous Limestone Series is seen at the margin of the coalfield. The Lower Carboniferous rocks proved in the Stall Street Inclined Borehole therefore represent a higher stratigraphical level than the highest rocks proved in the Kingsmead Borehole and many hundreds of feet higher in the Carboniferous succession than the Clifton Down Limestone proved in the Sports Centre Borehole.

From a practical point of view the discovery of open fissures carrying thermal water in an almost insoluble mass of hard quartzitic sandstone was important since it was possible to terminate the open-ended lining of the borehole within the sandstone. Drilling continued to about 84 m (downhole) when the hole penetrated another faulted contact and a short length of Blue Lias limestone was recovered. At this point it was decided to discontinue drilling and the hole was lined throughout with a continuous string of 100 mm diameter Aquiline lining tubes. The bottom of the hole, drilled in almost indestructible quartzitic sandstone, has been left open to ease the passage of the thermal water from the fractured and faulted Carboniferous rocks into the lined well (Fig. 7.8).

Continuous testing for more than a year showed that the thermal water, which has been deflected from the rising waters feeding the King's Spring, is free of *N. fowleri*. It is clear and carries very little sediment or granular material, has a temperature averaging about 46°C and a hydrostatic head of 27 m A.O.D. For reasons of safety and in order to maintain a satisfactory flow of thermal water into the Roman Reservoir and Great Roman Bath, the Stall Street Borehole is not permitted to yield its maximum possible artesian flow. It is thought that flows of up to 20 000 gallons per hour could be achieved at the Stall Street well head. This is almost twice the normal daily natural flow of the King's Spring and the recovery of thermal water on this scale over a long period might permanently damage the King's Spring and imperil the stability of its basin. Used wisely, however, the Stall Street Inclined Borehole can be employed to control the flow of the King's Spring and to supply sediment-free biologically clean thermal water to the Pump Room and other installations.

At a late stage in the drilling of the inclined borehole, it was observed that the water in the basin of the King's Spring became discoloured by muddy drill water which was entering the rising thermal water stream. This cloudiness in the King's Spring cleared fairly quickly once drilling ceased. When drilling and pumping were resumed the interval between resumption of pumping and the time of reappearance of the cloudiness gave a rough approximation of the time taken by the muddy water to rise from the drill bit to the spring basin above. This gave an average speed of vertical flow of more than 100 m per hour. Circulation of water by the drill pump may have accounted for part of this velocity but this factor was probably offset by the time taken to detect the onset of cloudiness in the spring basin. If the measured velocity is maintained at depth it would imply that the water is capable of rising through a vertical distance of about 4500 m (*c.* 15 000 ft) in two days. Unless the reservoir of thermal water is immeasurably greater in size than has yet been suspected, this is most unlikely. However, it is obvious that the upward movement of the thermal water is constrained by the presence of the Quaternary sediments (mainly gravel) that have accumulated in the spring pipe and, as will be seen, the grain size of the material in these natural "bore-holes" (as William Smith called them) does seem to be related, at least in part, to their diameter and yield. From this it follows that it is likely that the original breakthrough of the water rising from tectonically fissured and fractured rocks may well have been due to the formation of hot springs of much greater volume and erosive capacity than those which now exist. Constriction of the exit by infillings of sand and gravel has led to the creation of high-velocity streams or jets of thermal water with a reduction in the overall rate of emission.

The Cross Bath and Hetling Springs

Although the Cross Bath Spring and Hetling Spring are generally supposed to account for only 10 per cent of the total output of the hot springs at Bath, there has always been some doubt as to their relative strength and their relationship to the King's Spring. Old records are difficult to interpret, partly because of the obscure units (e.g. "Hogsheads") in which they are commonly expressed and partly because of uncertainty about water and ground surface levels which cannot always be precisely related to Ordnance Datum.

Both the Hetling Spring and the Cross Bath Spring were affected by the deepening of Pinch's Well (Armstrong 1838), but this does not necessarily indicate that there is any direct connection between the two, even though the springs are only about 35 m apart. The temperature of the Hetling Spring (48°C) is higher than the Cross Bath Spring (41–42°C) and marginally higher than that of the King's Spring (46 or 46.5°C). This indicates that, small though it is, the Hetling Spring may derive a higher proportion of its water from a deep fissure system, while the Cross Bath Spring (Plate 14) is fed by an indirect route or draws a higher proportion of its water from a shallow, cooler source connected with the recharge mound.

The timing of new investigations in the Cross Bath–Hetling Spring area has been largely controlled by plans to develop Bath Spa and by the fact that the Hetling Spring issues beneath the middle of a public highway and is, therefore, difficult of access. From a geological point of view it would have been preferable to have investigated the Hetling Spring first but this has not been possible and the extent to which leakage and other factors are affecting the flow of this spring cannot yet be assessed.

Cross Bath Inclined Borehole (1986)

Drilling of an inclined borehole beneath the Cross Bath (Fig. 7.9) commenced in February 1987; ground level is 22.17 m A.O.D. and the hole was carried to a depth of 109.2 m downhole. The methods used by Drillsure in constructing the borehole were very much the same as those used for the Stall Street Inclined Borehole as were the Aquiline linings. The evidence used to determine the orientation (175° Mag.) and inclination (74°) of the hole was less complete than for the Stall Street well but to have delayed sinking of the borehole could have held up progress in the restoration of the Cross Bath. It was hoped that the inclined borehole might be adapted to act as a control for the flow of the Cross Bath spring in much the same way as can be done by the Stall Street Inclined Borehole, but this was not possible, though the results indicated how this control could be achieved at relatively low cost during the course of routine trial drilling in the Cross Bath.

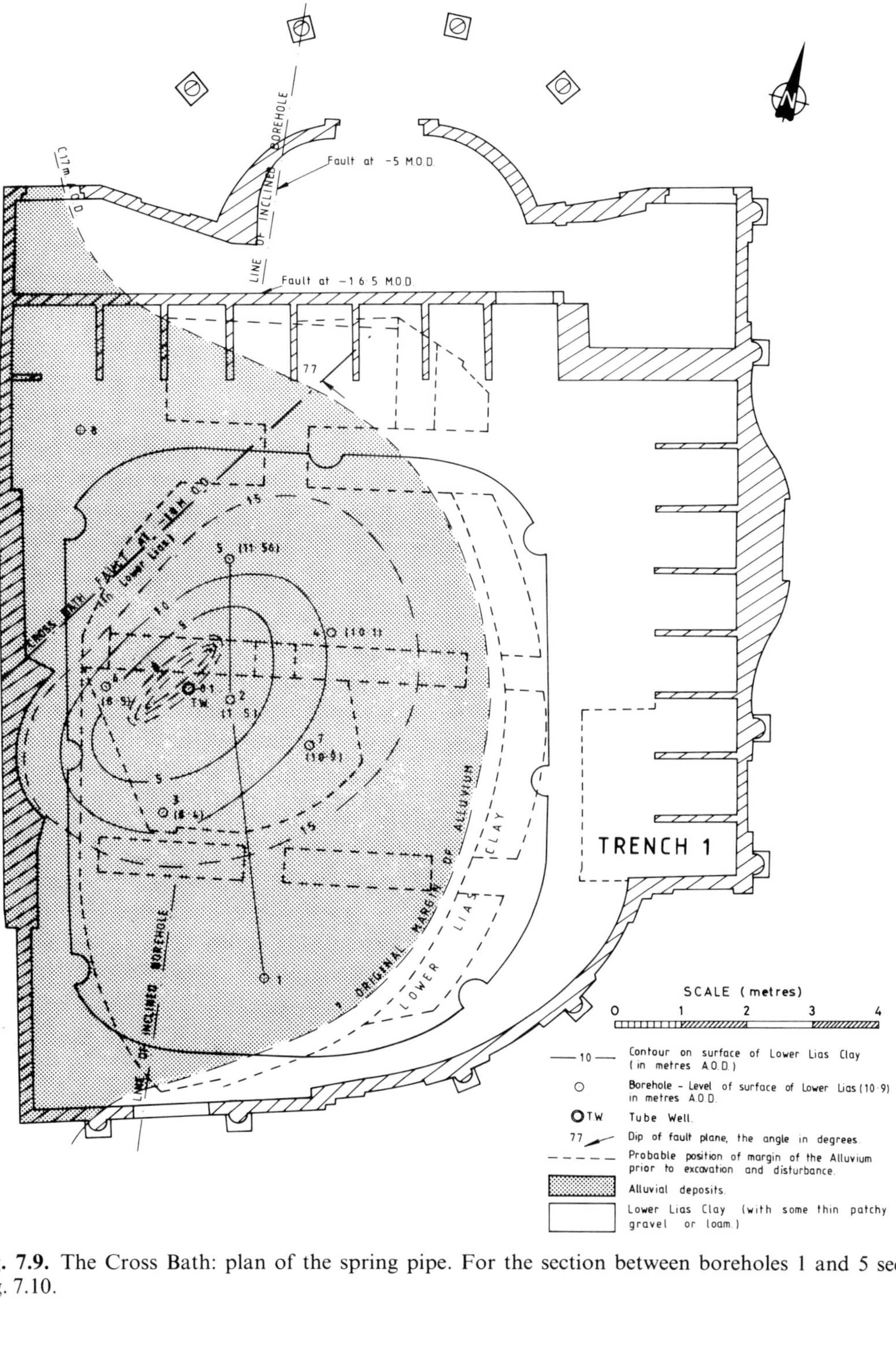

Fig. 7.9. The Cross Bath: plan of the spring pipe. For the section between boreholes 1 and 5 see Fig. 7.10.

In calculating the approximate depth to the top of the Blue Lias the average level for the Cross Bath area was taken as about −26 m A.O.D. This proved to be correct though there is some local variation due to small scale faulting (Fig. 7.10). At a downhole depth of 28.46 m the borehole entered faulted Lower Lias Clay in contact with the upper part of the Blue Lias. Here a small flow of thermal water was encountered with a temperature of 78°F (25.5°C). However, the clays were unusually weak and structurally unstable and it was essential to line out the hole in order to preserve it. It is highly probable that this thermal water is connected with that of the Cross Bath Spring but its temperature is too low for it to be the principal source. The presence of water in faulted and disturbed Lower Lias Clay immediately above the Blue Lias appears to be a feature of the Cross Bath area. Much of the Lower Lias Clay shows evidence of local disturbance, possibly due to Quaternary structural movements in the valley floor.

No other measurable influx of thermal water was detected in the upper part of the Blue Lias and at a downhole depth of 36.25 m the boring entered a series of sticky clays with small limestone nodules, representing the Saltford shales (Donovan & Kellaway 1984, p 12). At 40.54 m there was evidence of structural disturbance, though no thermal water was encountered. The thickness of the Saltford Shales as proved in the inclined borehole is almost double the normal thickness. It therefore appears that a fault with a displacement of about 3.5 m cuts the Saltford Shales which act as a minor aquiclude within the Blue Lias.

The aquicludes in this succession tend to be effective as long as any intersecting faults are mineralized or where the displacement of the fault is appreciably less than the vertical thickness of the aquiclude. The fault displacing the Saltford Shales is not sufficiently large to break the continuity of the aquiclude and no thermal water was recovered, though the structure may be one along which thermal water is travelling to the Cross Bath Spring. Only a very slight lateral variation in the amount of throw of the fault would be sufficient to allow the thermal water to pass (Fig. 7.10). Below 44.17 m, the boring proved a normal succession in the basal Jurassic and uppermost Triassic rocks. Almost 100% recovery was attained in the boring, showing that the strata are in regular order with low dips and no evidence of faulting or fissuring. The black shales of the Westbury Beds are firm and the harder shales retain their original fissility while the Tea Green Marl—a stratum from which thermal water issued with some force at Kingsmead—proved to be firm and dry. Red and green mudstone (Mercia mudstone) passing down into hard calcareous sandstone with a basal bed of calcareous breccia form a deposit of hard massive Triassic rocks resting on Carboniferous Limestone at 40 m below O.D.

The condition of these red rocks, which, if fissured, form a Triassic aquifer underlying the impervious black shales of the Westbury Beds demonstrates that the intensity and continuity of fracturing and fissuring and the absence or presence of mineralization of the joints and fissures which are the critical factors controlling the transmission of thermal water through these rocks. The waterless red Triassic rocks proved in the Cross Bath Inclined Borehole do not differ significantly from those yielding thermal water farther west in the Kingsmead and Bath Brewery Boreholes. The old Bath Brewery Borehole [72546520] is situated in a belt of fissured and faulted rocks including trough gulls and dip-and-fault structures of Quaternary age. The water-bearing Triassic strata proved in the well are almost certain to have been fissured even though this was not recorded in the log.

Beneath the Triassic rocks the Cross Bath Inclined Borehole continued into Hotwells Limestone in which a fissure was encountered at a downhole depth of 72.8 m. This produced a flow of about 240 gallons per hour at 100.5°F (38°C) and an increase to 325 gallons per hour at 104°F (40°C) at 72.8 m downhole on 4th February 1986. By April 11th the flow had increased to 351 gallons per hour with a temperature of 106°F (41°C). The

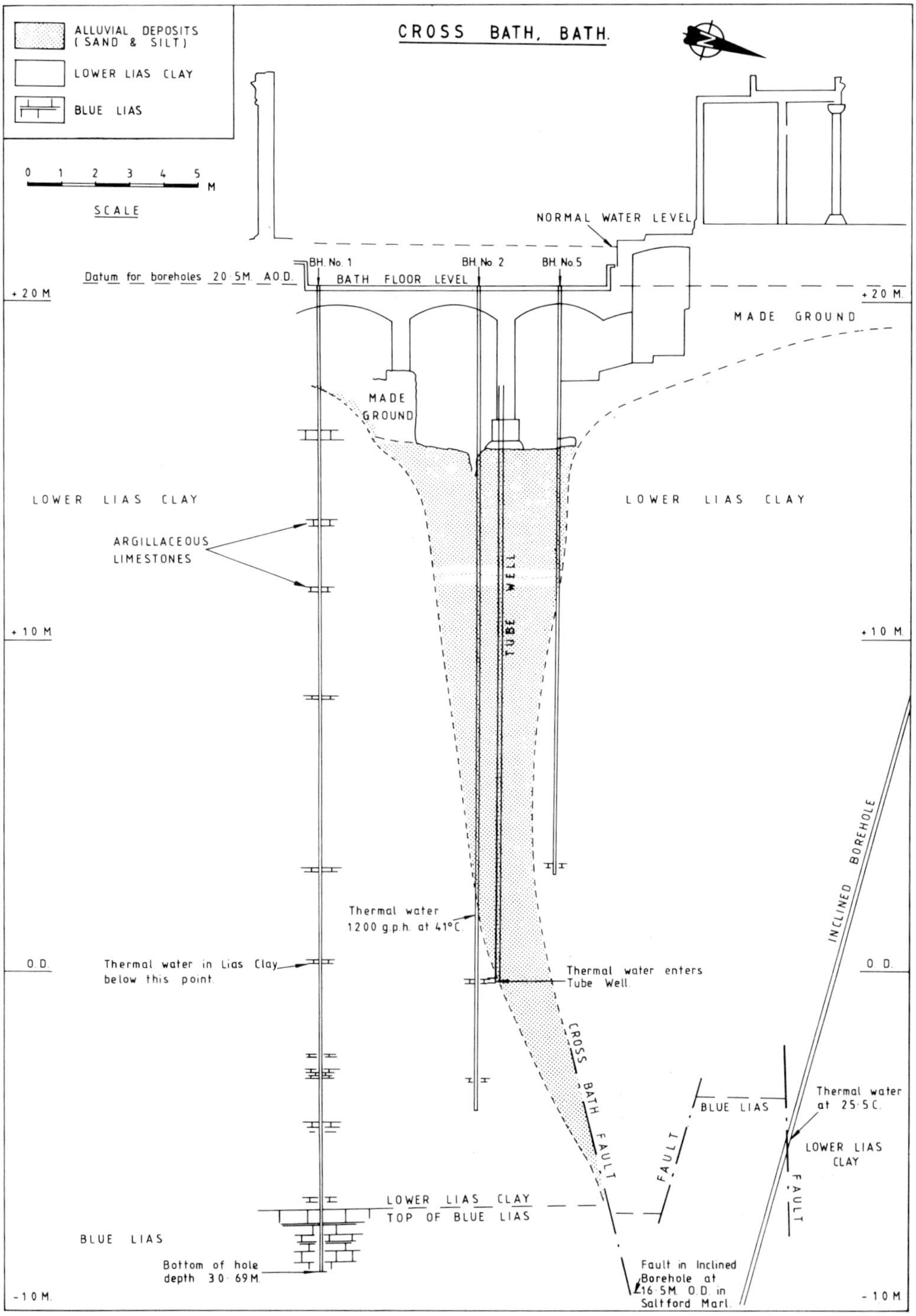

Fig. 7.10. Section through the Cross Bath Spring.

water was tested for a period of more than one year and was found to be free of *N. fowleri*. An analysis of the records of the flow from the inclined borehole and the Cross Bath Spring (prior to drilling of the spring basin in October–November 1987) has disclosed no evidence of a connection between the two sources. It is possible that the thermal water in this fissure in the Carboniferous Limestone is one of the sources of the Hetling Spring which has a temperature of 113°F (48°C), but there is at present no means of confirming this and further advances await the monitoring of the Hetling Spring.

While the Cross Bath Inclined Borehole cannot be used to control the flow of the Cross Bath Spring, the geological results of the boring indicated that it would be advisable to confine the rising thermal water by drilling in the centre of the spring pipe. Trial borings of the Cross Bath Spring basin were necessary for engineering purposes and the siting of these was therefore planned to provide both hydrogeological and engineering data at the same time.

Investigation of the Cross Bath Spring (1987)

A series of eight exploratory boreholes was sunk in the basin of the Cross Bath Spring by Geotechnical Services of Gloucester between 28th September and 30th October 1987. These proved the general form of the upper part of the spring pipe (Fig. 7.10) which is a conical or carrot-shaped structure as seen in cross section, apparently formed by water rising from the faults proved in the Cross Bath Inclined Borehole cutting the Saltford Shales at −16.5 m O.D. The fact that this fault was not carrying thermal water in the inclined borehole has already been referred to above. Of the trial boreholes shown on the plan (Fig. 7.10) Nos 1 and 8 were designed to test the Lower Lias Clay and Blue Lias in the wall of the spring pipe, the remaining holes were sited to prove the form of the spring pipe, the nature of its contents and the flow and temperature of the thermal water.

As will be seen, the conical spring pipe is oval or elliptical in plan with its long axis orientated in a NE–SW direction. The orientation of the faults proved in the Cross Bath Inclined Borehole could not be determined, but the main waterbearing fault is likely to have a NE–SW strike.

Trial Borehole No 1 encountered thermal water in Lower Lias Clay with thin limestone bands at about O.D. level and thermal water flows were encountered down to the top of the Blue Lias. No additional water was found in the upper part of the Blue Lias and the boring was ended at a depth of −30.69 m O.D.

The curious situation in the No 1 Borehole at the Cross Bath where thermal water is present in the lower part of the Lower Lias Clay is paralleled by the occurrence of thermal water in the inclined borehole in precisely the same strata. In Borehole No 1 difficulty was experienced in controlling the very weak waterlogged clay and the hole was eventually cemented up and abandoned. Borehole 2 was sited near the centre of the spring pipe (Fig. 7.10). The boring proved the south side of the spring pipe and encountered thermal water at 1.5 m below O.D. This has a temperature of 41°C, the normal temperature of the Cross Bath, but the initial flow was of the order of 1200 gallons per hour as opposed to 300–400 gallons per hour in the Cross Bath Spring prior to drilling. The flow later stabilized at about 800 gallons per hour, still more than twice the flow of the spring previous to drilling. Sinking of a permanent Aquiline-lined well at the Cross Bath was achieved in 1988, enabling restoration of the Cross Bath to be accomplished with a considerable saving in expense. The water is free of *Naegleria* and suitable for use in the restored bath or in other installations.

Preservation of the hot springs of Bath

Although thermal water from the Stall Street Inclined Borehole has been restored to the Pump Room fountain where it is drunk in small quantities by visitors who are curious to sample its flavour, its use for systematic medicinal treatment has virtually ceased. Restoration of the historic Cross Bath is now underway but even when this has been completed the diminutive size of the bath limits its usefulness.

The volume of clean thermal water now available for use at Stall Street is adequate to supply the Pump Room fountain and to enable restoration of the Spa to be carried out and the addition of the output of the Hetling Spring would provide further resources of clean thermal water. Nevertheless any environmental threat to the total yield of the hot springs is a serious matter and close attention has been given to this possibility, a monitoring programme being a vital element in conservation of the thermal water. It may, therefore, be useful to summarize the steps which have already been taken, or which will be taken in order to protect the hot springs in the future.

1. Long-term monitoring of the temperature and total yield of the hot springs and the boreholes of the centre of Bath. This awaits installation of the necessary equipment.
2. Control of sinking of shafts, wells, boreholes, tunnels and deep excavations within the geographical and depth limits defined in the Avon Act (1979).
3. Test boring and monitoring of trial borings sunk in peripheral areas in order to evaluate the conditions in the re-charge mound and in bodies of groundwater which balance the pressure of the thermal water now issuing from the hot springs.
4. Evaluation of changes in the standing water levels in those areas adjacent to Bath (e.g. Wick and the north eastern Mendips) where quarrying and similar activities could cause loss of pressure in the thermal water. These activities are also of concern to the Wessex Water Authority as they are related to regional groundwater resources affecting public water supplies, and include areas outside the geographical limits described in the Avon Act of 1979.

General theories of the origin of the hot springs

A number of hypotheses have been put forward to explain the origin of the thermal springs at Bath (Crickett 1869). William Smith seems to have favoured a northerly origin (Phillips 1844), deriving the water from the Wick–Hamswell area. Tomkins (1894) proposed a Mendip source, envisaging the Clandown Fault as a possible channel. Nearly all authors have looked to the Carboniferous Limestone (with or without the Dolomitic Conglomerate) as the most likely formation to be actively collecting and transmitting groundwater. The appearance of thermal water springs on the eastern side of the Bristol and Somerset Coalfield at Bath and on the western side at Hotwells, Bristol, gives much support to the view that the water is circulating in the Carboniferous Limestone at depth beneath a thick impervious cover of Millstone Grit (Namurian) and Coal Measures shales and mudstones. Andrews *et al.* (1982) therefore proposed the central Mendips near Emborough as the main source, the standing water level in the Carboniferous Limestone being higher in this area than it is farther east at Whatley and Mells.

The problem remains, however, of explaining the location of the thermal water springs at Hotwells and Clifton which are structurally isolated from the Mendips by the anticlinal structure of Broadfield Down and by the Yanley Fault. The Ashton Park Borehole, three-quarters of a mile south of Hotwells (Kellaway 1967) proved inflows of cold groundwaters at various depths down to 2340 ft below sea level and encountered no thermal water. This suggests an easterly or northerly origin rather than a direct southerly connection with the

Mendips for the Hotwells thermal spring. A revised model dealing with the complex hydrogeological problems is put forward below but whatever view is taken of the origin of the thermal water, continuous observation of water levels in the Carboniferous Limestone uplands will be required. The isotopic age of the thermal water is also relevant to an understanding of this problem (Andrews *et al.* 1982), though as will be shown later, the "age" of the springs in terms of their total life and the "age" of the water (in the sense of the date of its precipitation) may be very different.

Igneous rocks known to exist in the Bristol district and the Mendips include lavas and tuffs, mainly of andesitic or basaltic composition and of Silurian and Carboniferous age. Small quantities of volcanic dust are present in Fuller's Earth (Jurassic) at Bath but, apart from Lundy, there are no post-Variscan granites and nothing to indicate post-Jurassic vulcanicity. It is generally considered that the radioactivity of the Bath waters and their associated gases has originated from the decay of minerals in the Palaeozoic sedimentary rocks (Andrews *et al.* 1982).

In the Massif Central of France, numerous thermal springs exist in a region where the last manifestations of Plio-Pleistocene volcanic activity can still be detected, but many thermal springs are known to be situated on faults or fissures of tectonic origin, irrespective of the presence or absence of volcanic activity. Some are found on actively moving (or recently active) fissure and fault belts, commonly situated in the foothills of upland or mountainous areas. Others emerge in the floors of deep lakes, as at the southern end of Lake Garda in Italy, or beneath the sea and along the shoreline as at Tenakee in S.E. Alaska. Water falling on upland areas enters the rocks and descends to depths of many thousands of feet. Propelled by hydrostatic pressure generated by the difference between a high standing water level in the catchment area and the low level of the resurgence, the water abstracts heat from the rocks through which it migrates. Rising temperature causes expansion and consequently a reduction in viscosity and density which assists the passage of the water through voids and narrow fissure systems and causes it to rise to the surface by way of deeply penetrating open fractures and fissures which are of recent tectonic origin.

The presence of relatively impermeable formations composed primarily of mudstone and shale can exercise a restraining influence on the rise of the thermal water and this may be the reason for the location of the hot springs of Bath and Bristol at the margin of the coalfield (Fig. 7.2). In this case the principal aquiclude is the Coal Measures which achieves a maximum thickness of some 8000–9000 feet. In the Lower and Middle Coal Measures there are thick beds of mudstone and fireclay which soften up and rapidly become decompacted in the presence of water under pressure. Fissures formed in this material may have a relatively short life. On the other hand some of the largest inflows of cold groundwater met with in the Coal Measures at depths of up to 3000 feet below O.D. have appeared in mixed mudstone–sandstone successions in which faulting and fissuring have been the controlling factors.

When fissured, the Triassic sandstones and sandy mudstones which have been proved at Bath also transmit the thermal water but are dry and waterless when encountered in a massive and unbroken condition at depths of 50–80 m. In the same way, the small faults which cut the Blue Lias and Lower Lias Clay in the valley floor at Bath largely control the movement of the dilute thermal water in the recharge mound of the hot springs. This association of thermal water with open faults and fissure systems is equally noticeable in the Carboniferous Limestone. Thus in the bottom of the Avon Gorge the main flow of thermal water issues from strongly fissured Hotwells Limestone at sea level, whereas the Hotwells Limestone of the Ashton Park Borehole, about 1 mile south of the hot spring, yielded no thermal water at depths of 1714–1981 feet (528–603 m) below O.D. Indeed the highest temperature, recorded at a depth of about 2100 feet (640 m) below O.D. in the borehole, had not attained a value equal to that of the Hotwells spring, 76°F (24°C) at sea level, when the borehole was abandoned. Even if some allowance is made for subsequent thermal recovery in the Ashton Park Borehole, it is most unlikely that this contrast would have been greatly diminished.

Energy due to hydrostatic pressure is the prime mover in forcing the groundwater into the deeper levels of the fissure systems. When the rising thermal water reaches an open fissure system up which rapid ascent is possible, part of the pressure energy will be converted to kinetic energy, and if, as is usual at Bath, dissolved and exsolved gases are present, the liberation of the gases with the formation of rising and expanding gas bubbles at higher altitudes may accentuate the "lift" or acceleration of the moving stream of thermal water. The increase in flow which has resulted in almost every case where the damping effect of the drift deposits filling the spring pipes has been temporarily removed by drilling, indicates that the original flow of the thermal water may have been larger or the pressure much higher when the natural penetration of the Lower Lias Clay was effected. Once the Mesozoic cover had been removed no such check would have been exerted on the flow of the Hotwells thermal water and it is not impossible that the much greater dilution and lower temperature at Hotwells may be due in part to lack of restraint in the upper part of the fissure system in the Carboniferous Limestone.

From the evidence at present available, it would appear that flow along fissures and fracture planes is the dominant factor down to depths of at least 2000–3000 feet (600–900 m) below sea level. Conditions in Carboniferous uplands such as the Mendips present special features at levels from about 100 m below O.D. upwards, due to solution effects produced by Quaternary climatic conditions. Large underground channels permit the groundwater to circulate freely in this zone and in the southern and western parts of the Mendips much water issues as low level overflow springs. For this reason it is doubtful whether the western part of the Mendips contributes significantly to the emission of the thermal water rising at Hotwells and Bath.

At Bristol the open fissure system along the Avon Gorge is of post-Triassic age, post-dating Mesozoic mineralization. It almost certainly extends into the upper Old Red Sandstone and where the relatively impervious Lower Old Red Sandstone is missing, any underlying Silurian rocks are also likely to have been affected. In the central part of the coal basin the downward penetration of groundwater may, therefore, have continued locally to the surface of the Tremadoc Shale underlying the Silurian–Carboniferous succession.

Assuming a structural thickness of 15,000 feet (4600 m) of post-Tremadoc strata, a mean annual temperature of 50°F (10°C) and a temperature gradient of 1°F for 75 feet would give a rock temperature of 250°F (121°C) at the deepest levels east of Bristol, i.e. in the vicinity of Brislington. West of Bristol the temperature in the zone of effective thermal water circulation would be lower, as the depth to the Tremadoc Shale beneath the Trym valley is much less. In both areas groundwater circulation is likely to diminish significantly below the base of the Carboniferous Limestone, with a further substantial reduction at the base of the Upper Old Red Sandstone.

Measurements made in the Bristol and Somerset collieries between 1945 and 1950 suggested that the highest thermal gradients are found in the north and the lowest in the south of the coalfield but temperatures recorded in coalmines can be unreliable owing to the effects of ventilation and drainage and some of those made in the coalfield may also have been affected by unequal heating due to the oxidation of coal. The detailed results were not considered to be sufficiently accurate for publication, yet the apparent difference in the gradient in the south and the north may be not unrealistic (Downing *et al.* 1987,

Fig. 2). The temperatures measured in Ashton Park Borehole (Bullerwell in Kellaway 1967) give a maximum gradient of 1°F for 78 feet (1°C for 42.8 m), but owing to lack of recovery time, the correct figures could not be measured and the true gradient must be slightly steeper. Wethered (1882) gave a gradient of 1°F for every 76 feet at Kingswood but this was determined underground in mine workings. The average figure used by Andrews *et al.* (1982) is 20°C/km (1°C for 50 m). If the groundwater in the Carboniferous Limestone of the northeastern Mendips is migrating northwards towards the Avon Valley then the chief structural feature causing it to rise at Bristol and Bath is the tectonic belt extending from the Avon Thrust to the Wick Thrust in the east and the intermediate Kingswood Anticline. Although basically Variscan in origin this tectonic unit has also been strongly affected by Mesozoic and later movements, producing uninverted structural relief with a broad ridge of soft Middle Coal Measures shales extending from Eastville to Warmley. The complexity and mineralization of the Variscan structures between the Mendips and Bath are such that hydrological continuity of the Carboniferous Limestone is questionable (Kellaway & Welch 1991).

At Bath the rate of upward migration of the thermal water is such that there has been no possibility of measuring the geothermal gradient at the depths proved by the various boreholes. It seems likely, however, that the geothermal regime has been highly modified by the movement of the thermal water.

Acknowledgements. It will be clear from the nature of the work described above that the results given in the paper depended largely on the successful operation of the drilling programme. To Mr David Vine (formerly City Engineer, Bath), Mr M. Froggatt and Mr T. Bletso I am particularly indebted for help. Wessex Water Authority made two holes which were supervised by Dr W. I. Stanton. Mention should also be made of the drillers, particularly Mr Robin Meeks of Drillsure whose construction of the inclined Borehole in Stall Street was an outstanding achievement. To Mr David Beeton (former Chief Executive, Bath City Council) I am grateful above all for the moral support that he gave me at the outset of a difficult task.

References

Andrews, J. N., Burgess, W. G., Edmunds, W. M., May, R. L. F. & Lee, D. J., 1982. The Thermal Springs of Bath. *Nature*, **298**, 339–343.

Anstie, J. 1873. The Coalfields of Gloucestershire and Somersetshire and their resources. London.

Armstrong, W. 1838. *An Account of the Tapping and Closing the Springs of Hot Water at Mr Pinch's Brewery, Bath.* Bristol.

Cave, R. 1963. (The Hamswell Borehole) in Summary of Progress of the Geological Survey. (for 1962), 38–39.

Crickett, R. E. 1869 *Thermal Springs at Bath and Elsewhere.* Proceedings of the Bath Natural History and Antiquarian Field Club 1. (1867–69), 78–98.

Cunliffe, B. & Davenport, P. 1985. *The Temple of Sul Minerva at Bath.* Oxford University Committee for Archaeology: Monograph No. 7.

Donovan, D. T. & Kellaway, G. A. 1984. "Geology of the Bristol District: The Lower Jurassic Rocks". Memoirs of the British Geological Survey.

Downing, R. A., Edmunds, W. M. and Gale, I. N. 1987. *Regional ground water flow in sedimentary basins in the U.K.* Geological Society Special Publication, 4, 105–125.

Kellaway, G. A. 1967. The Geological Survey Ashton Park Borehole and its bearing on the geology of the Bristol District. *Bulletin of the Geological Survey of Great Britain*, **27**, 47–153.

—— & Welch, F. B. A. 1955. The Upper Old Red Sandstone and Lower Carboniferous Rocks of Bristol and the Mendips compared with those of Chepstow and the Forest of Dean. *Bulletin of the Geological Survey of Great Britain*, **9**, 1–21.

—— & —— 1991. *The Geology of the Bristol District.* Memoirs of the British Geological Survey.

Moore, C. 1867. On Abnormal Conditions of Secondary Deposits when connected with the Somersetshire and South Wales Coal-Basin: and on the age of the Sutton and Southerndown Series. *Quarterly Journal of the Geological Society, London*, **23**, 449–568.

Rastall, R. H. 1925. "Report on the Geology of the Mineral Springs in the City of Bath, with special reference to the Conservation of the Water Supply". (Report submitted to Bath City Council, March 24th 1925.)

—— 1926. Note on the Geology of Bath Springs. *Geological Magazine*, 98–104.

Richardson, L. 1928. *Wells and Springs of Somerset*, with a Bibliography of the Bath Thermal Waters by W. W. Whittaker. Memoirs of the Geological Survey of Great Britain.

Tomkins, N. 1895. *Bath Thermal Springs, their Supposed Origin and Source*. Proceedings of the Bath Natural History and Antiquarian Field Club.

Wethered, E. 1882 *Underground Temperatures*. Proceedings of the Bristol Naturalists Society, NS 3 (for 1879–82), 9–19.

Woodward, H. B. 1876. *Geology of the East Somerset and the Bristol Coal-fields*. Memoirs of the Geological Survey of Great Britain.

From Kellaway, G. A. (ed.) *Hot Springs of Bath*, pp 127–142

8. Hydrogeology of the hot springs of Bath

W. I. Stanton

The hot springs of Bath consist of three separate thermal springs, the large King's Spring, the small Hetling (or Hot Bath) Spring and the very small Cross Bath Spring. All three thermal sources were developed and used by the Roman inhabitants of Aquae Sulis but only the King's Spring, which lies 120 m ENE of the other two, has been studied in detail by archaeologists. However, it is clear from modern investigation and historical records that at each site the hot water rises through gravel deposits that, at the King's and Cross Bath Springs, fill funnel shaped hollows in the underlying Lower Lias Clay.

Past estimates of the output of thermal water have varied from 8 litres/second to 27 litres/second, but a commonly accepted round figure was a quarter of a million gallons in one day (13 litres/second). The water temperature was said to vary between 40°C and 49°C (104–120°F). Among the striking characteristics of the spring waters are the copious bubblings of exsolved gases and the rust-coloured precipitates of iron hydroxides that encrust the baths and water channels, but the dominant dissolved ions in the thermal waters are calcium and sulphate.

Early writers supposed that the heat of the springs was volcanic in origin. A more popular view, dating from about 1880, was that the source of the springs lay 20 km (12 miles) southwest of Bath in the Mendip Hills. Rain falling on the Mendips travelled through caves in the Carboniferous Limestone beneath the Radstock coalfield, where it reached 2.5 km depth and acquired the necessary heat, and rose up at Bath along a fault that fractured the confining beds of Lias and Rhaetic clay.

Wells and boreholes in and near Bath have sometimes encountered thermal water (Fig. 8.1). Small flows of tepid water have come from shallow excavations into the Lias, larger flows of warmer water from the Trias at greater depths. At Batheaston, 4 km (2.5 miles) northeast of the hot springs, a shaft sunk in search of coal in 1812 struck a strong artesian flow of tepid water from Dolomitic Conglomerate (Trias) at a depth of about 110 m. Within the city limits hot artesian water was met in the Trias at Pinch's Well, Kingsmead Square, in 1838, in probable Lias at Bathwick Brewery Borehole in 1880, and in Lias Clay in a well near Bath Spa Station in 1911. At all these sites the Trias and Lias overlies Carboniferous Limestone. The Twerton or Pennyquick colliery on the west side of Bath did not encounter thermal water.

Modern studies of the hot springs began with work by the Institute of Geological Sciences in 1977–1978 (Burgess *et al.* 1980). The hypothesis of a Mendip source for the water was supported and a recharge mound in Triassic strata overlying Carboniferous Limestone beneath Bath was proposed. The travel time from the Mendips to Bath was considered to be *c.* 3000 years. Boreholes drilled by Bath City Council in 1980 discovered the gravel-filled funnel below the King's Spring. In 1981–1982 the presence of Carboniferous Limestone beneath Bath, long suspected, was finally proved in two exploratory boreholes drilled for Bath City Council by Wessex Water Authority. Artesian hot water flowed from the Kingsmead Borehole, 112 m deep, and from the 90 m deep Sports Centre Borehole (Plate 11). When the borehole water was proven to be free of the amoebae *Naegleria* spp that had contaminated the hot springs, Bath City Council sank production boreholes to the Carboniferous Limestone at Stall Street (1983) and Hot Bath Street (1986).

The results of controlling and monitoring the flows from these boreholes, and measuring the reaction at the King's Spring, are presented in this paper.

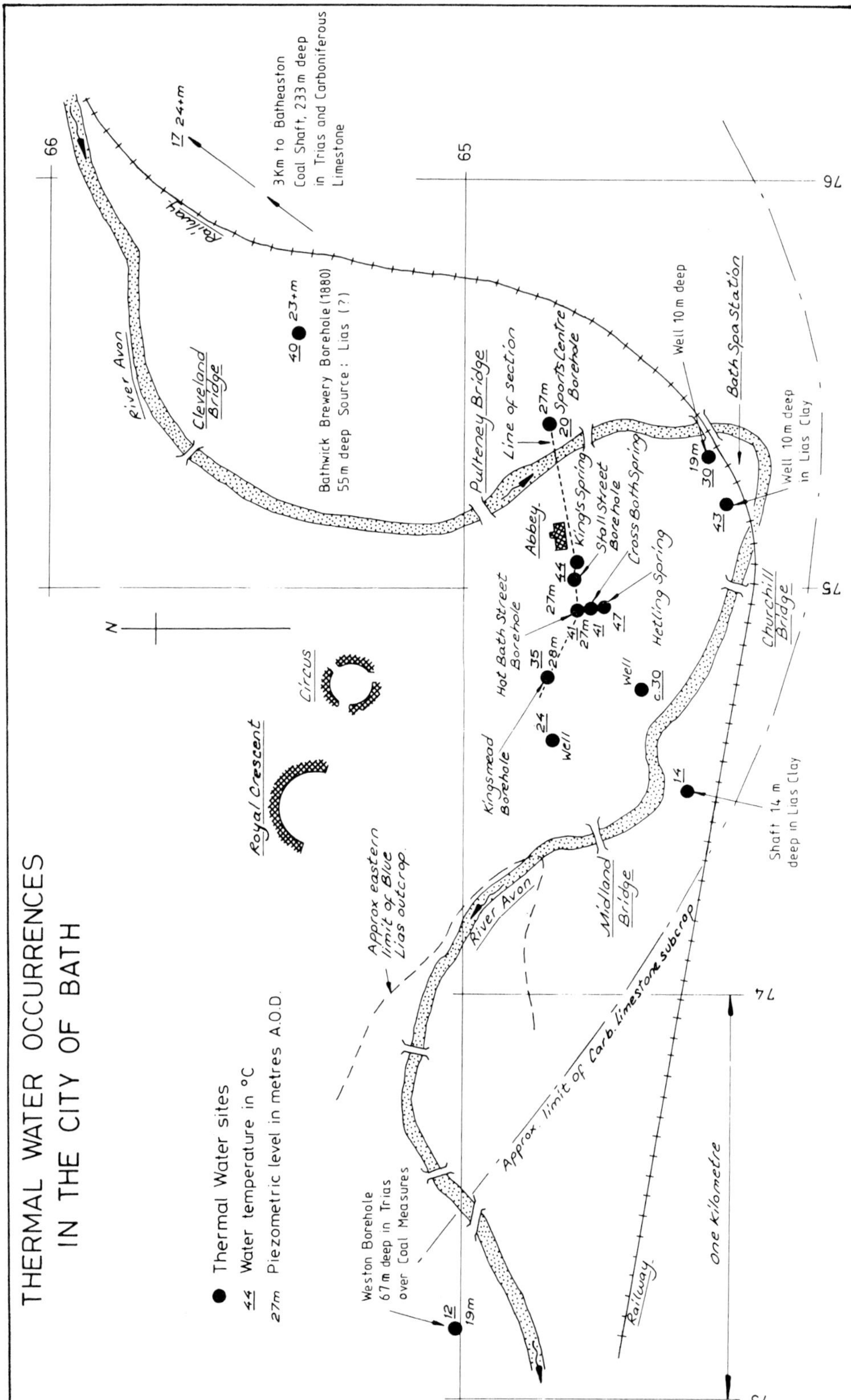

Fig. 8.1. Thermal water occurrences in the City of Bath.

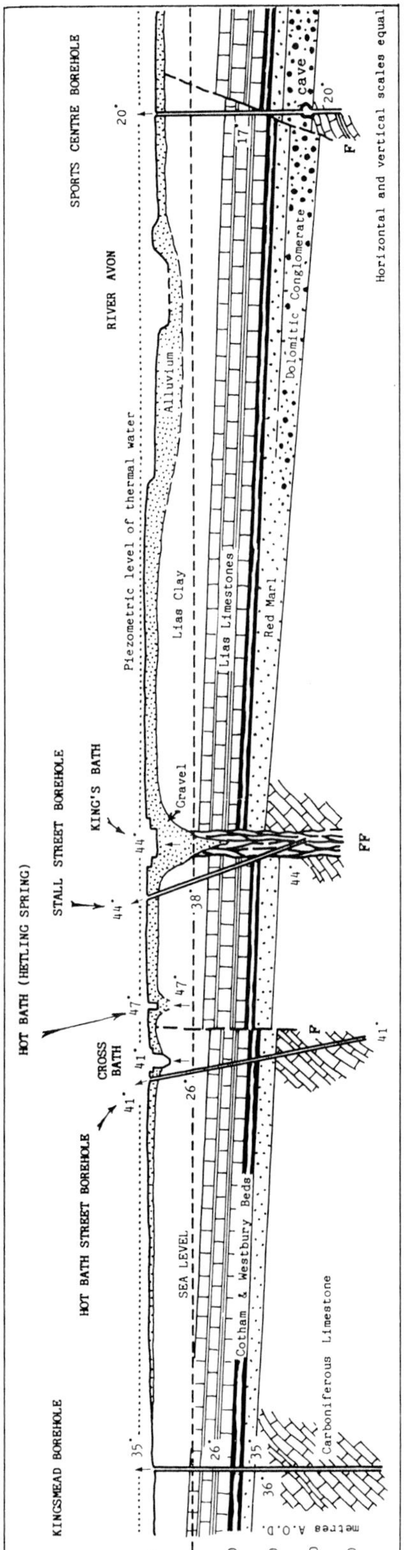

Fig. 8.2. Schematic geological section through the springs and boreholes showing groundwater temperatures in °C.

Hydrogeology

The thermal water rises from a great depth through fissures in Carboniferous Limestone, which appear to be few in number and confined to the hot springs area in central Bath. The Stall Street Borehole proved the existence of a complex subvertical fault zone beneath the King's Spring (Fig. 8.2). The fault cuts through the full geological sequence including the confining Jurassic and Triassic strata and is without doubt a major conduit used by the hot waters. The Hot Bath Street Borehole indicated that the Cross Bath Spring is supplied by a different, smaller, fault. Geological conditions beneath the Hetling Spring have not been investigated by drilling.

The Jurassic/Triassic confining layer, 50–70 metres (165–230 ft) thick, consists of mudstones and marls (Lower Lias Clay, Penarth Beds, Tea Green and Red Marls) interbedded with carbonate rocks (Blue and White Lias, Dolomitic Conglomerate). The topmost mudstone layer, the Lower Lias Clay, has disintegrated where the thermal water rises up fractures in it, forming deep funnels now filled with riverborne gravel.

The resistance to free flow exerted by the massive gravel chokes occupying the funnels is the principal cause of the artesian head of 5–10 metres that characterizes the thermal water in the vicinity of the springs. It also causes thermal water to migrate laterally through fissured limestone beds in the confining layer, notably the Dolomitic Conglomerate. This water is present beneath large areas of the city, under artesian head, becoming cooler as it moves out from the central zone but retaining its chemical signature as it mixes with meteoric water that has percolated down into the aquifer through fractures in the Lias Clay, through the Blue Lias from outcrops at Weston, 2 km northwest of the King's Spring, or through the Trias from faults and distant outcrops north of Bath.

Lateral flow in the Dolomitic Conglomerate is in widely spaced fissures and caves. In the Sports Centre Borehole, artesian flow came from a cavity 3 m high in the conglomerate, 71 m below ground (Fig. 8.2). In the Kingsmead Borehole, flow was from small fissures in conglomerate and Red Marl. In the Stall Street Borehole, large voids in the fault zone provided the flow. The fissures and caves form a well integrated groundwater reservoir, as was proved by the fast reaction at some boreholes and springs to flow changes at the others.

The volume of thermal water stored under pressure as a recharge mound greatly exceeds 100 million litres (Ml). This was the approximate volume removed from storage during the period of artesian flow testing at the Kingsmead and Sports Centre Boreholes, making allowance for the entry of local meteoric water as the recharge mound shrank.

If the recharge mound were absent, abstraction of thermal groundwater from any borehole near the hot springs would cause a rapid and equal reduction in spring output. The recharge mound forms a valuable buffer to the effects of short-lived over-abstraction. It is also an effective barrier to the free flow of local meteoric water to the hot springs.

Groundwater hydrology

The total 'normal' output of the three thermal springs is shown to be about 15 litres/second (Fig. 8.3). Output can be more than doubled by abstracting from boreholes, but this is achieved by drawing on storage in the recharge mound. Lowering the overflow level of the King's Spring increases the artesian head on the spring, causing an increase in output (Fig. 8.4). The maximum head gain of 2 m augmented the output by 3 litres/second. However, it was found that if the level was lowered to minimum and then raised, the outputs at given water levels were not necessarily repeated. The same unpredictability was noted at the Cross Bath Spring (Fig. 8.4). It is explained by assuming that unstable

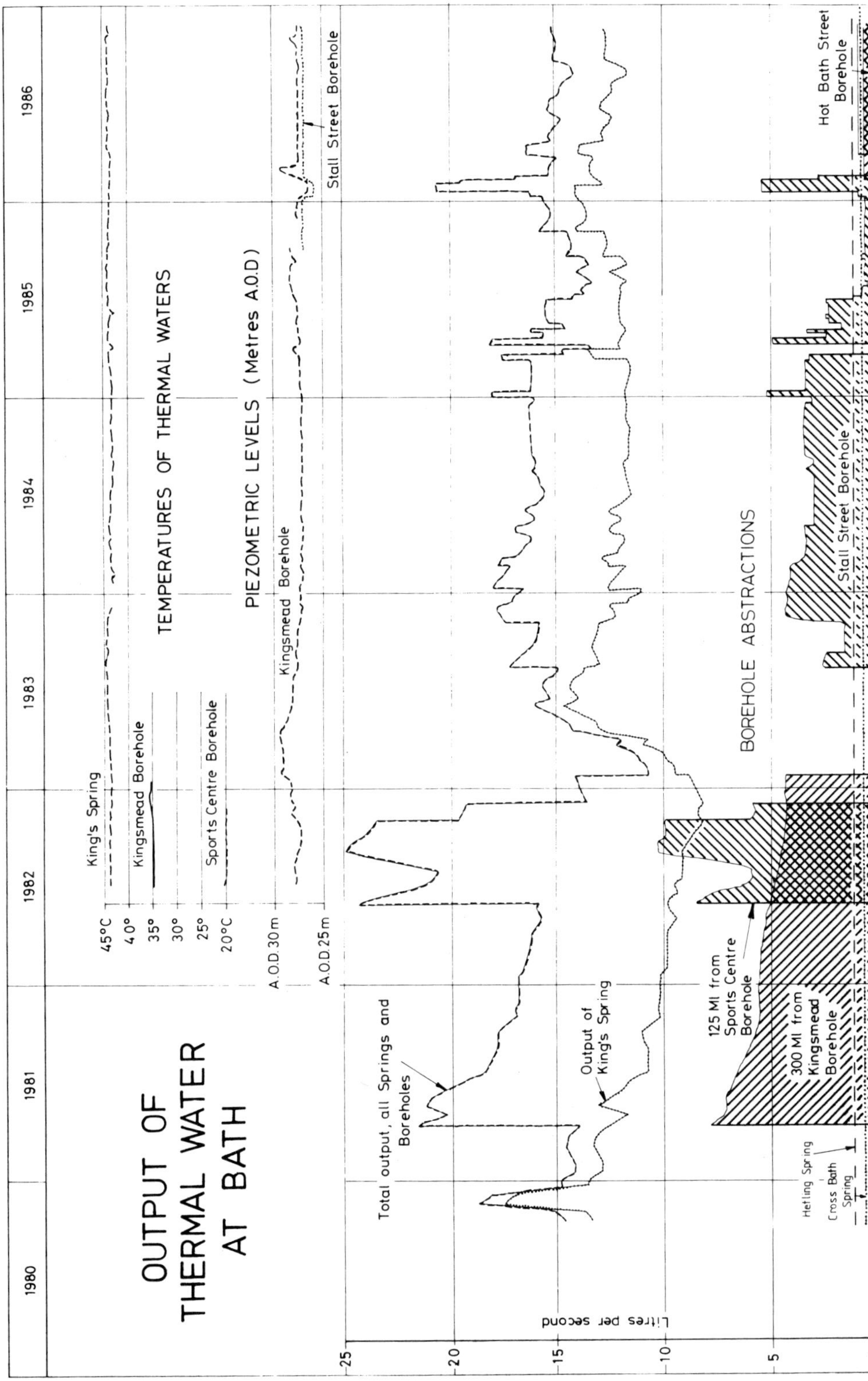

Fig. 8.3. Output of thermal water at Bath.

channels exist through the gravels in the spring pipe, kept open by the steady flow of water. When the flow velocity changes, the channels suffer erosion, deposition and settlement, changing their configuration and their resistance to flow. Unpredictable settlements of the gravel are thought to be responsible for the sudden small changes in spring output that occasionally figure on the King's Spring flow charts.

In the past, measurements of spring output were usually based on the time taken for the King's Bath or the Cross Bath to fill or partly fill. They were inaccurate because the flow reduced as the water level rose.

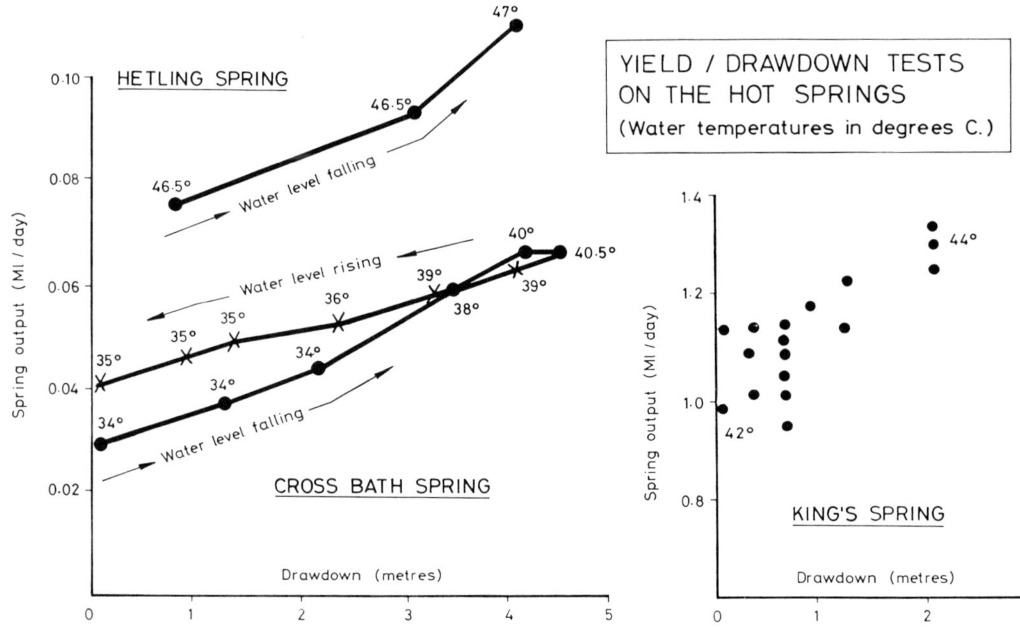

Fig. 8.4. Yield/drawdown tests on the three springs.

Prolonged flow testing of the artesian boreholes demonstrated that the recharge mound represents a well integrated groundwater reservoir. There is a close hydraulic link between the King's Spring and the Kingsmead Borehole, 300 m distant. When the borehole valve is opened, flow at the spring decreases within hours by the equivalent of *c.* 5% of the borehole output (Fig. 8.5). The 24-month flow test of spring and borehole together showed an initial 40% increase over the output of the King's Spring alone (Fig. 8.3), but this apparent gain reduced to nearly zero after 14 months as the output of both sources diminished.

Fourteen months into the test, flow from the Sports Centre Borehole was added to that from the Kingsmead Borehole and the King's Spring. Again there was an initial increase in total output followed by diminishing individual outputs (complicated at this stage by the need to restrict flow from the Sports Centre source to avoid damage to the meter by waterborne rock fragments). During the six months of Sports Centre Borehole testing the output of the other two sources diminished faster than before.

When, 20 months into the test, flow from the Sports Centre Borehole was stopped, the combined output of the other two sources was 12 litres/second, significantly less than the 'normal' output of the King's Spring alone (13 litres/second). Over the succeeding 1.5

months the combined two-source output rose gradually to 'normal'. At this point flow from the Kingsmead Borehole was stopped. There followed a rapid recovery in the King's Spring output, which reached 'normal' in only three months.

Over the 24-month period of over-abstraction followed by recovery, the total abstraction from the three sources was 1056 Ml. The 'normal' output of the King's Spring over the same period would have been 821 Ml. Thus to restore the *status quo*, recharge to the recharge mound from local sources was 244 Ml over the whole period, or 0.33 Ml/day (3.8 litres/second) on average.

At the start of the 24-month flow test in April 1981 (Fig. 8.3), water temperature at the King's Spring is thought to have been *c.* 45°C (113°F). Burgess *et al.* (1980) used a calibrated thermistor installed in the spring chamber to consistently record 46.5°C (115.7°F) in 1977. The same thermistor, used in the Kingsmead Borehole in 1981, proved to have become unreliable. Bristol Avon River Authority divers in 1968 and 1969 recorded temperatures between 45.0°C and 45.4°C in the spring chamber using a mercury in glass certified thermometer. Bath City Council's boreholes in the King's Spring, drilled in 1980, encountered water at 45°C and this figure is consistent with temperatures recorded since 1983, from the King's Spring feeder pipe at 50 m depth, in water intercepted by the Stall Street Borehole.

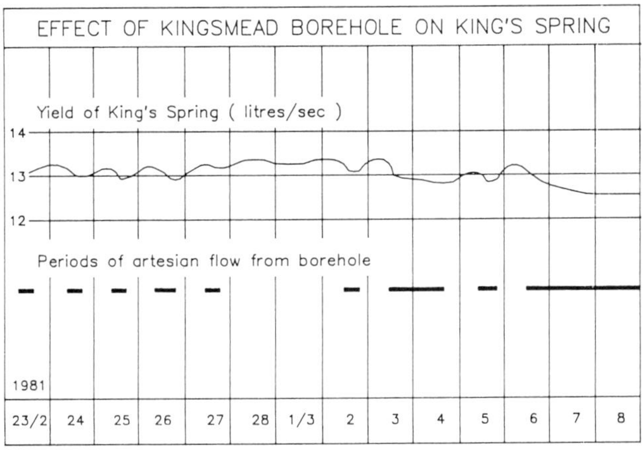

Fig. 8.5. The effect of Kingsmead Borehole on flow from the King's Spring.

When regular monitoring of spring and borehole temperatures began in July 1982 (Fig. 8.3) the temperature of the King's Spring had fallen to 44°C, and it had declined further to 43.5°C when all borehole abstraction ceased in January 1983. During the recovery stage of the flow test, the King's Spring temperature rose to 44.5°C in May 1983. In contrast, the water temperature at the Kingsmead Borehole rose from 35°C to 36°C over the 21 month abstraction period. The cause of these temperature fluctuations is thought to be the temporary establishment of new flow routes through the recharge mound, leading to diversion of thermal water from the King's Spring towards the Kingsmead and Sports Centre Boreholes, and its replacement by cooler water from elsewhere in the mound. That the replacement water was stored thermal water is demonstrated by the fact that there was no significant change in water chemistry at any source during the test period.

Artesian head measured at Kingsmead Borehole fell during the 21 months when the Kingsmead and Sports Centre Boreholes were flowing and rose when their flows ceased

(Fig. 8.3). It varied between 7 and 9 m above ground level (27–29 m AOD) and appeared to be inversely proportional to the total output of thermal water. The same relationship was seen when the Stall Street Borehole was tested from August 1983 to June 1985. It is an indication of the extent to which the recharge mound is depleted by abstraction in excess of the 'normal' spring output of 15 litres/second.

There can now be little doubt that any attempt to extract more thermal water than is naturally produced by the hot springs will fail in the long term, although an initial excess would be produced as the recharge mound beneath Bath was drained.

Groundwater chemistry and biology

The thermal waters from the three springs and the four boreholes all contain sodium, calcium, chloride, and sulphate ions in high concentrations. The waters from the Weston Borehole and the Batheaston Coal Shaft, respectively 2 km and 4 km distant from the hot springs (Fig. 8.1), were clearly of thermal origin (Table 8.1).

Table 8.1. Basic chemistry of Bath thermal waters.

	King's Spring and Stall St. Borehole	Hetling Spring	Cross Bath Spring	Kingsmead Borehole	Sports Centre Borehole	Hot Bath Street Borehole	Weston Borehole	Batheaston Coal Shaft
Temp (°C)	44°	47°	41°	35°	20°	41°	13°	17°
Na (mg/l)	187	195	183	300	550	200	420	476
Ca (mg/l)	390	358	380	360	296	367	110	261
SO_4 (mg/l)	1010	1015	1050	1060	1000	1020	795	940
Cl (mg/l)	286	340	288	440	770	350	235	964
HCO_3 (mg/l)	199	193	189	154	198	—	245	81
Mg (mg/l)	53	57	54	56	71	—	35	65
Si (mg/l)	21	21	—	17	7	13	—	10
Fe (mg/l)	1	0.5	0.2	1	—	2	0.2	1.5

The main chemical difference between the waters is in their NaCl content, which is lowest in the central springs and boreholes and increases away from them. Apparently, then, the bulk of the NaCl is acquired locally from the Mesozoic cover. West of Bath, groundwater with high NaCl content is known from boreholes at Keynsham (528 mg/l chlorides) and Bitton (278 mg/l NaCl). In both cases the saline water comes from the Triassic aquifer, whereas the overlying Lias is low in chlorides. It is assumed, therefore, that the thermal springwater acquires additional NaCl from the Triassic strata during its long residence in the recharge mound beneath Bath.

The thermal water contains *c.* 1 mg/l of Fe, and ochreous iron staining and iron hydroxide sludges and encrustations are among the most prominent features of the Hot Springs. In the spring chamber beneath the floor of the Cross Bath the sludge deposit, formed in about 100 years, was 12 cm thick. Hard encrustations in the channel leading from the King's Spring to the Great Bath, and on the waterfall in the Museum, reproduce many of the forms typical of lead veins in the Mendips and Pennines. They are perhaps the best British examples of metalliferous mineral deposits in active growth. Deposition of the iron appears to be dependent on cooling, oxygenation and agitation of the thermal water, thus samples of water taken where it leaves the baths are likely to be reduced in Fe content.

Water samples taken from the four boreholes have proved to be microbiologically

Plate 11. Thermal water issuing under artesian pressure from the borehole at the Sports Centre (1982). Photo – W. I. Stanton.

Plate 12. The King's Bath after removal of the mass concrete floor.

Plate 13. The flow recorder in the Roman Drain. This monitors the flow of the water issuing from the King's Spring.

Plate 14. Measuring the flow of the Cross Bath Spring prior to drilling of the spring-pipe (1986).

sterile. The original goal of the groundwater investigation, which was to secure a reliable supply of spa water free from pathogenic organisms, has thus been achieved.

Monitoring of groundwater yield and heat output

An editorial in the *Bath Chronicle* of 21st October 1897 called for regular gauging of the hot springs, in view of their importance to the city: "Apparently, the yield is as regular as the chemical ingredients are constant. Still, there may be changes in the remote laboratory where they are compounded, and also by the works of man."

There was, however, no systematic monitoring of the springs until 1980, when Wessex Water Authority and Bath City Council jointly installed a 90° V-notch weir and stage recorder on the 'Main Roman Drain' which brings together the various outflows, neglecting possible leakages, from the King's Spring. Subsequently, all borehole outputs have been continuously measured by flow meters except in the case of the Hot Bath Street Borehole, from which the small artesian flow (0.45 litres/second) is gauged at intervals and appears constant. The small yields of the Hetling and Cross Bath Springs (*c.* 0.9 litres/second and *c.* 0.4 litres/second respectively) are known from occasional gauged pumpings (Fig. 8.4) and are assumed to be constant.

This monitoring has confirmed that the 'normal' yield of the King's Spring before 1980 was about 13 litres/second but is now reduced to about 12 litres/second due to continuous minor abstraction from the Stall Street and Hot Bath Street Boreholes. On the same basis the total yield of the three springs was *c.* 15 litres/second, now reduced to *c.* 14 litres/second. All measurements are considered to be 95% reliable.

In 1987, to comply with the conditions of their abstraction licence, Bath City Council installed integrator flow meters on all three springs and on those boreholes that are regularly used in order to produce more accurate and complete records of thermal water yield.

How constant is the spring flow? Such records as exist, going back to the 17th Century, cannot be relied upon. Thus in 1897 C. E. Davis reported to Bath City Council that the Cross Bath was leaking. In evidence he said that the effective flow of the Cross Bath Spring reduced from 1.1 litres/second to 0.4 litres/second as the bath filled, and he quoted records showing that in 1693 the effective spring yield had been 1.3 litres/second and in 1748 0.9 litres/second. Although the Cross Bath may have leaked to some extent, changes in artesian head (i.e. overflow or filling levels) at the Cross Bath itself can adequately explain the variable flows (Fig. 8.4). It is significant that in 1897, when the mean output of the Cross Bath Spring was about double that of today, the King's Bath was in regular use. Thus in 1897, the overflow level of the King's Spring was *c.* 1.7 m higher than at present, and its yield would have been less (Fig. 8.4). In compensation, flow from the other two springs would have been greater.

Meaningful interpretation of old records of spring flow is impossible unless contemporary overflow levels are also known, but the available evidence strongly suggests that, overall, there has been no major change in total spring output over the last few centuries.

To the visible and measurable output of the Hot Springs must be added an unknown quantity of thermal water that finds its way to the ground surface through many minor leaks in the strata confining the recharge mound. The fact that groundwater in the mound is warm, but becomes cooler with increasing distance from the main springs, points to there being a slow outward movement from the centre. The proportion of the thermal water thus diverted is not known. Strikes of hot thermal water have been recorded well away from the hot springs (Fig. 8.1), so it is not impossible that there exist other, minor, sources of deep-seated water discharging mainly into the recharge mound. In particular the well near Bath Spa Station that encountered water at 43°C in the Lias Clay in 1911, could repay the trouble of relocating it.

If it is assumed, for the sake of argument, that 20% of the deep-seated thermal water is dispersed into the recharge mound, the full heat output of the hot springs, above the natural groundwater temperature of 10°C, can be calculated for any day on which full measurements are made. See Table 8.2 for measurements made on 1 November 1980, when only the three main springs were flowing.

Table 8.2. Total heat output of the hot springs (1 November 1980).

Source	Water output (m^3/day)	Temperature excess (above 10°C)	Heat output (Mcal/day)
King's Spring	1180	35°	41300
Hetling Spring	75	37°	2775
Cross Bath Spring	35	31°	1085
Total	1290		45160

Adding 20% for losses gives a total heat output of 54192 Mcal/day.

On 15 October 1982, late in the 24-month flow test, the groundwater yield had increased by 58% but the heat output was virtually unchanged (Table 8.3).

Table 8.3. Total heat output of the hot springs (15 October 1982).

Source	Water output (m^3/day)	Temperature excess (above 10°C)	Heat output (Mcal/day)
King's Spring	720	34°	24480
Hetling Spring	75	37°	2775
Cross Bath Spring	35	31°	1085
Kingsmead Borehole	360	25°	9000
Sports Centre Borehole	850	10°	8500
Total	2040		45840

Adding 20% for losses gives a total heat output of 55008 Mcal/day.

Fig. 8.6, constructed on this basis and within the limits of the available data, shows that the heat output of the whole thermal source varies much less than the output of thermal water. Following a change in water output, whether positive or negative, the heat output trends back towards the figure of 5.5×10^{10} calories per day. This is taken to be the steady heat output of the deep-seated thermal water. It is unaffected by temporary changes due to depletion or recovery of the recharge mound.

Real changes in the performance of the hot springs must therefore be measured by changes in the heat output. Variations in the yield of thermal water are not directly relevant. Satisfactory monitoring of the thermal source will depend on frequent measurements of water temperature as well as yield at all springs and boreholes and a steady heat output chart will be the indication that all is well with the hot springs.

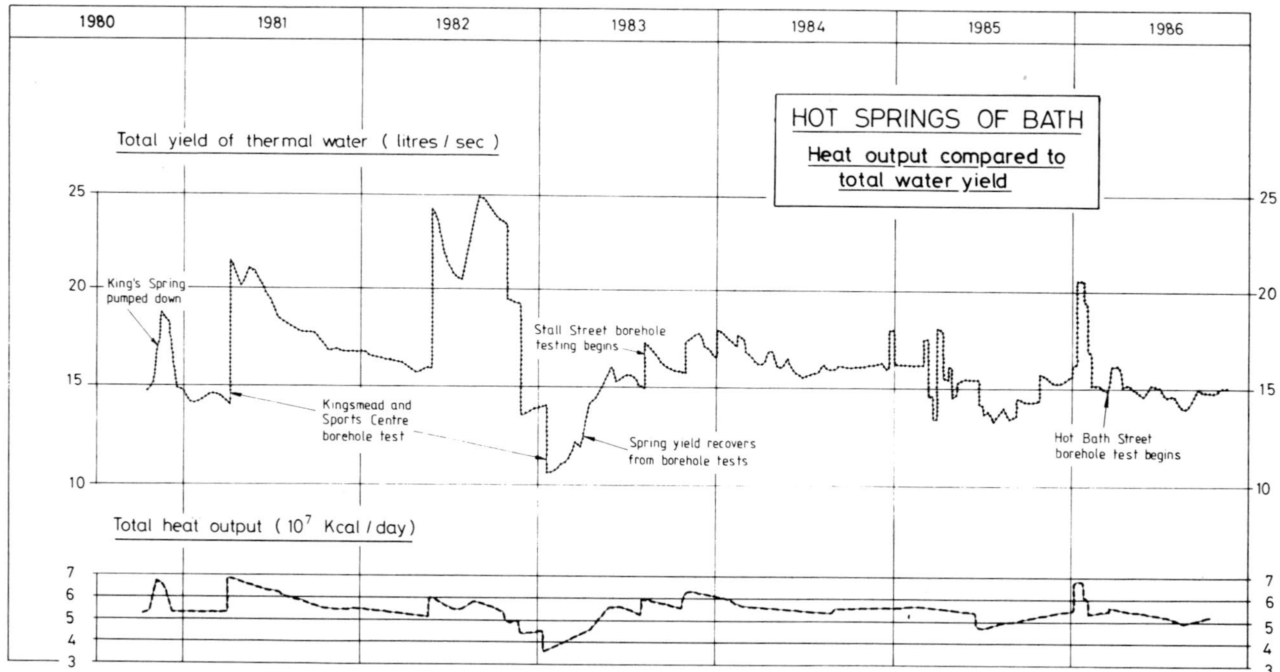

Fig. 8.6. Heat output compared to total water yield.

Possible sources of the hot springs

If the thermal water does indeed derive from the Mendips, and the circulation is driven by the high water table at the Mendip end of the hypothetical U-tube in Carboniferous Limestone beneath the Radstock coalfield, the water output of the hot springs will increase or decrease in step with increases or decreases in the hydrostatic head. The actual head, depending on the source area, is not known. A Mendip source area will presumably be in limestone outcrops on the north flanks of one or more of the Old Red Sandstone periclines (Fig. 8.7). The maximum possible head is 170 m, north of Shepton Mallet. To east and west the possible heads decline to 50 m near Frome and zero west of Blackdown.

Limestone outcrops north of Shepton Mallet, between Green Ore and Stoke St. Michael, may therefore be the most likely catchment area for the hot springs. The areas to east and west cannot, however, be ruled out. The 'likely' area comprises the catchments of the Gurney Slade, Ashwick Grove and St. Dunstan's Well spring groups, together with a small part of the Wookey Hole Caves catchment. Flow from these catchments towards the Hot Springs may be regarded as an early stage in the capture of relatively high level springs by a distant spring at a much lower level.

Within these catchments are several large limestone quarries (Fig. 8.7) that have lowered the water table in their immediate vicinity by up to 20 m. As they develop, the water table over a wide surrounding area may be depressed by 100 m or more. No reaction has been detected, so far, at the hot springs. It is expected that the more precise monitoring of the heat output that will begin in 1987 will detect any such reaction at an early stage.

The local cold meteoric water that joins the hot springs when conditions in the recharge mound permit appears to move mainly through the Triassic aquifer. Favourable conditions for Triassic recharge exist at Hamswell, 7 km north of Bath in the Swainswick valley, and at Grandmother's Rock, 3 km west of Hamswell. Both outcrops are at a level *c.* 60 m higher than the hot springs. If a hydraulic connection does exist, the water movement must be slow, because the lack of tritium in the hot springs water (Burgess *et al.* 1980) indicates that none of it is post-1953 recharge.

The hypothetical U-tube in Carboniferous Limestone linking the Mendips to Bath is *c.* 20 km long and *c.* 2.5 km deep (Andrews *et al.* 1982). In the upwelling limb, beneath Bath, the mean water temperature must be at least 30°C higher than in the descending limb, and the lower density of the warmer water will cause a head difference between the two limbs of 15–20 m. As a driving force for the thermal circulation this is small compared to the potential head difference of 170 m between the Mendips and Bath, but as it is acting in the same direction it increases the potential head to *c.* 190 m.

Age of the thermal water

Andrews *et al.* (1982) showed on isotopic and hydraulic grounds that the bulk of the thermal water resurging at Bath is less than 10,000 years old and probably less than 4000 years old. Volumetric considerations suggest that the age is unlikely to exceed a few hundred years. If the output of deep-seated thermal water is taken to be 18 litres/second on average (15 litres/second + 20%) the annual storage requirement for water in transit from the Mendips to Bath is 5.7×10^5 m^3. If the transmitting segment of Carboniferous Limestone 20 km long, 1 km wide and 700 m thick postulated by Andrews *et al.* has an effective porosity of 1% (a high figure for this rock at depth) its capacity will be reached in *c.* 250 years. Thus for a residence time of 1000 years the segment would have to be 4 km wide and fully occupied in transmitting water to Bath, and if the Carboniferous Limestone is so uniformly permeable one must ask why the main flow is not directed towards the

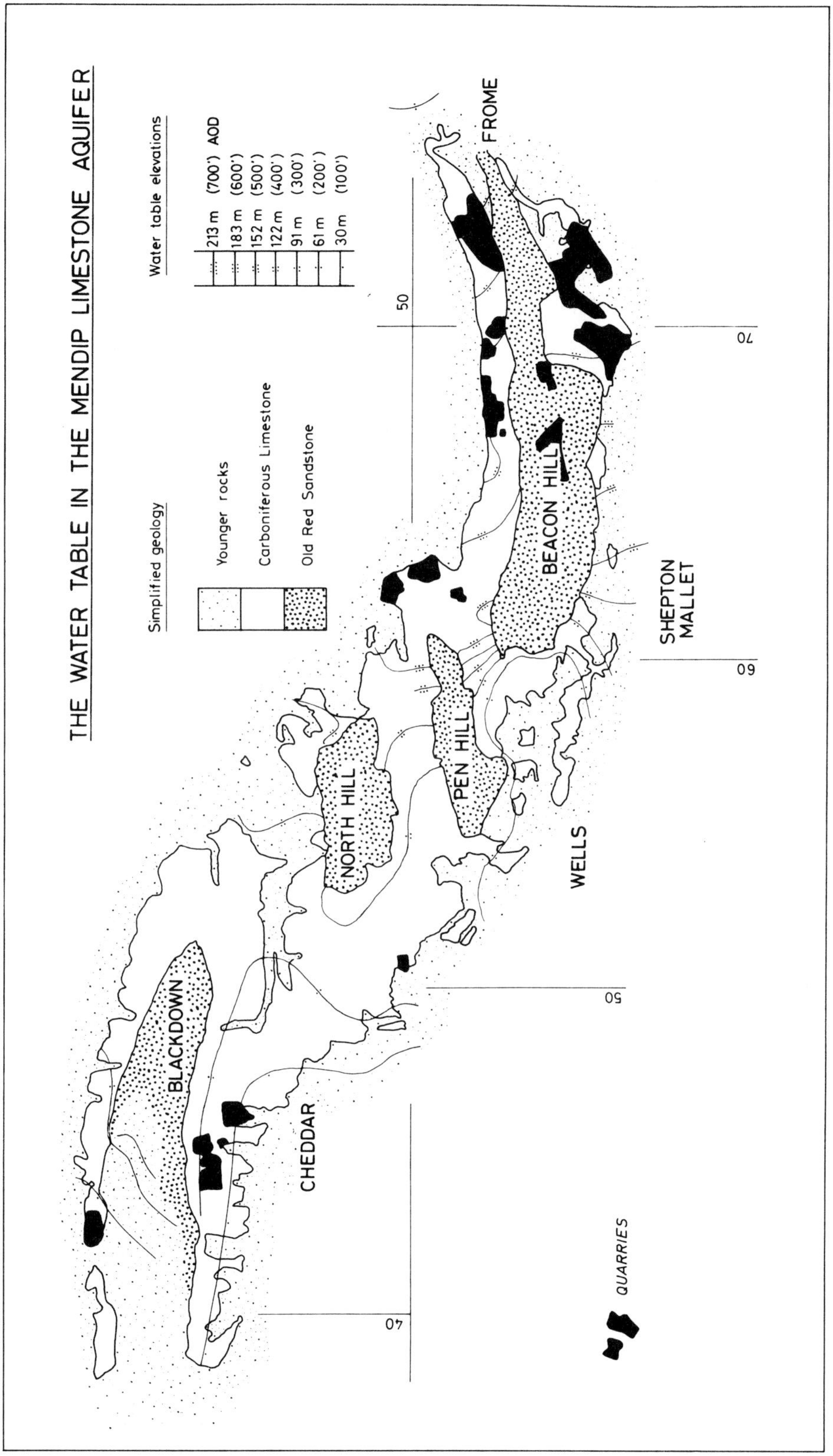

Fig. 8.7. The water table in the Mendip Limestone Aquifer.

Hotwells Spring at Bristol, which at 1.4 m A.O.D. is the lowest outcrop of Carboniferous Limestone in the Avon Valley.

All of the British and most of the European thermal springs (except those in volcanic regions) are associated with limestone strata in which caves can form. It is therefore a reasonable assumption that the waters flowing from the Mendips to Bath follow fissures that have been widened by solution and are incipient caves. Within the network of small tubes that are the first stage of cave development (Smith & Drew 1972) a few preferred routes enlarge at the expense of the remainder, which become inactive. In this model the route from the Mendips to Bath is an immature cave system, carrying conduit, not diffuse, flow. Because the storage capacity of the cave system is not great the water residence time is likely to be tens or hundreds, rather than thousands, of years.

Acknowledgements. The research here described was carried out by Wessex Water Authority staff in the course of routine duties. Beside the author, Gordon Hobbs and David Palmer were particularly closely involved. The Authority's collaboration with Terry Bletso and Mike Froggatt of Bath City Council, and with Geoffrey Kellaway, the Council's geological consultant, has been much appreciated. This paper appears by permission of the Chief Executive, Wessex Water Authority.

References

Andrews, J. N., Burgess, W. G., Edmunds, W. M., Kay, R. L. F. & Lee, D. J. 1982. The thermal springs of Bath. *Nature*, **298**, 339–343.

Burgess, W. G., Edmunds, W. M., Andrews, J. N., Kay, R. L. F. & Lee, D. J. 1980. *The Hydrogeology and Hydrochemistry of the Thermal Water in the Bath–Bristol Basin.* Institute of Geological Sciences Report to the EEC.

Smith, D. I. & Drew, D. P. 1972. *Limestones and Caves of the Mendip Hills.* David and Charles, Newton Abbot.

From Kellaway, G. A. (ed.) *Hot Springs of Bath*, pp 143–156

9. The geochemistry of Bath thermal waters

W. M. Edmunds & D. L. Miles

The subject of fascination by physicians, philosophers and artists for several centuries, the thermal waters of Bath have also been a focus for the curiosity of scientists, especially in the last 175 years. In particular they have been a field laboratory for testing developments in analytical chemistry. No fewer than ten analyses of the springs have been published since 1823 and, although they may lack a little in accuracy, they testify to the progress of chemistry in the 19th and 20th Centuries. Whereas the analysis was often the goal of the scientific study, highlighting the discovery of an element, gas or radioactivity (Dewar & Strutt 1904; Munro 1928) or new analytical techniques such as colorimetric analysis (Riley 1961), the present geochemical investigation seeks to use the chemistry to answer questions concerning the origin of the water.

The geochemical investigations described in this paper were initiated as part of the search for geothermal energy in the UK. Despite being located in a region of low heat flow (Whieldon & Rollin 1986), the Bath thermal springs at 46°C (115°F) are the warmest geothermal manifestation in the United Kingdom although springs with temperatures up to 55°C (131°F) have been encountered underground in the tin mines of Cornwall (Edmunds *et al.* 1984).

It was considered that a thorough investigation of the Bath thermal waters would provide clues as to the presence of other geothermal provinces in the UK and also serve as an experimental site to evaluate new geochemical techniques, following a long tradition. As a result of eight years of survey the geothermal resource base for the country has now been defined (Downing & Gray 1986), but the most promising areas are not at Bath but in certain Mesozoic sedimentary basins such as the Wessex Basin as well as in granitic rocks, providing new 'Hot Dry Rock' technology can be developed.

A hydrogeological model for the Bath thermal waters has already been presented (Burgess *et al.* 1980; Andrews *et al.* 1982). Mainly on the basis of geochemistry it was concluded that the thermal springs originate from rainfall on the Mendip Hills and by circulation in the Carboniferous Limestone to a depth of 2.7–4.3 km where they reach a temperature of 80 ± 16°C (176 ± 29°F). The objective of the present paper is to review the existing geochemical data and to present new results from: (a) the monitoring of the springs since the drilling of the Stall Street Borehole, (b) further isotopic analysis and (c) additional inorganic analysis notably for trace elements. These data will then be reviewed to provide an updated geochemical model and especially to give revised estimates of the subsurface temperature, source aquifers, residence time and near-surface mixing. A more detailed account of the radioelements and gases in the geochemical story is given in Chapter 10 where our conclusions on the age of the water are combined.

Methods of investigation

Detailed investigations of the King's Spring and its environs were undertaken during 1977 and 1978 (Burgess *et al.* 1980). As well as monitoring the discharge of the thermal spring, samples of water were collected from the other accessible thermal sources in Bath, from the Hotwells Spring at Bristol and from representative non-thermal groundwaters in the vicinity. The analytical and sampling methods are described in the original report. Representative data have been used in the present study.

Intermittent sampling of the thermal springs and the exploration boreholes (Sports Centre and Kingsmead Square) was carried out between 1978 and 1983. Following the completion of the Stall Street Borehole, monthly monitoring of the discharge was carried out between late 1983 and early 1986. At the same time, monitoring of the Hetling Spring was carried out on a less regular basis, using a submersible pump to sample the spring chamber beneath York Street.

Two comprehensive inorganic and isotopic analyses of the King's Spring were carried out in 1977 and 1986, continuing the tradition established for almost 200 years. Chemical analysis was carried out using inductively coupled plasma optical emission spectrometry (ICP–OES) and ICP mass spectrometry to provide a reliable and detailed analysis of cations and a range of trace elements. Automated colorimetric analysis was used for anion analysis.

Historical perspective

Joseph Glanvill FRS, Rector of the Abbey Church in Bath presented an account (Glanvill 1669) at the Royal Society entitled 'Observations concerning the Bath Springs' which is still worth quoting some 320 years later. He writes:

> "The town and baths are of great Antiquity. Besides what I find in very ancient Chronicles to that purpose, one of our great Antiquaries (Mr. P.) asserts that these Baths *were* 800 years before Christ. Which, if so, would give occasion to enquire, how consistent with it that hypothesis (concerning the heat of these waters) would be which makes it to be the fermentation of minerals [*in fieri*]; and, whether it be likely that the minerals through which these waters should pass, should be in that state of imperfection so many hundred years, and that the whole disposed matter in those places should not be perfectly concreted in so great a tract of time."

Glanvill was therefore the first to write on the subject of geothermics, in Britain at least, proposing an exothermic hypothesis for the origin of the temperature of the waters. He also provided the earliest geochemical account of the rock–water interaction. He continues:

> "You doubtless know the other conjecture, which supposeth the cause of this heat to be,—that two streams having run through and imbided certain sorts of different minerals, meet at last, after they have been deeply impregnated, and mingle their liquors, from which commixture arises a great fermentation that causes heat [like as we see it is in vitriol and tartar, which though separately they are not hot, yet when mingled beget an intense heat and ebullition between them]. This seems to me a probable cause of the lastingness of the heat of these waters."

Thus, he also deserves credit as the first to propose a mixing model for the spring even though one can no longer accept mixing as a cause of the elevated temperature.

Results of the 1977–1986 investigations

The present series of studies can be considered in three phases.

1. Regional setting

The Bath thermal springs are not the only thermal source in the area and it is worthwhile to compare the geochemical characteristics of Bath with other thermal and non-thermal groundwaters in the region. The chemistry of groundwaters in the wider Bath–Bristol region has been described by Burgess *et al.* (1980). Of interest in the present study are the thermal waters intercepted within a few miles radius of Bath, either in springs or in boreholes (Chapter 8). The recent analyses of these sources are given in Table 9.1, with Hotwells and Bath shown for reference.

The results of the regional studies show that no other groundwater samples except Hotwells have chemical or thermal characteristics similar to those at Bath. Topographically and geologically the Avon Valley could be the location of other thermal springs or thermal groundwaters (see Chapter 12). The Hotwells source at Bristol has a composition equivalent to a mixture of Bath-type thermal water and shallow Carboniferous Limestone water in the ratio 1:2:3. This lends support to the hypothesis that both thermal sources have a similar structural control over their flow and discharge.

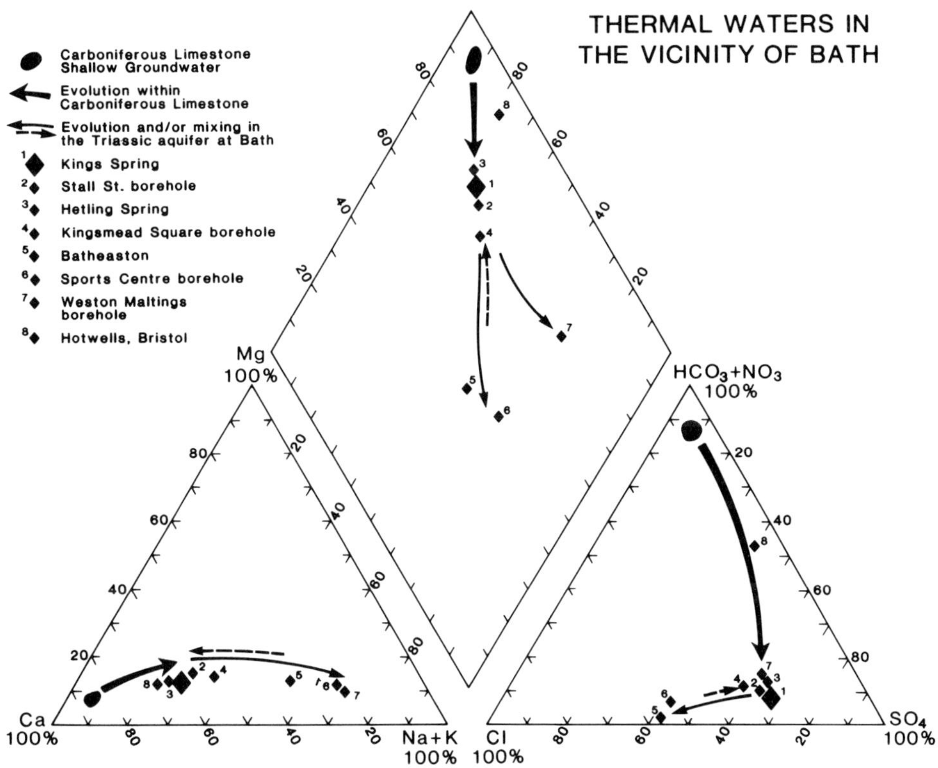

Fig. 9.1. Trilinear diagram showing the chemical composition of thermal waters in the vicinity of Bath. Lines of possible evolution are also shown.

Within the vicinity of Bath itself a number of thermal waters have been discovered, mainly as a result of drilling. The interrelationships between these and the main King's Spring are illustrated in Fig. 9.1 where the major ionic compositions are plotted. The close affinities of the Hetling Spring, Stall Street Borehole and Kingsmead Square Borehole can be seen, demonstrating that these are all related to the King's Spring discharge (as is Hotwells). Three non-thermal or slightly warm sources have been described from up to 1 km away from the King's Spring to both the east and west and occurring in the confined Triassic succession. It therefore appears that the rising thermal water under pressure is recharging the Trias sequence, particularly in the basal conglomerates. In the course of its

Table 9.1. Representative chemical analyses of thermal waters in the vicinity of Bath. Analyses by BGS except for Batheaston. An asterisk denotes laboratory analysis.

	King's Spring Bath	Hetling Spring	Stall St Borehole	Kingsmead Square	Cross Bath	Sports Centre Borehole	Weston Maltings	Batheaston (Heriot 1877)
Date of analysis	1979	1985	1986	1981	1986	1982	1977	1875
Temperature (°C)	45.3	44.8	44.4	34.8	36.1	20.2	13.2	16.7
pH (field)	6.65	6.5	6.70	7.70*	6.9	6.7	7.95*	–
Na	183	179	228	282	220	486	420	476
K	17.4	16.8	18.1	19.1	18.0	23.0	22.8	28
Ca	382	385	390	405	393	296	110	261
Mg	53	54	58	62	58	71	35.0	65
HCO_3	192	188	187	198	170	198	245	81
SO_4	1032	1000	1030	1040	1080	898	795	938
Cl	287	284	335	465	325	820	235	964
NO_3–N	<1.0	<1	<0.05	<0.5	0.35	–	<0.5	–
Sr	5.9	5.5	6.3	6.2	6.3	7.4	10.5	–
Ba	0.024	0.024	0.030	0.026	0.030	0.010	<0.2	–
B	0.59	0.47	0.59	0.95	0.50	1.4	–	–
Si	20.6	19.3	19.5	15.8	19.5	6.5	4.7	–
Fe_T	0.88	<0.010	1.00	1.11	<0.05	0.88	0.17	–
Mn	0.068	0.045	0.050	0.054	0.030	0.040	0.022	–
Fe	2.08	2.00	2.0	–	2.0	2.3	3.1	–
Br	2.02	1.70	2.0	2.8	1.8	–	–	–
I	0.043	0.028	0.042	0.056	0.041	–	–	–
Total mineralization	2179	2136	2278	2498	2295	2811	1864	2815

flow over 2 km, away from the point of ascent, the thermal water cools down to near normal temperature and its composition is modified. In the case of Weston Maltings to the west of Bath there is a change in Na/Ca ratio, but elsewhere there is evidence of mixing with a chloride-rich groundwater. The modification also leads to a very high content of dissolved helium derived from the higher radioactivity of the Triassic succession (see Chapter 10).

2. *Composition of the thermal springs: changes with time*

Some of the historical analyses of the King's spring are summarized in Table 9.2, quoting only the principal constituents analysed. Allowing for changes in analytical methods over the past century it is concluded that the composition has remained constant over the period. However, a significant recent change in temperature and chemistry occurred in 1983 following the drilling of the Stall Street Borehole (Chapter 7). There was a fall in the mean temperature of the spring from 45.3 to 44.4 in 1983 following the abstraction of water from the borehole, but over the same period the temperature measured on the Hetling Spring remained constant (Fig. 9.2). Changes in the chemistry in this period are illustrated in Figs 9.3 and 9.4 The most significant change is in the chloride which increases by about 60 mg l^{-1} (an increase of some 20%). There is also a complementary fall in sulphate. Considering the complete analysis it is apparent that mixing of cold Na, Cl water is taking place which is 'diluting' the Ca, SO_4 water of the thermal source. The percentage mixing must be below 5%. This 'contamination' probably is the same as that modifying the composition of the Sports Centre Borehole and Batheaston (Fig. 9.1) where the mixing proportion is slightly greater. Access to this water has been denied to the main spring conduit, possibly having been sealed off over the period of flow during geological timescale by mineral precipitation.

Table 9.2. Some historical analyses of the King's Spring compared with results of present study.

Analysis No	1	2	3	4	5	6
Year of analysis	1874	1888	1936	1961	1979	1986
Temperature	46	47	49	48	46.5	43.5
Total solids	1920	2360	2162	2166	2203	2278
Na	129	135	177	174	183	228
K	39	31	15.4	15.7	17.4	18.1
Ca	377	402	392	392	382	390
Mg	47	52	51	54	53	58
HCO_3	86	88	193	216	192	187
SO_4	869	1061	1001	1021	1032	1030
Cl	280	277	274	276	287	335

1. Heriot (1875)
2. Ekin (1888)
3. Judd-Lewis (1938)
4. Riley (1961)
5. King's Spring, this study
6. Stall Street Borehole (King's Spring), this study

Another feature of the King's Spring over the ten year period to 1986 has been the variable redox potential (Eh) of the water (Fig. 9.4). It is unusual for a deep groundwater to vary, as the King's Spring has done, over a range of some 300 mV. This unsettled Eh has been accompanied by fluctuations in the concentrations of dissolved total iron (present in solution as Fe^{2+}). This high total iron is clearly related to the presence of anaerobic conditions at depth allowing dissolution, yet the water is rather below saturation with siderite ($FeCO_3$) which is a likely control on solubility in the thermal water. The ubiquitous iron deposition of the Bath Spring upon aeration is well known both to those who have 'taken the waters' over the years (see Glanvill 1841, for example), and to the visitor surveying the Baths. The deposition of iron also occurs within the spring chamber by the mixing with shallow oxidizing groundwater. This is substantiated by the presence of traces of dissolved oxygen, measured over the same period and this is the likely cause of the fluctuation in Eh, in itself a very sensitive indicator of dissolved oxygen.

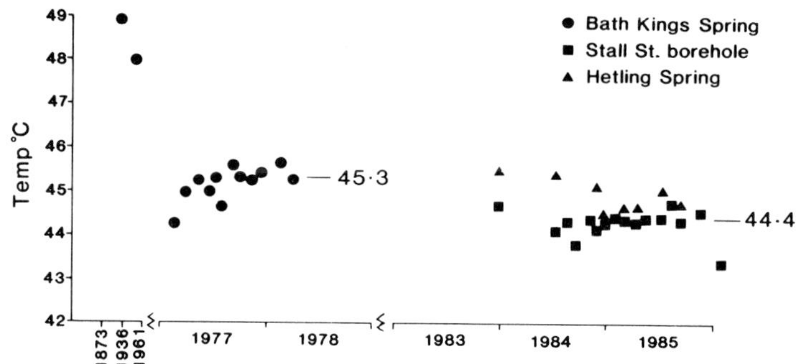

Fig. 9.2. Temperature (°C) of the Bath King's Spring, Hetling Spring and Stall Street Borehole measured between 1977 and 1985 by BGS.

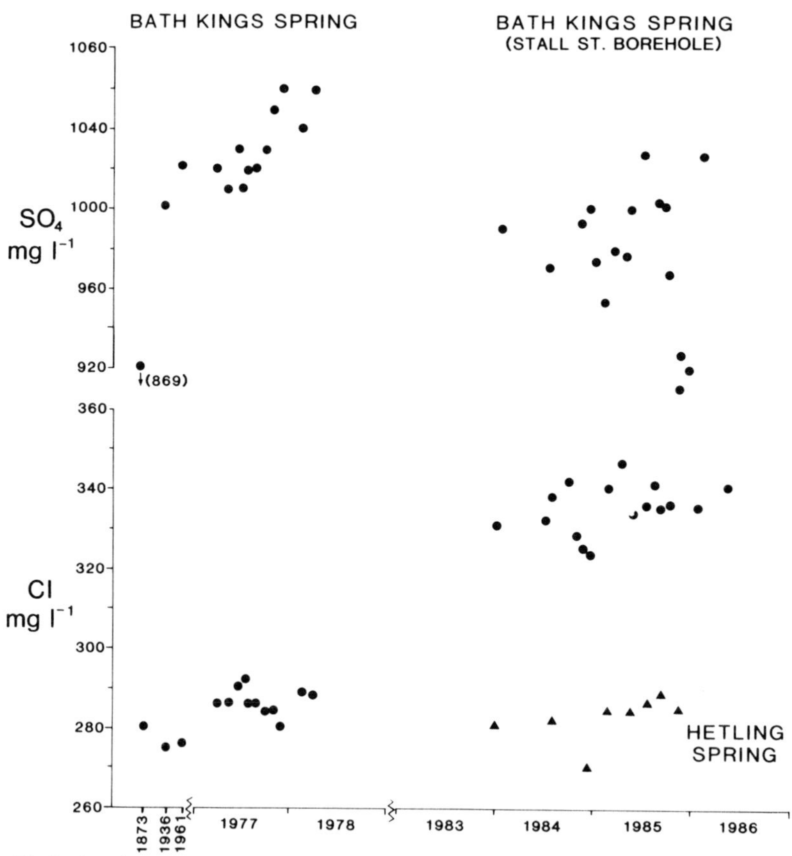

Fig. 9.3. Variation in sulphate and chloride in groundwater from Bath King's Spring, Stall Street Borehole and Hetling Spring measured between 1977 and 1985.

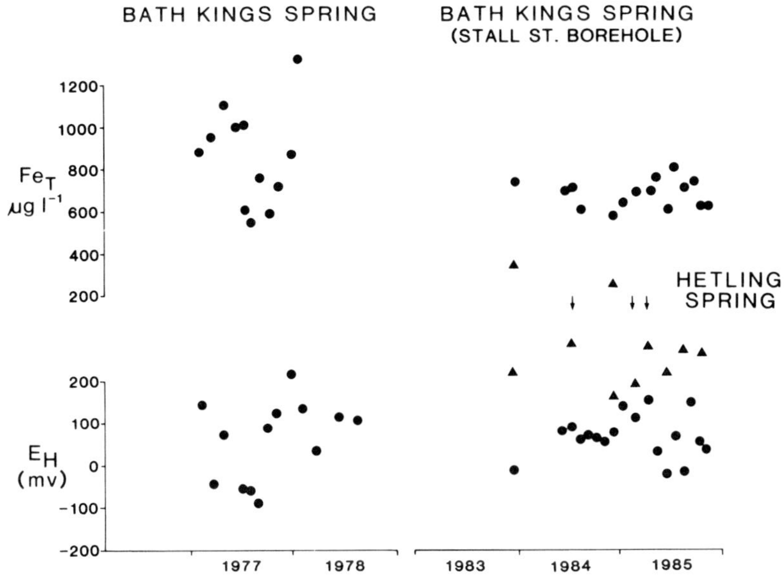

Fig. 9.4. Variation in Fe_T and redox potential (Eh) in groundwater from Bath King's Spring, Stall Street Borehole and Hetling Spring measured between 1977 and 1985.

Tritium (derived from thermonuclear testing) was also detected from time to time and the balance of evidence would suggest that a small quantity of shallow groundwater (so small as hardly to affect temperature) is seasonally entrained in the groundwater discharge. Reduced sulphur was measured but was below the limit of detection (1 $\mu g\,l^{-1}$). From time to time a smell of sulphur may just be detected—but dissolved sulphide is clearly unimportant in quantity. There is, however, a record of marcasite being found as encrustation on the deposits of the deep spring chamber (Kellaway, pers. comm.) and this may be evidence of sulphide deposition over a geological timescale and the reason why H_2S is not readily detected.

Following the construction of the Stall Street Borehole (1983) the Fe^{2+} and, to some extent, the Eh have been rather more constant, suggesting that any shallow seasonal influence might have been eliminated. It is still puzzling why the warmer Hetling Spring has a higher Eh than the King's Spring. It is likely that the Hetling Spring represents the similar conditions of discharge as the former King's Spring (same conduit but with shallow mixing). The Stall Street Borehole taps thermal water below any shallow oxygenated water, but its lower temperature and higher Cl attests to mixing with some groundwater in the aquifer system above the unconformity.

3. Geochemical characteristics of the King's Spring

The detailed geochemical analysis of the King's Spring (1979) and King's Spring Borehole (1986) are given in Table 9.3. The significance of these various properties may be considered.

Oxygen and hydrogen isotopes. The ratio of the stable isotopes of oxygen ^{18}O, ^{16}O ($\delta^{18}O$) and of hydrogen ^{2}H, ^{1}H (δD) provide clues to the evolution and source of the water. Enrichment in heavy oxygen (^{18}O) or deuterium (^{2}H) relative to the lighter isotopes can indicate whether the water molecule has undergone evaporation or similar changes and

Table 9.3. Chemical and isotopic composition of the King's Spring and King's Spring Borehole.

	King's Spring (1979)	King's Spring Borehole (1986)
Temp (°C)	45.3	44.4
pH	6.65	6.70
Eh (mV)	variable	variable
	mg/l	mg/l
Diss O_2	<0.2	<0.1
Na	183	228
K	17.4	18.1
Ca	382	390
Mg	53	58
HCO_3	192	187
SO_4	1032	1030
Cl	287	335
NO_3	<1	<0.05
Si	20.6	19.5
Sr	5.92	6.3
Ba	0.024	0.030
Li	0.242	0.230
Rb	<0.02	0.011
F	2.08	2.0
Br	2.02	2.0
I	0.043	0.042
B	0.59	0.59
Se	<0.004	<0.002
HS	<0.001	–
Total Fe	0.88	1.0
Total Mn	0.068	0.050
Cd	<0.0025	<0.001
Co	<0.02	<0.001
Cr	<0.03	<0.002
Al	<0.05	<0.001
Cu	0.002	<0.001
La	<0.01	<0.001
Hg	<0.0001	<0.001
Mo	<0.1	<0.002
Ni	0.022	<0.003
Pb	<0.2	<0.0002
V	<0.2	<0.001
Zn	<0.02	0.036
Sc	<0.001	<0.0005
Y	<0.001	<0.001
Zr	<0.015	<0.002
Sb	<0.002	<0.001
U	0.00005	0.00005
Total mineralization	2179 mg/l	2278 mg/l
$\delta^{18}O$ (SMOW)	−7.4	−7.4
$\delta^{2}H$ (SMOW)	−47	−47
^{3}H (TU)	<3	–
$\delta^{13}C$ (PDB)	−1.5	–
^{14}C (PMC)	4.5	–
$\delta^{34}S$ (CD)	–	19.1
$S^{18}O_4$ (SMOW)	–	16.4
$^{87}Sr/^{86}Sr$	0.71075	

also whether it is derived from climatic conditions much different to the present day. The $\delta^{18}O$ and δ^2H of the spring have remained constant during the period of measurement. The isotopic compositions lie close to the worldwide meteoric line (Fig. 9.5) and the ratios are similar to modern shallow groundwaters in the region. This signifies that the thermal water has been derived from rainfall very similar to that of the present day. There is no evidence of any water with a lighter (more negative) isotopic composition, which might signify late Pleistocene groundwater, similar to that found in some other UK groundwaters (Bath *et al.* 1979).

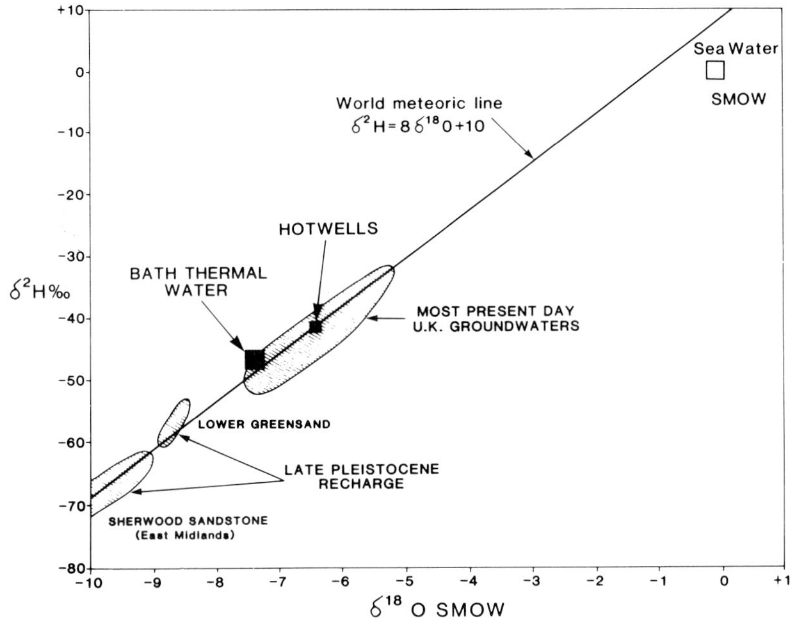

Fig. 9.5. The isotopic composition (δ^2H against $\delta^{18}O$) of the Bath thermal water shown in relation to the world meteoric line and to other present-day groundwaters; the fields of Pleistocene groundwaters, identified in the Sherwood Sandstone (Trias) and Lower Greensand (Cretaceous) aquifers is also shown.

Dissolved constituents: major, minor and trace elements. The dominant chemical characteristic of the Bath thermal water is its calcium and sulphate content which together make up some 65% by weight of the total mineralization of 2.18 g l^{-1}. Sodium and chloride are next in order of importance. The minor and trace element composition has also been determined to the limits of analysis and it is found that with the exception of iron, the water is very low in dissolved metals and any other harmful elements. Thus, it is suggested there is no reason why the water should not be taken (once aerated to remove radon) and providing the tast of 'warm flat irons' can be tolerated. Whether this has any beneficial effect has to be left to a more subjective judgement.

The chemical analyses have been investigated further using the geochemical computer program WATEQ-F, the results of which are summarized diagrammatically in Fig. 9.6. This program takes into account the temperature of the spring and also any speciation and ion interaction which may be occurring at the higher salinity. The composition of the King's Spring is determined almost exactly by the equilibration with the minerals indicated. Most elements lie within limits of saturation of those minerals shown but the waters are supersaturated with respect to barite, celestite and quartz. The water therefore has a mature composition consistent with prolonged residence time in the aquifer and does not have the capacity to dissolve any further the minerals shown except for compositional adjustments within the constraints of equilibrium. In other words, the concentrations of Ca, HCO_3, Mg, SO_4, Ba, Sr, F, SiO_2 are all adequately explained by interaction between water and rock. The concentration of lithium is not controlled by mineral solubility and its relatively high value also indicates a considerable residence time in the aquifer (Edmunds *et al.* 1986). The Br/Cl ratio may also be used as a diagnostic parameter. In the King's Spring this is enriched (7.0×10^{-3}) compared with the sea water ratio of 3.4×10^{-3}. This implies that the chloride cannot be derived either from dissolution of evaporites, which are deficient in Br, or from unmodified sea water. It is most likely that the water has been in contact with sediments rich in organic carbon; this is a feature of waters which have evolved within the Carboniferous sequence (Edmunds 1975).

The inorganic analysis has also been used (Andrews *et al.* 1982) to derive rock-water equilibration temperatures (geothermometry), using the amounts of dissolved Na, K, Ca and SiO_2. Although the cation thermometers were not valid for the conditions of the sedimentary formations of the Bath–Bristol region where feldspar dissolution was not the only solubility control, the SiO_2 composition has been used. The thermal groundwaters are closer to equilibrium with chalcedony than quartz (Fig. 9.6). It is not possible to be certain which mineral acts as a control in this case, but using quartz, an upper temperature of equilibration at depth of 96°C (205°F) is calculated (Fournier 1973) and using chalcedony, a lower limit of 64°C (147°F) is derived. On the assumption that the regional geothermal gradient is about 20°C/km this would place the range of likely circulation depths of the Bath thermal water as 2700–4300 m.

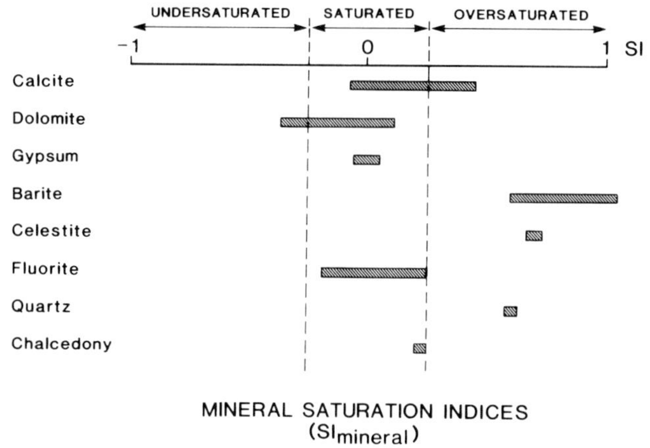

Fig. 9.6. Saturation indices for various minerals computed from analyses of the King's Spring.

Sources of solutes: isotopic evidence. Sulphate is the dominant anion and an attempt has been made to determine the geological source using the sulphur and oxygen isotope ratios of sulphate. The $\delta^{34}S$ isotope ratio of evaporite minerals has varied in composition through geological time, mainly as a consequence of the evolution of the oceans, more enrichment in the heavier isotope occurring during the periods of global desertification such as the Permo-Trias. The $\delta^{34}S$ composition of the Bath thermal spring is plotted on Fig. 9.7a and that of sulphate oxygen in Fig. 9.7b. The sulphur ratio fits with a Carboniferous/Devonian source rather than Triassic, when considered against the global mean; however there is the possibility of atypical values occurring in the Trias of Europe (Claypool *et al.* 1980) and therefore, in the absence of analyses of local sulphate minerals, any interpretation must remain tentative. The sulphate oxygen results also point to a more likely source being pre-Permian.

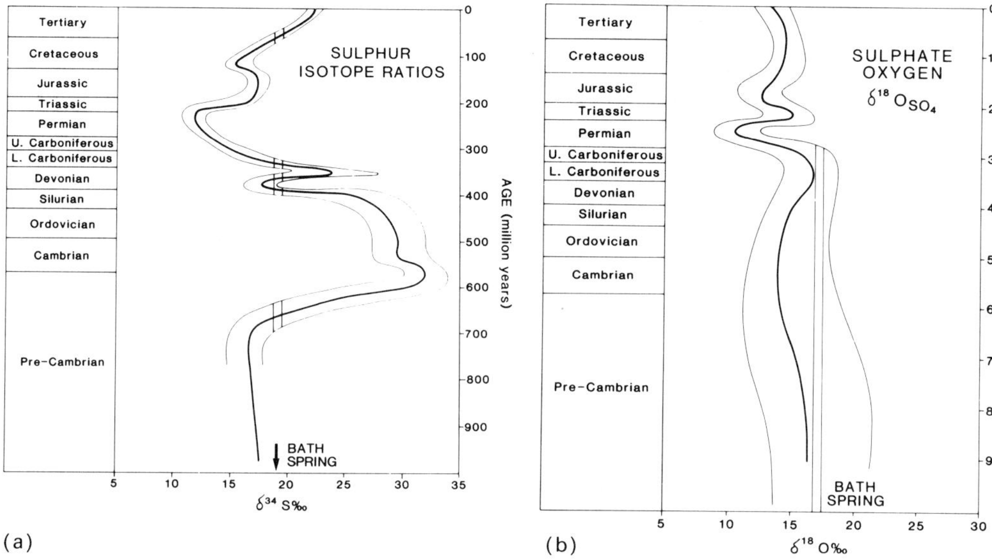

Fig. 9.7(a). Sulphur isotope age curve for sulphate. The bold curve is the best estimate for $\delta^{34}S$ of sulphate minerals in equilibrium with the world ocean surface sulphate of that date. The lighter curves show the uncertainty. The composition of the King's Spring is superimposed. **(b)** Oxygen isotope age curve for sulphate, key as for (a).

Discussion and conclusions

The geochemical evidence presented here and in earlier papers can be used to give the most likely explanation for the origin of the waters.

The King's Spring originated as meteoric water which entered the aquifer as rainfall under very similar conditions to the present day. The geochemical evidence cannot identify where the recharge area, or areas, were, but it is constrained by other evidence which points to the aquifer of storage and evolution as the Carboniferous Limestone. The hydrogeological evidence has been in favour of the Mendip Hills as a recharge source.

The silica geothermometer provides a maximum (96°C, 205°F) and a minimum (64°C, 147°F) temperature for equilibration with rock. If mixing with shallow water (around 5%) has occurred then these temperatures would be a few degrees higher.

The most probable aquifer for transmission and storage of the thermal water is the Carboniferous Limestone series. The likely depth of this aquifer in the Bristol Basin is in the range 2700–4300 m which is consistent with the geothermometry. The strongest evidence for the Carboniferous Limestone as the main thermal aquifer lies in the carbon isotope results. The sulphur and strontium isotope results also indicate that the Carboniferous succession is a likely source of sulphate, although it is also possible that a Triassic signature exists in which case secondary sulphate vein mineralization within the Carboniferous is the likely sulphate source.

The residence time of the thermal water is one of the more difficult parameters to determine. In a previous paper (Andrews *et al.* 1982) it was concluded that, based on the carbon-14 results (Table 9.2) a maximum age of about 20,000 years can be derived. A further reduction in the maximum age to 10,000 years can be made when one considers the oxygen and hydrogen stable isotope results which indicate that no component of Pleistocene water is found. The carbon stable isotope results provide an additional qualitative indicator that the water is likely to be thousands (as opposed to tens or hundreds) of years old.

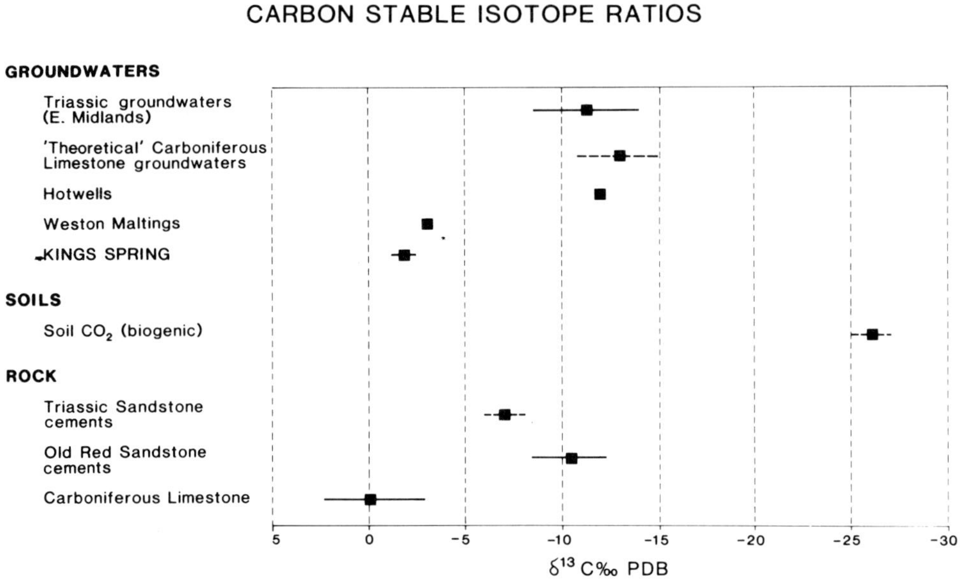

Fig. 9.8. Carbon stable isotope ratios for rocks considered to be possible participants in the flow path of the Bath thermal waters. Soil gas composition and groundwater compositions are also shown in relation to the $\delta^{13}C$ of the King's Spring. Solution of carbonate minerals in the soil zone probably takes place according to the reaction:

$$Ca\underline{\underline{C}}O_3 + \underline{\underline{C}}O_2 + H_2O \longrightarrow 2H\underline{\underline{C}}O_3 + Ca^{2+}.$$

Thus one mole of carbonate carbon combines with one mole of CO_2 gas to give two moles of dissolved bicarbonate in groundwater.

The $\delta^{13}C$ of -1.5 indicates that the water closely approaches isotopic equilibrium with the marine limestone. Even at temperatures of around 60°C (140°F) this is unlikely to be due to simple isotopic exchange (Pearson & Hanshaw 1970) and is instead attributed to incongruent reaction along the aquifer flow path following achievement of carbonate saturation near the point of recharge. Similar results from the English Chalk are generated within a period of 10,000 years (Edmunds *et al.* 1987) but this is a rather special case where the reaction is facilitated by the very high specific surface area presented by the rock. Such rapid reaction would not be anticipated in a massive carbonate aquifer such as the Carboniferous Limestone. It is therefore suggested that while elevated temperatures might tend to compensate for a lack of surface area in the Carboniferous Limestone aquifer system, considerable time measured in thousands of years is still required for $\delta^{13}C$ enrichment by incongruent carbonate reactions. The same process would of course affect the ^{14}C and lead to an age considerably younger than the apparent age of 20,000 years. There is a slight possibility that radiocarbon is being added from a shallow source complicating the use of ^{14}C still further. The ^{13}C, being a conservative parameter, is considered to be more reliable than ^{14}C as an indicator of residence time in the present case. The best estimate for the age of the King's Spring therefore, using carbon isotopes and the stable isotopes $\delta^{18}O$, $\delta^{2}H$ as palaeoclimatic evidence, would be 6000 $\pm$ 2000 years. This question is considered further in Chapter 10.

Leaving aside the possibility of mixing of deep groundwaters (a subject raised in earlier papers) there is the strong possibility that mixing with shallow groundwaters derived from the Mesozoic succession is taking place. The quantities involved must be small (5% or less), but they contain small amounts of oxygen which are responsible for the extensive oxidation observed in the ascending thermal water.

The thermal spring gives rise to a recharge mound which spreads laterally; groundwater with chemical characteristics the same as the King's Spring can be recognized up to 2 km from the point of discharge.

Acknowledgements. We wish to express our thanks to colleagues in the British Geological Survey who have been involved either with the field work or the chemical analyses presented here. The co-operation of the City of Bath Engineers Department in affording access to the Springs is gratefully acknowledged. This paper is published by permission of the Director, British Geological Survey (Natural Environment Research Council).

References

Andrews, J. N., Burgess, W. G., Edmunds, W. M., Kay, R. L. F. & Lee, D. J. 1982. The thermal springs of Bath. *Nature*, **298**, 339–343.

Bath, A. H., Edmunds, W. M. & Andrews, J. N. 1979. Palaeoclimatic trends deduced from the hydrochemistry of a Triassic sandstone aquifer, United Kingdom. *In*: *Isotope Hydrology 1978*. Vol. II. IAEA, Vienna. pp 545–568.

Burgess, W. G., Edmunds, W. M., Andrews, J. N., Kay, R. L. F. & Lee, D. J. 1980. *The Hydrogeology and Hydrochemistry of the Thermal Water in the Bath–Bristol Basin*. Report of the Institute of Geological Sciences. London.

Claypool, G. E., Holser, W. T., Kaplan, I. R., Sakai, H. & Zak, I. 1980. The age curves of sulphur and oxygen isotopes in marine sulphate and their mutual interpretation. *Chemical Geology*, **28**, 199–260.

Dewar, J. & Strutt, R. J. 1904. Reports to the Corporation of Bath.

Downing, R. A. & Gray, D. A. 1986. *Geothermal Energy—the Potential in the United Kingdom*. HMSO, London.

Edmunds, W. M. 1971. *Hydrogeochemistry of Groundwaters in the Derbyshire Dome with Special Reference to Trace Constituents*. Report 71/7. Institute of Geological Sciences, HMSO, London.

—— 1975. Geochemistry of brines in the Coal Measures of northeast England. *Transactions of the Institute of Mining and Metallurgy*, **84**, B39–B52.

——, Andrews, J. N., Burgess, W. G., Kay, R. L. F. & Lee, D. J. 1984. The evolution of saline and thermal groundwaters in the Carnmenellis granite. *Mineralogical Magazine*, **48**, 407–424.

Ekin, C. 1888. *In*: Freeman, H. W. (ed.) *The Thermal Baths of Bath*. Hamilton Adams, London.

Fournier, R. O. 1973. Silica in thermal waters: laboratory and field investigations. *In*: Ingeson, E. (ed.) *Proceedings of Symposium on Hydrogeochemistry and Biogeochemistry*. Vol. 1. Clarke, Washington. 122–139.

Glanvill, J. 1669. Observations concerning the Bath springs. *Philosophical Transactions*, **4**, 977–982.

Granville, H. 1841. *The Spas of England*. Colburn, London.

Heriot, M. 1877. The mineral spring at Batheaston. *Proceedings of the Bath Natural History and Antiquarian Field Club*, **3**, 171–177.

Judd-Lewis, S. 1938. *In*: *Bath Official Handbook*. Mendip, Bath.

Munro, J. M. H. 1928. Report to the Corporation of Bath.

Pearson, F. J. & Hanshaw, B. B. 1970. Sources of dissolved carbonate species in groundwater and their effects on Carbon-14 dating. *In*: *Isotope Hydrology*. 1970. IAEA Vienna. 271–286.

Riley, J. P. 1961. Composition of mineral water from the hot spring at Bath. *Journal of Applied Chemistry*, **11**, 190–192.

Whieldon, J. & Rollin, K. E. 1986. Heat flow. *In*: Downing, R. A. & Gray, D. A. (eds) *Geothermal Energy: the Potential in the United Kingdom*. HMSO, London. 8–20.
Wood, M. W. & Shaw, H. F. 1976. The geochemistry of celestites from the Yate area near Bristol (U.K.). *Chemical Geology*, **17**, 179–193.

From Kellaway, G. A. (ed.) *Hot Springs of Bath*, pp 157–170

10. Radioactivity and dissolved gases in the thermal waters of Bath

J. N. Andrews

Thermal waters emerge naturally from Mesozoic strata in Bath at the King's Spring, the Hetling Spring (the Hot Bath), the Cross Bath Spring and directly from the Carboniferous Limestone below the high-water mark in the tidal part of the River Avon at Hotwells in Bristol. A number of boreholes have been drilled in Bath since 1978, in search of a bacteriologically uncontaminated source of thermal water (Chapter 7). Boreholes at Kingsmead Square, North Parade (Sports Centre Borehole) and at the site of the King's Spring (Stall Street Borehole) yielded artesian flows of mineralized water. The hydrological interdependence of these flows and the geological relationship of the boreholes is described by Stanton (Chapter 8). Waters from the Stall Street Borehole and from the naturally flowing Hetling Spring have been monitored over a period of almost two years. The significance of their chemical composition is described by Edmunds and Miles (Chapter 9). The use of radioelement and dissolved gas contents of the waters to constrain their origin and circulation time is discussed in this paper. Some previous chemical and radiochemical analyses of the King's Spring in the period 1977–1979 (Andrews *et al.* 1982a) were compared with the composition of groundwaters from aquifers in both the Palaeozoic and Mesozoic rocks of the region (Burgess *et al.* 1979).

The artesian head for all the boreholes was similar at about 9.5 m (31 ft) suggesting that they had the same hydraulic control, possibly the piezometric level in the Carboniferous Limestone of the Mendip Hills about 20 km (12 miles) south of Bath (Andrews *et al.* 1982a). The thermal water is thus the result of geothermal heating during transit through the Carboniferous Limestone beneath the Radstock coalfield. There is also the possibility of some admixture of shallow circulation local groundwaters with these deep circulation thermal waters. Up to 200 m of oolitic limestones of Jurassic age lie south, east and particularly to the north and northwest of Bath.

These limestones have high porosities and high permeabilities and spring catchments from them provide significant public water supply sources near Bath. Although they are hydraulically separated from the Carboniferous by impermeable clays (Liassic) and marls (Triassic), it is possible that the transmissivity of such impermeable strata over a wide area could permit some recharge to the Carboniferous Limestone and contribute to the thermal water flows at Bath and Bristol (Chapter 12). Although some leakage from Jurassic strata may occur, a proportion of direct recharge through Triassic strata from north of Bath is a more likely source of shallow circulation groundwater (Chapter 8).

Radioactivity of the Bath and Bristol thermal waters

An extended series of sampling and analyses for the radioelement and noble gas contents of the thermal waters from the King's Spring (Stall Street) Borehole and from the Hetling Spring were carried out over the period June 1984 to January 1986. The parameters determined were the natural U-content, $^{234}U/^{238}U$ activity ratio, ^{226}Ra-content, ^{222}Rn-content and dissolved noble gas contents (He, Ne, Ar, Kr, Xe). Occasional determinations of the $^{228}Ra/^{226}Ra$ activity ratio were also made. Samples for major element and some trace element analyses (Chapter 9) were collected simultaneously with samples for radioelement and noble gas analysis.

The objectives of these analyses were to identify any seasonal changes in the thermal water composition and hence to place limits on the extent of thermal water mixing with the non-thermal waters of the superficial aquifers in the Mesozoic sediments of the Bath district. The mean contents of those radioelements (^{226}Ra and ^{222}Rn) which may be significant radiological health hazards have also been established.

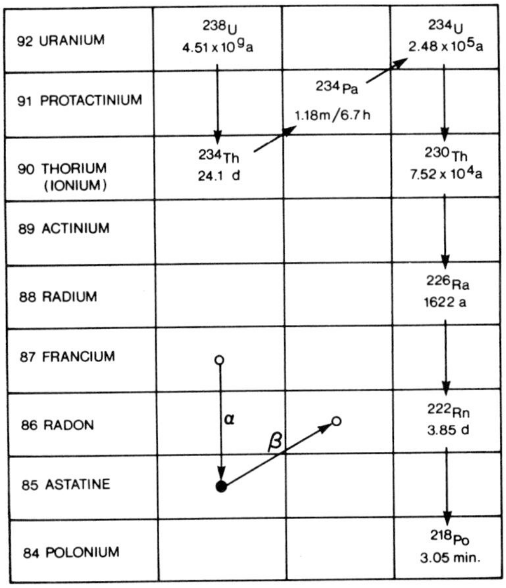

Fig. 10.1. Part of the ^{238}U decay series showing the decay relationships between ^{238}U, ^{234}U, ^{226}Ra and ^{222}Rn.

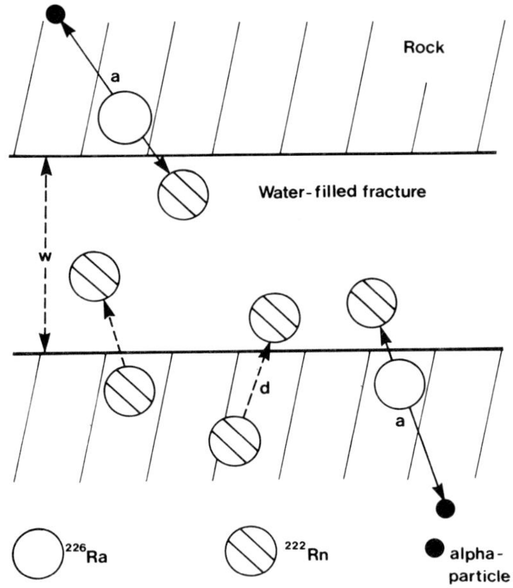

Fig. 10.2. The dissolution of ^{222}Rn by the alpha-recoil process (a) and diffusive release (d) from the rock surface. (a) The ^{222}Rn atom may be ejected into solution by the recoil-energy imparted by the alpha-particle decay of its parent ^{226}Ra if the latter is situated close to the rock/water interface. (d) ^{222}Rn atoms formed by ^{226}Ra decay within the body of the rock must diffuse to the rock/water interface before solution.

Table 10.1. Radioelement contents of the thermal waters from the Stall Street Borehole, the Hetling Spring and Cross Bath Borehole in the period from June 1984 to January 1986.

		Stall Street	Hetling Spring	Cross Bath
Uranium,	μg/litre			
	range	0.025–0.051	0.027–0.053	0.055
	mean (*n*)	0.038 (21)	0.035 (9)	0.055 (1)
$^{234}U/^{238}U$	activity ratio			
	range	2.86–3.64	2.85–3.53	2.77
	mean (*n*)	3.37 (21)	3.03 (9)	2.77 (1)
Radium:	^{226}Ra, pCi/litre			
	range	7.2–16.0	8.5–14.9	10.2
	mean (*n*)	11.8 (76)	10.8 (35)	10.2 (1)
Radon:	^{222}Rn, pCi/litre			
	range	1650–4800	3300–8600	750–850
	mean (*n*)	2380 (46)	5375 (20)	800 (2)
Helium:	^{4}He, $10^{-8}cm^3STP/cm^3H_2O$			
	range	1160–2600	590–1460	700
	mean (*n*)	1800 (21)	1000 (9)	700 (1)

1 picocurie (pCi) = 0.037 becquerels (Bq) = 0.037 disintegrations per second
n = number of measurements.

The radioelements which are present in the thermal waters, and also to a lesser extent in the local shallow groundwaters, belong to the ^{238}U decay series (Fig. 10.1). In general the solution of radioelements by a groundwater is dependent upon its chemical character, that is its pH, redox potential, salinity and the particular dissolved species present. The radiogenic ^{4}He, ^{222}Rn, ^{226}Ra and ^{238}U contents of the thermal springs are summarized in Table 10.1 and are compared with data for various local groundwaters in Table 10.2. Analyses for the thermal waters of Derbyshire and South Wales have been included for comparison. It may be noted that high ^{222}Rn and ^{226}Ra contents are characteristic of

Table 10.2. Helium-4 and radioelement contents of groundwaters from the region surrounding Bath and Bristol.

Formation and site	^{4}He *	^{222}Rn pCi/l	^{226}Ra pCi/l	^{238}U μg/l	$^{234}U/^{238}U$ A.R.	$^{226}Ra/^{238}U$ A.R.
Jurassic						
Inferior Oolite/Midford Sand						
Monkswood	—	470	0.10	1.00	1.28	0.30
Lias						
Norton Malreward	62	1330	0.54	2.12	1.19	0.77
Triassic						
Keuper						
Keynsham	29	590	0.40	3.99	1.25	0.30
Weston, Bath	1430	1600	0.21	1.12	2.35	0.37
Carboniferous						
Pennant Series						
Hanham	6	440	0.31	0.88	1.19	1.07
Coal Measures						
Edford	8	65	0.15	0.42	2.04	1.07
Carboniferous Limestone						
Gurney Slade	4	160	0.20	0.42	1.36	1.43
Hapsford	4	160	—	0.48	1.42	—
Mells	4	30	—	0.60	1.48	—
Shepton Mallet	—	60	0.39	1.36	1.53	0.87
Devonian						
Old Red Sandstone						
Three Ashes	6	740	0.33	0.12	2.86	8.25
Waldegrave	6	350	0.13	0.21	1.96	1.86
Thermal springs						
a) Bristol						
Hotwells, 24.3°C	410	1170	0.99	0.85	1.99	3.5
b) Bath						
King's Spring, 45.5°C	1080	2320	10.3	0.06	2.62	515.0
Hetling Spring, 46.5°C	1000	5370	10.8	0.03	3.03	935.0
Cross Bath, 45.3°C	700	800	10.2	0.06	2.77	562.0
Stall St. b/h, 45.5°C	1840	2380	11.8	0.03	3.37	941.0
Kingsmead b/h						
Lias inflow	640	290	—	0.94	2.78	—
Trias/Carb., 36°C	2300	880	1.77	0.06	4.98	88.5
Sports Centre b/h, 20°C	3525	120	1.04	0.18	7.62	17.3
c) Derbyshire						
Buxton, 28°C	50	2090	0.93	2.78	1.69	1.0
Matlock, 20°C	75	2400	—	1.78	2.56	—
d) South Wales						
Taff's Well, 19°C	540	120	3.7	0.70	2.10	16.1

* in units of $10^{-8}cm^3STP/cm^3H_2O$. A.R. = activity ratio.
1 picocuire (pCi) = 0.037 becquerels (Bq) = 0.037 disintegrations per second

thermal waters. Also, because of their reducing character, these waters have relatively low ^{238}U contents and high $^{234}U/^{238}U$ activity ratios. The combination of high ^{226}Ra contents with low ^{238}U contents results in characteristically high $^{226}Ra/^{238}U$ activity ratios in thermal waters.

The natural radioelement contents of possible source rocks are important parameters for explaining the radioelement content of the thermal waters. Core samples from the Kingsmead Square Borehole were therefore analysed for their radioelement contents by gamma-ray spectrometry. These results show that the highest U, Th and K contents occur in the clays which overlie the Blue Lias Limestones and in the Rhaetic and Triassic formations (Table 10.3). The mean radioelement contents of the Carboniferous Limestone are 1.7 ppm U, 6.3 ppm Th and <0.1% K.

Table 10.3. Radioelement and radiogenic helium contents of local rocks.

Formation	Age (Ma)	U (ppm)	Th (ppm)	Li (ppm)	Gt *	$^{3}He/^{4}He$ $\times 10^{-8}$
Lias clays	180	4.5	20.0	180	52,000	5.66
Blue Lias	180	2.5	7.0	—	23,000	—
Trias	225	6.0	18.0	100–139	72,000	3.20–4.36
Carboniferous Limestone	280–345	1.75	6.0	100–150	280,000–340,000	3.99–5.60

* accumulated helium in units of 10^{-8} cm^3 STP/cm^3 H_2O
Stored helium (Gt) = t $\{1.19 \times 10^{-13}[U] + 2.88 \times 10^{-14}[Th]\}$
Density = 2.6 g cm^{-3} for all cases

Radon solution and the fracture flow system

Radon (^{222}Rn; half-life 3.8 days) has long been known to occur in thermal springs throughout the world. It was first discovered in the King's Spring by Strutt (1904) just a few years after the discovery of radioactivity. Later measurements were made by Ramsey (1912), Munro (1928) and Jacobi (1949). During the early part of the twentieth century, the healing properties of the water were attributed to its radioactivity and radon gas was collected by a large lead-lined wooden funnel placed over the spring. This gas was used for inhalation by patients up until the 1930s. The first attempts to use the radioactivity of the water to constrain its geochemical origin were made by Andrews and Wood (1972).

The ^{222}Rn content of Bath thermal springs contributes the largest proportion of their total radioactivity. Most of the water movement through the Carboniferous Limestone must occur through fractures because the interstitial permeability of the limestone is very low. As discussed below, the radon content of groundwaters in a fracture flow system can be related to the rock radioactivity and the consequent ^{222}Rn flux from fracture surfaces. In this way, an estimate of the fracture width in the deep flow system has been obtained and used in conjunction with Darcy's law for fracture flow to model the thermal spring flow system.

The activity of ^{222}Rn released into a surrounding air or water phase for a constant flux of Rn from rock surfaces, increases with time, t, according to the equation:

$$A_t = A_e\,[1 - \exp(-^{222}\lambda t)] \quad (1)$$

where A_t and A_e are the ^{222}Rn activities in the air or water phase at time t and $^{222}\lambda$ is the decay constant of ^{222}Rn. The solution mechanisms for radon are illustrated in Fig. 10.2. Radon is the immediate decay product of ^{226}Ra (formed by ^{238}U decay) and may be released from a planar rock surface by:

(1) ^{222}Rn recoil following alpha-decay of ^{226}Ra atoms close to the surface (within the ^{222}Rn recoil range of 0.036 μm);
(2) ^{222}Rn diffusion through the crystalline lattice from production sites below the rock surface;
(3) ^{222}Rn diffusion along crystal defects, grain boundaries or microfractures from greater depths below the rock surface.

The recoil mechanism alone cannot account for the observed ^{222}Rn contents of groundwaters and the extent of ^{222}Rn release into water is generally greatest for rocks with more than one mineral phase present (Andrews & Wood 1972). Intergranular diffusion is unimportant for the Carboniferous Limestone for which the experimentally measured flux of ^{222}Rn from plane surfaces (for specimens cut from Whatley Quarry) is 3 atoms $s^{-1}\,m^{-2}$.

The ^{222}Rn content, [Rn] of water within a rock fracture may be calculated from the Rn flux, F ($s^{-1}\,m^{-2}$), from rock surfaces and the fracture dimensions using the equation (Andrews *et al.* 1986):

$$[\mathrm{Rn}] = (2F/w)\,[1 - \exp(-{}^{222}\lambda t)]\ (\mathrm{s}^{-1}/\mathrm{m}^3) \tag{2}$$

where t is the fluid residence time in the fracture and w is the width (m) of the fracture. If the residence time, t, is greater than a few half-lives of ^{222}Rn, the equilibrium Rn content of the fluids becomes:

$$\begin{aligned}[\mathrm{Rn}] &= (2F/w)\ (\mathrm{s}^{-1}/\mathrm{m}^3)\\ &= (2F/w)\ (\mathrm{Bq}/\mathrm{m}^3)\\ &= (2F/37w)\ (\mathrm{pCi/litre}).\end{aligned} \tag{3}$$

Most groundwaters, even those which penetrate to relatively shallow depth, have residence times greater than 20 days so that equilibrium ingrowth of Rn is certain to have occurred in the case of deeply circulating thermal waters.

The fracture width required for the Bath thermal water to attain its mean ^{222}Rn-content (2380 pCi/litre, Table 10.1) within the Carboniferous Limestone, calculated from equation (3), is 0.07 mm. This fracture width estimate may then be used in conjunction with Darcy's law for flow in a fractured system to calculate the time required for flow between recharge on the Mendips to Bath, if this fracture width is applicable along all of the flow path. Darcy's law relates the fracture width, fracture length and fluid transit time by the equation:

$$t = \frac{12\eta}{\Delta P}\left(\frac{l}{w}\right)^2 \tag{4}$$

where ΔP is the pressure differential, l is the length of the flow system, η is the fluid viscosity and w is the mean fracture width. For the piezometric head between Mendip and Bath (170 m) and a flow path length of 30 km (somewhat greater than the linear distance to allow for tortuosity), the calculated flow time from Mendip to Bath (age of the thermal water) is 44 years. This is increased to 58 years if the effective piezometric head is reduced to 120 m for the reason discussed below. The minimum groundwater age is therefore about 58 years. Static storage in palaeokarst cavities (which cannot be estimated) could increase the possible thermal water age to some hundreds of years or perhaps more.

Radium solution and the thermal water age

Early measurements of radium (^{226}Ra, half-life 1620 years) were made by Ramsey (1904) and Munro (1928). The measurement techniques used at this time, however, grossly overestimated ^{226}Ra which is only 0.5% of the ^{222}Rn activity (Table 10.1). The solution of ^{226}Ra by groundwater is caused by chemical dissolution or rock-etch processes and/or by alpha-recoil of ^{226}Ra on decay of ^{230}Th in the rock surface. The chemical solution of ^{226}Ra should increase in importance as the salinity of the groundwater increases, provided that sufficient barium or calcium ions are present in solution. The residence time of ^{226}Ra in solution depends upon barium and calcium solution/deposition as well as upon its half-life (1620 years). The activity of ^{226}Ra dissolved by the recoil mechanism increases with time according to an equation similar to that for radon:

$$^{226}A_t = {}^{226}A_e[1 - \exp(-{}^{226}\lambda t)] \tag{5}$$

where $^{226}A_t$ and $^{226}A_e$ are the ^{226}Ra activities at time, t, and at equilibrium, respectively, and $^{226}\lambda$ is the decay constant for ^{226}Ra. The ^{226}Ra activity in solution should, if alpha-recoil is the dominant mechanism, become constant after about 8000 years. However, geochemical processes in which ^{226}Ra participates, such as calcium carbonate solution/deposition or adsorption of ^{226}Ra on hydroxy complexes of oxidized Fe or Mn, usually prevent the ^{226}Ra activity in groundwaters from ever attaining this equilibrium value. In very reducing groundwaters such adsorption processes are absent and high ^{226}Ra contents are therefore generally found in evolved groundwaters such as geothermal waters and oilfield brines (Bloch & Key 1981; Andrews *et al.* 1987). Because of these chemical effects, quantitative estimation of groundwater age using ^{226}Ra is not possible, but in general high ^{226}Ra contents occur only in older, more evolved, groundwaters. The high ^{226}Ra contents of the Bath thermal waters, compared with young shallow goundwaters, may be attributed to their deep circulation, relatively high salinity and age.

Radium-226 in the thermal waters is the most significant radiological health hazard for ingestion of the water (Andrews & Wood 1974). It is present in amounts which are about 100 times those found in shallow groundwaters or in public water supplies of the Bath/Mendip area. Although the ^{226}Ra content of the thermal water is above the permissible level for drinking water, there is no hazard from the quantities taken in occasional 'tastings'. Although the ^{222}Rn content is much higher than that of ^{226}Ra, radon is only hazardous if it is released in significant quantities into confined air spaces where its decay products (radon daughters) can be ingested by respiration.

The ^{226}Ra/^{228}Ra ratio in the thermal water

The first few members of the ^{232}Th decay series are:

$$\begin{array}{cccc} {}^{232}_{90}\mathrm{Th} \xrightarrow{\alpha} & {}^{228}_{88}\mathrm{Ra} \xrightarrow{\beta} & {}^{228}_{89}\mathrm{Ac} \xrightarrow{\beta} & {}^{228}_{90}\mathrm{Th} \xrightarrow{\alpha} \\ 1.4 \times 10^{10}\ \text{years} & 5.7\ \text{years} & 6.13\ \text{hours} & 1.91\ \text{years} \end{array}$$

The ^{232}Th content of the thermal waters of Bath is below the detection limit of 0.01 μg/litre and the ^{238}U-series isotope, ^{230}Th, is also absent. However, ^{228}Ra, the decay product of ^{232}Th, has identical chemical properties to ^{226}Ra and is dissolved by similar mechanisms. The $^{226}Ra/^{228}Ra$ ratio in solution is therefore controlled by the relative availability of ^{238}U and ^{232}Th in the aquifer rock. The value expected for the Bath thermal water, based on the U/Th ratio in the Carboniferous Limestone is 0.9 compared with the mean measured value of 13.4. This suggests that uranium deposition has occurred on fracture surfaces along the flow path and enriched uranium on surfaces to about 15 times its concentration in the bulk rock. Figure 10.3 shows how such uranium deposition increases the ^{226}Ra solution rate. Consequently, the ^{234}U recoil rate and the ^{222}Rn flux should be proportionately increased, at least in the deeper and more reducing parts of the flow system. For such uranium deposition, the increased ^{222}Rn flux would contribute the observed ^{222}Rn concentrations to water within fractures with widths up to 1.0 mm. However, the measured ^{222}Rn content of the thermal water must, because of its short half-life, be acquired at the end of its flow path where uranium deposition is less likely to occur.

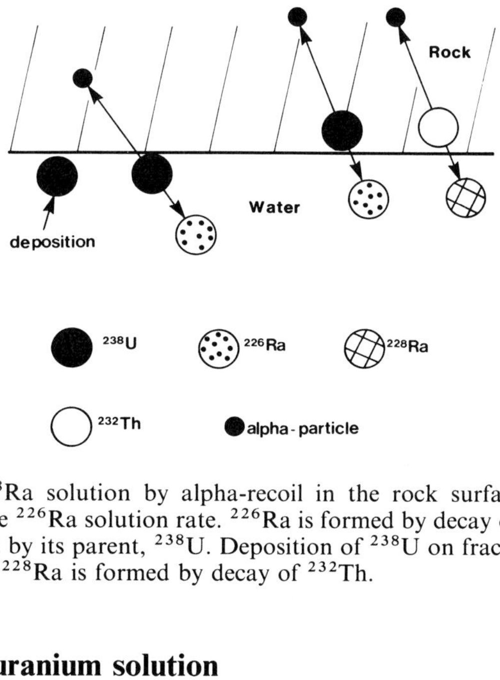

Fig. 10.3. ^{226}Ra and ^{228}Ra solution by alpha-recoil in the rock surface showing the effect of uranium deposition on the ^{226}Ra solution rate. ^{226}Ra is formed by decay of ^{230}Th which is situated in sites formerly occupied by its parent, ^{238}U. Deposition of ^{238}U on fracture surfaces can increase the dissolution of ^{226}Ra. ^{228}Ra is formed by decay of ^{232}Th.

The mechanism of uranium solution

The uranium content of the King's Spring water is so low that it has only become possible to determine it with the development of the modern technique of alpha-particle spectrometry. This has also enabled the ratio (in activity terms) of the two uranium isotopes, ^{238}U and ^{234}U, to be determined. Uranium in solution is stabilized as carbonate complexes of both oxidized uranium, U^{VI}, and reduced uranium, U^{IV}, and is dependent upon the pH, Eh and bicarbonate content of the water (Langmuir 1978). Radioactive equilibrium is established throughout the ^{238}U decay series within 1.25×10^6 years in closed systems containing ^{238}U and at equilibrium the ^{234}U and ^{238}U activities are equal. Very frequently, however, the $^{234}U/^{238}U$ activity ratio for uranium dissolved by groundwaters is not in equilibrium, generally being greater than unity. Such disequilibrium may be a consequence of either preferential chemical dissolution of ^{234}U or of alpha-recoil induced solution of ^{234}U as illustrated in Fig. 10.4. Preferential solution of ^{234}U rather than ^{238}U

atoms may occur because the former are the decay products of ^{238}U and must, therefore, be present in crystalline regions which have suffered damage during the decay process. It is also probable that the ^{234}U atoms are oxidized to the more soluble U^{VI} oxidation state during the decay process (Rosholt *et al.* 1963). Decay of ^{238}U atoms in the rock surface close to the rock–water interface may result in the direct ejection of alpha-recoil ^{234}Th atoms into the solution (Kigoshi 1971). These short-lived ^{234}Th atoms either decay in solution or are deposited on the rock surface where, on decay, the ^{234}U formed may readily be dissolved.

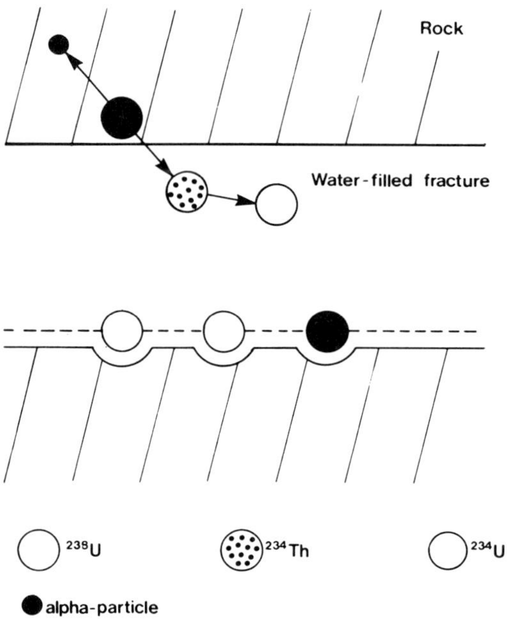

Fig. 10.4. Uranium solution processes: alpha-recoil solution of ^{234}U (top) and chemical etch solution of ^{238}U and ^{234}U (bottom). Decay of ^{238}U may *either* eject ^{234}Th directly into solution where it decays to ^{234}U after being adsorbed at the rock surface *or* cause the ^{234}U atom formed within the rock to be more easily dissolved because of lattice damage or oxidation. Slow chemical dissolution (etch) of the surface preferentially dissolves ^{234}U because of its oxidation state.

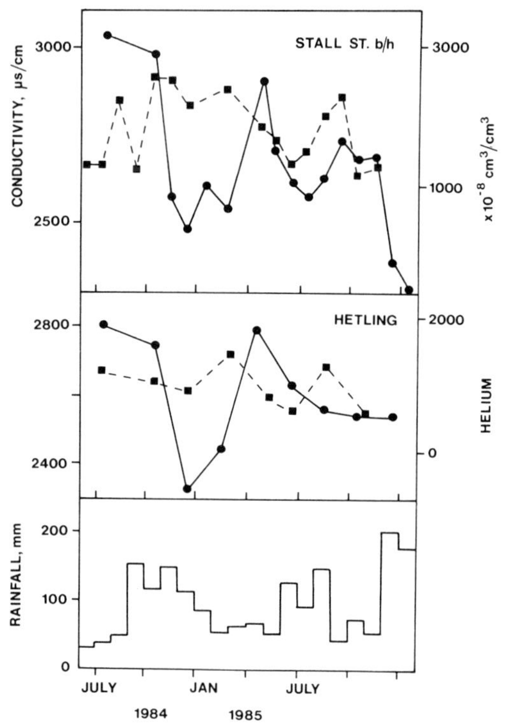

Fig. 10.5. Temporal variations of the conductivity (circles) and ^4He-content (squares) in thermal waters from the Stall Street Borehole and the Hetling Spring for June 1984 to January 1986. The local rainfall record is shown for comparison.

The $^{234}U/^{238}U$ activity ratio in solution, AR_t, is dependent upon the extent of rock surface from which ^{234}Th recoil can occur as well as upon the U-contents of the rock surface, $[U]_r$, and the solution, $[U]_s$. The extent of available rock surface may be expressed as the surface area, S, per unit volume of water. This may be calculated from the fracture width which was estimated from the radon content of the thermal water. For a groundwater in which uranium dissolution is no longer active (reducing conditions) the activity ratio for dissolved uranium, AR_t, changes with time according to the equation (Andrews *et al.* 1982b):

$$AR_t = 1 + (AR_i - 1)\exp(-{}^{234}\lambda t) + 0.25\,\rho SR[1 - \exp(-{}^{234}\lambda t)][U]_r/[U]_s \quad (6)$$

where AR_i is the initial activity ratio of the dissolved uranium as the groundwater enters the reducing zone ($t = 0$); $^{234}\lambda$ is the decay constant of ^{234}U; R is the recoil range of ^{234}Th and ρ is the rock density. An initially enhanced activity ratio may be a consequence of either preferential solution of ^{234}U or of alpha-recoil induced solution of ^{234}Th. However, the recoil process requires time for subsequent ^{234}U ingrowth to become significant and enhanced activity ratios for groundwaters close to recharge must be the result of preferential ^{234}U solution. The recoil becomes more significant as the residence time of the groundwater increases, or if uranium deposition occurs as the groundwater evolves towards more reducing conditions. It is evident from equation (6) that groundwaters with very low uranium contents, $[U]_s$, are most likely to undergo activity ratio increase due to alpha-recoil. This will tend to counterbalance the decay of any excess ^{234}U which was dissolved by preferential solution closer to recharge. If reducing conditions prevail so that uranium solution by etch processes has ceased, equation (6) may be used to calculate groundwater residence time from the observed $^{234}U/^{238}U$ activity ratio, the initial characteristic activity ratio at recharge and the specific surface in contact with the water phase.

Dissolved uranium and the age of the Bath thermal water

The uranium contents of the Bath thermal waters are in the range 0.01 to 0.02 pCi/litre (0.02–0.06 μg/litre) and the $^{234}U/^{238}U$ activity ratio ranges from 2.8 to 3.6 (Table 10.1). These low uranium contents are due to the reducing character of the thermal water. Shallow circulation groundwaters of recent origin within the Carboniferous Limestone have uranium contents in the range 0.42 to 1.36 μg/litre with activity ratios of 1.36 to 1.53. This shows that dilute solutions of CO_2 (from the soil zone) must preferentially dissolve ^{234}U in these groundwaters which are too young and too oxidizing for the alpha-recoil process to be operative. Laboratory etching of Carboniferous Limestone has conclusively proved that chemical etch processes can produce activity ratios up to 1.70 in the resulting solution.

The mean activity ratio for the King's Spring water may be the result of additional ^{234}U solution by the recoil process. Using the estimated specific surface with which the water is in contact (calculated from fracture width estimates based on ^{222}Rn contents) and theoretical ^{234}U recoil rates for new fracture surfaces in the Carboniferous Limestone, a minimum residence time of 55,000 years is necessary to produce an activity ratio of 3.0. Such a long residence time is not compatible with hydrogeological data and it must be concluded that either the enhanced activity ratio is entirely due to preferential chemical solution of ^{234}U, or that there is considerable enhancement of ^{234}U recoil due to uranium deposition on fracture surfaces as the recharge waters deposit uranium on entering the reducing zone.

Uranium which is dissolved by oxidizing recharge waters must subsequently be deposited on the surfaces of the flow system as the groundwater becomes reducing during migration. If the flow system has been active throughout the Holocene, at least 4.7 tonnes of uranium must have been mobilized and re-deposited in this manner. This estimate is based upon the difference between the uranium contents of the thermal water and recently recharged waters in the Carboniferous Limestone of the Mendip Hills and the present flow rate of the springs. Deposition of this amount of uranium in the flow system would increase alpha-recoil solution of ^{234}U. The $^{226}Ra/^{228}Ra$ ratio suggests that the ^{234}U recoil rate has been increased by a factor of 15 and the calculated groundwater residence time to generate the observed $^{234}U/^{238}U$ activity ratio in the King's Spring water under these conditions is reduced to 3,000 years.

Noble gas contents and recharge temperatures

Groundwater recharge temperatures may be estimated from the amounts of the atmospheric noble gases (Ne, Ar, Kr and Xe) which have been dissolved (Andrews & Lee 1979; Bath *et al.* 1979). It has been possible by this method to show, for example, that some groundwaters in the East Midlands Triassic Sandstone were recharged during the cooler climatic conditions that prevailed in the interstadial prior to the last (Devensian) glaciation. The stable isotope ratios ($^{18}O/^{16}O$ and $^{2}H/^{1}H$) of the water itself also reflect such climatic changes (Bath *et al.* 1979; Bath 1983). For the King's Spring thermal water both the recharge temperature and the stable isotopic composition (Chapter 9) show that atmospheric equilibration must have taken place during a similar climatic period to the present. The recharge time must therefore have occurred either during the Holocene (groundwater age less than 10,000 years) or more than 100,000 years ago. The other age considerations discussed in this paper show that it is the first of these alternatives that is the correct interpretation.

Dissolved helium and its isotopic ratio

Helium was dissolved in gases from the King's Spring by Sir James Dewar during the course of his investigations into gas liquefaction. In his report to the Corporation of Bath (1904), however, he recorded that the amount present was too small to be of commercial value. Although this remains true, the helium content of the thermal waters is much greater than that found in other groundwaters of the region and it is an indication of a deep origin for the water. Helium is formed by alpha-particle decay of uranium and thorium and their decay products and it is lost from geological formations of uniform composition by diffusion (Andrews 1985). For a diffusion coefficient typical of rock matrices, helium should be totally stored over the age of the rock at depths greater than a few tens of metres whereas, in an aqueous medium, total storage should occur at depths below a few hundred metres. In a fractured rock, helium may be transported from depth by water circulating in the fracture system. However, in narrow width fractures such as are likely in the thermal spring system, equilibration between the concentrations of helium stored in the rock and that dissolved by the water should occur within a few days.

The rate, G, of He generation in a rock depends upon its radioelement content and may be calculated from the equation

$$G = \rho(1.17 \times 10^{-13}[U] + 2.88 \times 10^{-14}[Th]) \ (cm^3/cm^3) \quad (7)$$

where ρ is the rock density (g/cm^3) and [U], [Th] are its U and Th contents (ppm). The total amount of stored helium in a formation of age t years and from which there has been no helium loss may then be obtained from the product Gt. Table 10.4 lists the calculated amounts of stored helium for various formations at depths where there is no helium loss. Helium analyses for the various thermal sources (Table 10.3) show that the ^{4}He contents of the thermal waters are much less than the possible stored helium in the Lias, Trias and Carboniferous Limestone. If the thermal water has equilibrated with Carboniferous Limestone at depth, then either it has not retained all its dissolved ^{4}He during its ascent or the rock must have retained much less helium close to the fracture surfaces than that estimated in Table 10.3. Such depletion of helium at fracture surfaces would be explained if they have been continuously flushed with fresh meteoric water for a considerable period of time. This process of He removal would be more efficient for a relatively low thermal water age.

Table 10.4. Helium-4 contents and $^3He/^4He$ ratios for dissolved He in the Bath thermal waters.

	^4He content $\times 10^{-8}$ cm^3/cm^3 H_2O	$^3He/^4He$* $\times 10^{-8}$	Li required (ppm)
Sports Centre b/h	2900–3700 (3350)	3.16 ± 0.25	101
		3.46 ± 0.25	101
Kingsmead Square b/h			
Lias inflow	530–810 (670)	—	
Carboniferous	1900–3200 (2450)	4.07 ± 0.24	
		4.17 ± 0.25	
King's Spring	1080	4.58 ± 0.23	119
Stall Street b/h	1160–2600 (1800)	—	
Hetling Spring	1000	—	
Cross Bath	700	—	

* determined in the Department of Earth Sciences, University of Cambridge.
( ) mean value

The isotopic ratio, $^3He/^4He$, of the dissolved helium can indicate the probable source of helium in the various waters. This ratio is determined by the Li, U and Th contents of a rock matrix and calculated values are given in Table 10.3. These may be compared with the measurements for helium in the thermal waters (Table 10.4). It may be noted that the Sports Centre Borehole which produces water from the Dolomitic Conglomerate (Trias) contains more helium than the thermal waters and that its isotopic ratio is compatible with its helium having an origin in the Trias. The King's Spring waters have a higher isotopic ratio which suggests that they probably have a Carboniferous Limestone source. The waters at Kingsmead Square appear to be a mixture of these two types.

Temporal variations in radioelement and helium contents

The results of radioelement measurements on the thermal waters are summarized in Table 10.1. Temporal variations of the uranium content, $^{234}U/^{238}U$ activity ratio and ^{222}Rn content occur for the Stall Street Borehole and for the Hetling Spring. The variations of conductivity and helium content are compared with the mean monthly rainfall record in Fig. 10.5. The radioelement content changes could generally be explained by admixture of up to 10% of groundwaters from the more superficial Triassic and/or Jurassic aquifers with a thermal water originating in the Carboniferous Limestone. Increased helium contents correlate with decreases in the conductivity of the thermal waters and the observed changes in the Stall Street Borehole and in the Hetling Spring are synchronous. The changes follow periods of increased rainfall after a delay of months. These responses may be the result of increased contributions from the Jurassic and Triassic sequences respectively causing dilution and the increase in helium content. A series of $^3He/^4He$ isotopic measurements is needed to confirm the origin of the increased helium.

Flow and storage considerations

It has been suggested in Chapter 8 that groundwater storage in the Carboniferous Limestone is unlikely to be sufficient to allow an age greater than a few hundred years for the thermal water if the present spring flow has been constant. The fracture width for flow in the Carboniferous Limestone, estimated on the basis of ^{222}Rn measurements on water and rock, has been used to calculate the transit time (equation (4)) from Mendip to Bath for Darcian flow. It is probable that the fracture width for fault system along which the thermal waters ascend is somewhat larger than that over most of the flow path in the deep Carboniferous Limestone. Table 10.5 gives the results of such calculations for a model which attributes about one third of the head loss to driving a fault system with a fracture width of 1.0 mm and about two thirds of the head loss to driving the Mendip/Bath flow through a system with fracture widths of 0.07 mm. With reasonable assumptions for path lengths, the thermal water rise time is estimated to be about 40 hours which is sufficiently short to avoid much cooling during ascent. The transit time (groundwater age) from Mendip is about 58 years. Longer effective transit times are possible if there is groundwater storage in palaeokarst regions within the limestone at depth, as indicated by Table 10.5. For a groundwater age of 5,000 years, the required karst storage is about 4% of the Carboniferous Limestone volume within a 30° segment from Bath to Mendip. This is about 150 times the fracture system volume required to support the thermal water flow. A palaeokarst with 4% of void space within the limestone is highly improbable and the thermal water age is therefore not likely to be more than a few hundred years (Table 10.5) if storage is restricted to this sector of the Carboniferous Limestone. If the storage involved the entire Bath–Bristol basin, longer groundwater ages would be possible.

Table 10.5. A fracture flow model for the thermal springs of Bath.

	Mendip–Bath flow path	Fault-aided ascent from 2500 m depth
Head loss (m)	120	50
Pressure differential (MPa)	1.2	0.5
Fracture length (m)	30,000	2,500
Fracture width (mm)	0.07	1.00
Fracture volume (Mm^3)	27	0.0045
Fracture transit time	58 years	42 hours
Karst storage required for total flow of 18 litres/s		
Thermal water age, years	200	5,000
Storage required* (%)	0.18	4.0

* Percentage of void space required in a 30° sector of uniformly thick (700 m) Carboniferous Limestone between Bath and the Mendips to support the stated groundwater age at constant flow rate.

Conclusions on the thermal water flow system

An overall synthesis of the conclusions from the geochemical results presented in Chapter 9 and of the radiochemical evidence presented in this paper is outlined below. This places some overall constraints on the origin and age of the thermal water.

(1) The isotopic composition and noble gas recharge temperatures of the thermal water show that it is of meteoric origin and that it was recharged within the last 10,000 years.

(2) Its composition indicates that it has chemically equilibrated with the Carboniferous Limestone. The similarity of the carbon isotope composition of dissolved bicarbonate in the thermal water to that of the limestone is conclusive evidence that this equilibration has occurred.

(3) Chemical geothermometers suggest that the maximum temperature attained was 64–96°C; probably closer to the lower estimate. For a geothermal gradient of about 20°C/km, water circulation to a depth of at least 2500 m is indicated.

(4) The fracture width in the flow system through the Carboniferous Limestone, deduced from the ^{222}Rn content of the thermal waters, is 0.07 mm. Although fracture widths up to 1.0 mm are possible where uranium deposition has occurred, this is only probable over limited parts of the flow path.

(5) Considerations of fracture flow based upon Darcy's law, show that fracture transit times from Mendip to Bath are about 58 years. The thermal water age can only be more than a few hundred years if there is an extensive palaeokarst system in the deep Carboniferous Limestone.

(6) The carbon-14 content of the thermal water is equivalent to a decay time of about 20,000 years (Chapter 9). However, removal of ^{14}C due to bicarbonate/carbonate exchange with the Carboniferous Limestone would also reduce the ^{14}C content of the water. Equilibration of the stable isotope, ^{13}C, with the rock has occurred and suggests that such ^{14}C-removal would be significant. Equilibration of the stable carbon isotopes between solution and rock matrix generally requires several thousand years in other environments. At the high temperatures ($>60°C$) reached by the thermal water, chemical reactions proceed much more rapidly and it is likely that isotopic equilibration could be attained in a shorter time.

(7) The uranium isotopes ($^{234}U/^{238}U$ activity ratio) suggest that the thermal water has resided for a considerable time under reducing conditions deep in the Carboniferous Limestone. The estimated age of 3,000 years is very dependent upon the extent of uranium deposition at depth and younger ages are possible if such uranium deposition has been underestimated.

(8) The high ^{226}Ra content is typical of geochemically evolved waters and suggests that the thermal water is old, probably at least several hundreds of years.

The balance of evidence suggests that the Bath thermal waters were probably recharged between a few hundred and a few thousand years ago. These ages require some karstic storage in the limestone at depth and the longer ages would only be possible if the storage is dispersed throughout the Bristol–Bath basin. The geochemistry of the thermal springs is controlled by equilibration within the Carboniferous Limestone and there is evidence of a response to seasonal rainfall patterns. Some admixture with minor amounts of more superficial groundwaters could explain their response. The radioelement contents, although greater than for shallow groundwaters, are an insignificant hazard for public tasting of the thermal waters.

Acknowledgements. The collaboration and interest of W. M. Edmunds, G. A. Kellaway and W. S. Stanton over many years of investigations on the Bath Thermal Springs is gratefully acknowledged. N. Hussain, D. J. Ford and M. J. Youngman are thanked for their contributions by experimental work on radioelement and noble gas measurements and for useful discussions. The Bath City Council kindly granted access to the spring sources and the Natural Environment Research Council are thanked for their support by research grant GR3/2622.

References

Andrews, J. N. 1985. The isotopic composition of radigenic helium and its use to study groundwater movement in confined aquifers. *Chemical Geology*, **49**, 339–351.

——, Burgess, W. G., Edmunds, W. M., Kay, R. L. F. & Lee, D. J. 1982a. The thermal springs of Bath. *Nature*, **298**, 339–343.

——, Giles, I. S., Kay, R. L. F., Lee, D. J., Osmond, J. K., Cowart, J. B., Fritz, P., Barker, J. F. & Gale, J. 1982b. Radioelements, radiogenic helium and age relationships for groundwaters from the granites at Stripa, Sweden. *Geochimica Cosmochimica Acta*, **46**, 1533–1543.

——, Hussain, N., Batchelor, A. S. & Kwakwa, K. 1986a. ^{222}Rn solution by circulating groundwaters in a "hot dry rock" geothermal reservoir. *Applied Geochemistry*, **1**, 647–657.

—— & Lee, D. J. 1979. Inert gases in groundwater from the Bunter Sandstone of England as indicators of age and palaeoclimatic trends. *Journal of Hydrology*, **41**, 233–252.

—— & Wood, D. F. 1972. Mechanism of radon release in rock matrices and entry into groundwaters. *Transactions of the Institute of Mining and Metallurgy*, **B82**, 198–209.

—— & ——, 1974. Radium-226, radon-222 and lead-210 in Bath Thermal Springs compared with some environmental waters. *Health Physics*, **27**, 307–310.

——, Youngman, M. J., Goldbrunner, J. E. & Darling, W. G. 1987. The geochemistry of formation waters in the Molasse basin of Upper Austria. *Environmental Geology and Water Sciences*, **10**, 43–57.

Bath, A. H. 1983. Stable isotopic evidence for palaeo-recharge conditions of groundwater. *In*: *Palaeoclimates and palaeowaters: a collection of environmental isotope studies*. I.A.E.A., Vienna, 169–186.

——, Edmunds, W. M. & Andrews, J. N. 1979. Palaeoclimatic trends deduced from the hydrochemistry of a Triassic Sandstone aquifer, United Kingdom. *In*: *Isotope Hydrology 1978, II*. I.A.E.A., Vienna, 545–568.

Bloch, S. & Key, R. M. 1981. Modes of formation of anomalously high radioactivity in oil-field brines. *Bulletin American Association Petroleum Geologists*, **65**, 154–159.

Burgess, W. G., Edmunds, W. M., Andrews, J. N., Kay, R. L. F. & Lee, D. J. 1979. The hydrogeology and hydrochemistry of the thermal water in the Bath–Bristol basin. Report EEC contract 084-76-EG (UK), British Geological Survey.

Jacobi, R. B. 1949. The determination of radon and radium in water. *Journal of the Chemical Society*, S314–S318.

Kigoshi, K. 1971. Alpha-recoil ^{234}Th: dissolution into water and the ^{234}U/^{238}U disequilibrium in nature. *Science*, **173**, 47–48.

Langmuir, D. 1978. Uranium solution mineral equilibria at low temperatures with applications to sedimentary ore deposits. *Geochimica Cosmochimica Acta*, **42**, 547–569.

Munro, J. M. H. 1928. Report to the Hot Mineral Baths Committee on "Radioactivity in the Bath Thermal Waters", Fyson, Bath.

Ramsey, Sir W. 1912. The mineral waters of Bath. *Chemical News*, **105**, 133–135.

Rosholt, J. N., Shields, W. R. & Garner, E. L. 1963. Isotopic fractionation of uranium in sandstone. *Science*, **139**, 224–226.

Strutt, R. J. 1904. Radioactivity of certain minerals and mineral waters. *Proceedings of the Royal Society of London*, **73**, 191–197.

From Kellaway, G. A. (ed.) *Hot Springs of Bath*, pp 171–177

11. Regional hydrodynamic influences on the Bath–Bristol springs

W. G. Burgess, J. H. Black & A. J. Cook

Introduction

Background

Thermal springs are uncommon in stable continental regions and those at Bath and Bristol have been the subject of conflicting hypotheses. Rastall (1926) invoked a deep-seated volcanic origin for the water and heat. Donovan (1957) argued for a meteoric source of water, and suggested the Carboniferous sandstones of the Coal Measures as the most likely aquifer. More recently, on the basis of an integrated hydrochemical and isotopic study, a fuller conceptual model of the hydrogeological system that gives rise to the springs has been proposed (Andrews *et al.* 1982), which identifies the Carboniferous Limestone as the principal storage aquifer. The essential features of this 'Mendips' model (Fig. 11.1) are as follows.

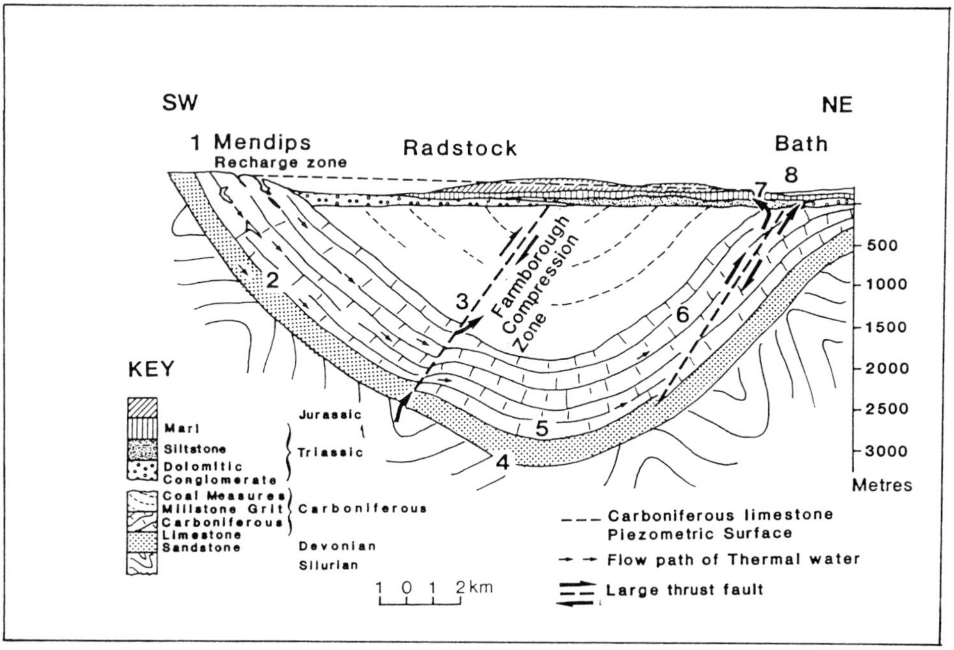

Fig. 11.1. A conceptual hydrogeological scheme for the Bath thermal springs: the 'Mendips' model. *Notes* 1. Recharge at the Carboniferous Limestone outcrop of the Mendips 2. Down gradient flow beyond the mature, phreatic, karstic system 3,4. Possible intercommunication between the Lower Palaeozoics, the Carboniferous Limestone, and/or the Upper Carboniferous at depth 5. Storage and chemical equilibration within the Carboniferous Limestone, at minimum temperature 64°C, (49°C minimum for Hotwells flowpath) 6. Relatively rapid upward flow in the thrust fault system associated with Hercynian tectonism 7. Recharge of Triassic siltstones by thermal water at Bath; at Hotwells, where there is no Triassic cover, the thermal water mixes with shallow groundwater of the Carboniferous Limestone 8. Discharge at the thermal springs, 46.5°C at Bath (24°C at Hotwells)

The thermal water is demonstrably meteoric. Under the prevailing geothermal gradient meteoric water circulates to a depth sufficient to be heated to a maximum temperature of 80°C (176°F) (range of uncertainty 64°C to 96°C) before discharging at 46.5°C (116°F) at Bath, cooling on ascent but not mixing appreciably with other groundwater. A separate but related cell of the flow system is heated to a maximum temperature of 60°C (140°F) before mixing with ambient groundwater of the Carboniferous Limestone and discharging at 24.3°C (75.7°F) at Hotwells, Bristol. Uranium is mobilized and redeposited in the flow system but uranium concentration and heat generation by radioactive decay is not responsible for the high temperature of the water. Rainfall on the Mendips acts as recharge to the system, and circulating groundwater equilibrates chemically within the Carboniferous Limestone during its northerly transit under a hydraulic gradient defined by the head difference between the water table on the Mendips and the pressure head of the thermal springs. The high temperature is due to circulation to a depth of at least 2700 m (Bath) and 1950 m (Hotwells) in a region where the conductive heat flow (about 40 mW/m^2) is lower than the average for the UK landmass. The bulk of the flow has a circulation time of less than 10,000 years but the residence time of a small component of the system is much longer and may involve transit in the Old Red Sandstone. Fractures associated with the Avon thrust fault allow flow of the thermal water to the surface. A proportion of the rising thermal water below Bath is released to the Triassic siltstones, which unconformably overlie the pre-Permian surface, as an approximately circular recharge mound. The thermal water cools and undergoes minor chemical re-equilibration within the Triassic.

Hydraulic basis for the regional model

Observed features of the thermal groundwater flow system must be satisfied by any hypothesis that seeks to explain them. Thus the thermal, chemical, and hydraulic aspects must all be in accord. The 'Mendips' model outlined above satisfies the majority of the chemical data, and accounts for observations and interpretations of the water's thermal history, but is supported by a hydraulic analysis that is necessarily limited by the paucity of the available hydrogeological data. This simplified hydraulic analysis has suggested a transit time of thermal water flow from the Mendips to Bath of 4,000 years (Andrews *et al.* 1982), but even the best estimates of the hydrogeological parameter values may be in error by orders of magnitude. The scale of fissuring and fracturing at depth, which would determine both the effective porosity and the gross transmissivity of the system, for example, is wholly undetermined. Quantitative hydrogeological data for the area (Burgess *et al.* 1979), particularly from the central parts of the basin, are too few adequately to support the current conceptual model. In view of this it is reasonable to subject the hypothesis to certain questions, and thence to explore the limits of its validity.

In particular, there is doubt as to whether northerly flow in the Carboniferous Limestone as far as Bath should reasonably be expected against the opposing piezometric head due to the Cotswold scarp and/or the Jurassic outliers to the south and west. The Mendips are an attractive proposal as the recharge area because of the piezometric head they offer. However, this is without consideration of the hydraulic benefit due to temperature induced density differences between the cold 'recharge' limb and the heated 'discharge' limb of the flow system, which may be equivalent to a head difference of as much as 40 m.

A major element of the 'Mendips' model is the $\delta^{13}C$ character of the springwater (average value −1.6‰), which is definitive of a marine rather than terrestrial origin for the source of dissolved bicarbonate, and hence favours the Carboniferous Limestone over the Devonian siltstones, Coal Measures sandstones or Triassic sandstones as the major storage aquifer at depth. The chemical data, recently supported by results from the drilling investigations in Bath City centre, are incontrovertible evidence for the Carboniferous Limestone as the storage aquifer at depth. The several marine limestones in the Lower Jurassic succession of the Cotswold scarp in close proximity to Bath, which also supply a considerable piezometric head, can justifiably be discounted as a source of the springs on the simple grounds of the temperature of the thermal water. Also, certain strata, for example the Midford Sands of the Middle Lias, have permeability that might be orders of magnitude higher than the mudstones of the Lower Lias; in which case they would be expected to limit the scale of vertical flow by allowing flow to springs near the base of the scarps. Perrenial springs do indeed flow from the base of the Midford Sands. Deep groundwater flow must also be inferred because it can be shown that uranium concentration in the Triassic cannot be at a level sufficient to act as the sole source of heat (Burgess *et al.* 1979). However, the influence of the Jurassic scarp and of Jurassic outliers on the potential for groundwater flow in the Carboniferous Limestone, not fully considered before, is the main subject of this chapter. The implications of the study are discussed with reference to the original 'Mendips' model.

Analytical description of regional groundwater flow

It is well documented that topographic relief has a marked effect on regional groundwater flow (Toth 1963, Freeze & Witherspoon 1967). Thick clays, although impermeable in the context of groundwater resources, have a significant long-term influence on regional flow systems. The problem can be investigated by the use of analytical models developed recently for application to long-term groundwater movement in respect of contaminant migration (Black *et al.* 1986), which demonstrate the manner and circumstances in which topographically related piezometric controls have influence over groundwater flow in subjacent aquifers across thick intervening argillaceous strata.

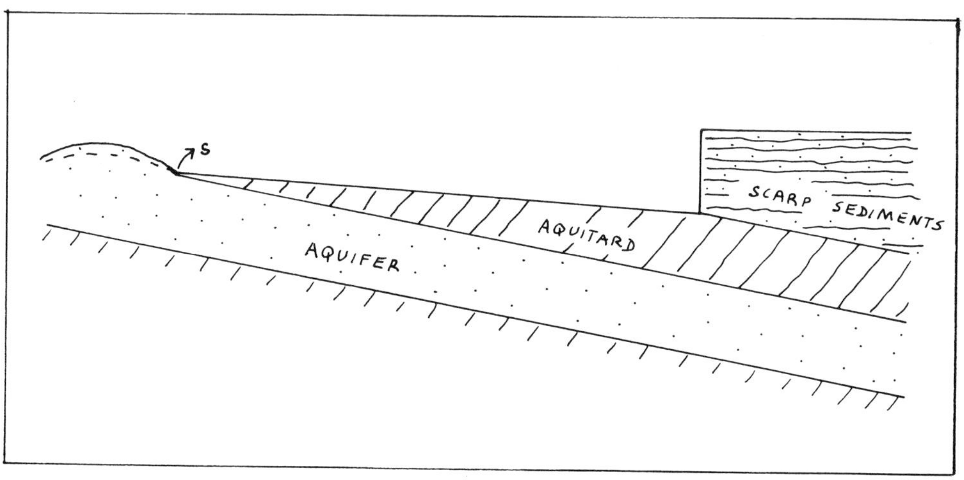

Fig. 11.2. The wedge approximation to a scarp/vale cross section (after Black & Barker 1981).

The analytical expressions are principally derived from solutions describing two-dimensional regional flow (Black & Barker 1981) for which the basic geometry is outlined in Fig. 11.2. An aquifer dips from its outcrop area below an argillaceous (leaky) wedge, beyond which rises an escarpment. Boundary conditions are imposed by the elevation of springs at the feather edge of the aquitard, by the water table in the aquitard which is assumed to follow the ground surface, and by the elevated water table on the escarpment. Expressions describing the head in the aquifer, $h(x)$, the horizontal gradient in the aquifer, dh/dx, and the vertical hydraulic gradient in the aquitard, dh/dz, are derived by Black & Barker (1981). Among the features of the flow geometry that can be investigated are:

(i) the direction of flow in the aquifer, under the competing influences of the aquifer outcrop and the scarp;
(ii) the points of flow balance in the aquitard and aquifer, if they exist;
(iii) the relative contribution of leakage through the aquitard to flow in the aquifer.

Additional expressions have been derived (J. A. Barker, pers. comm.) to describe groundwater potentials in an aquitard and aquifer caused by an overlying symmetrical mound-like head anomaly, such as might be caused by an isolated hill of low-permeability material.

The analytical expressions have been applied to the 'Mendips' model below. The results, their implications, and the requirements for field data to quantify the regional flow system are discussed.

Hydraulic and structural constraints on a regional model

The geology as related to the hydrogeology of the Bath–Bristol basin has been outlined elsewhere (Burgess *et al.* 1979; Andrews *et al.* 1982). Interpretation of the geological structure in detail is still a question of debate (Kellaway & Hancock 1983; Williams & Chapman 1986), but there is concensus that southerly dipping Variscan thrust faults disrupt the pre-Triassic strata at least to the base of the Carboniferous Limestone and probably deeper.

The hydrogeological features (piezometry, hydraulic conductivity) are listed in Table 11.1; constraints relate more to reasonable expectations of the limits to parameters that are largely unquantified, than to measured data. In practice the analytical expressions employ ratios and differences rather than absolute values; Table 11.1 indicates the basis for the chosen ranges of parameters used.

Table 11.1. Hydraulic constraints on a regional model.

Hydraulic conductivity		
Jurassic mudstones	(K')	10^{-10}–10^{-12} m/s
Carboniferous Limestone	(K)	10^{-6}–10^{-10} m/s
$\log (K/K')$		2–6
Piezometry		
Mendips	90–180 m	
Cotswold scarp	60–170 m	
Jurassic outlier	175 m	
Dundry Down	40–180 m	
Bath Spring	30–40 m	

The structural geometry must be simplified to make it amenable to analytical treatment. Three simplified geometries are proposed (Fig. 11.3) by which to investigate certain features of the regional flow system. The geometry illustrated in Fig. 11.3a has been used to investigate the range of conditions under which the Cotswold scarp might be expected to balance flow in the Carboniferous Limestone and prevent flow to Bath. The geometry illustrated in Fig. 11.3b has been used to investigate the influence of Jurassic outliers on flow in the aquifer at greater depths in the basin. Following this, a more realistic geometry (Fig. 11.3c) has been used to investigate the balance of flow in the aquifer and the overlying aquitard in the central regions of the basin.

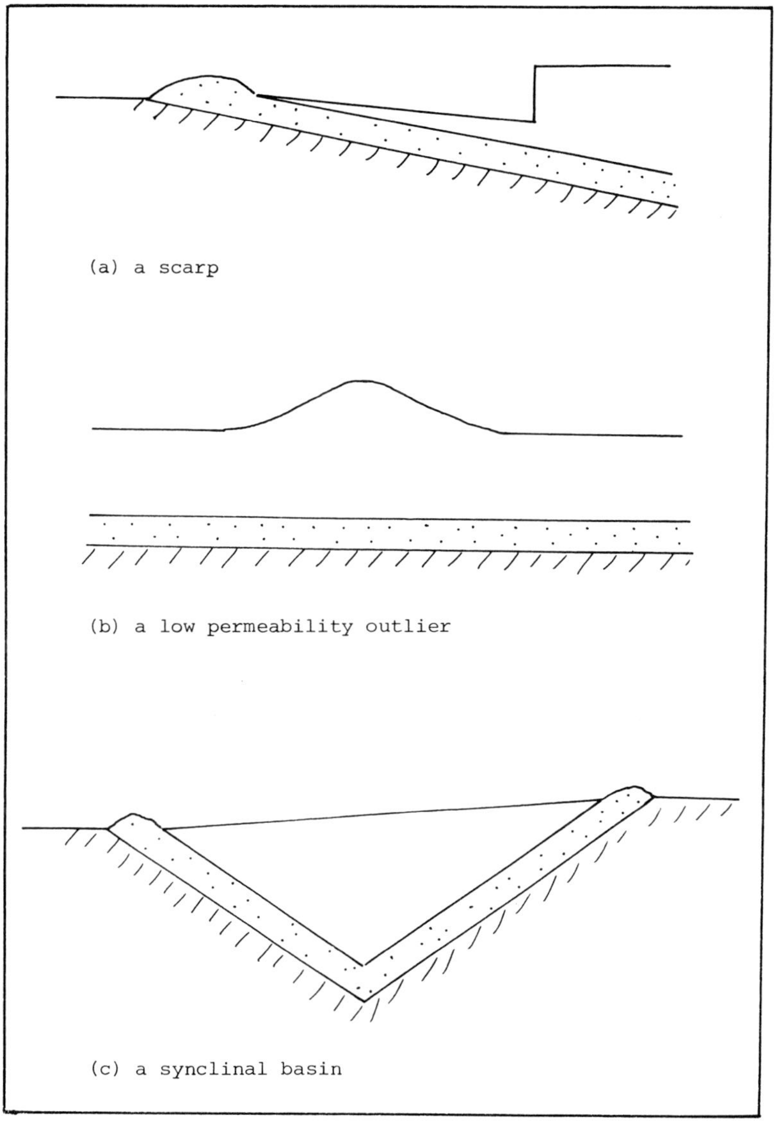

Fig. 11.3. The simplified geometries used to test influences on the groundwater flow system: (a) scarp; (b) a low permeability outlier; (c) a synclinal basin.

Results and discussion

Effects of topography

The model results show that groundwater in the Carboniferous Limestone can be expected to flow to Bath under all conditions envisaged in Table 11.1 for which $\log(K/K')$ is greater than 2.5 (geometry 3a and 3b). Surface topography has a significant effect on vertical flow in the aquitard, but flow to Bath is not impeded and the contribution of leakage to flow in the aquifer has only a minor influence on horizontal gradients in the aquifer.

Balance of flow in the aquitard

Water in the semi-confining layer consistently flows vertically upwards as leakage from the downflow limb of the subjacent aquifer, and subsequently downwards as leakage augmenting flow in the aquifer. The Carboniferous Limestone is thus recharged along part of its length by vertical leakage through the Jurassic and Triassic cover, and the Pennant Series and Coal Measures. While the model is not valid for the conditions local to Bath where thermal water recharges the Triassic strata, upwards leakage from the Carboniferous Limestone at depth under the Radstock coalfield along the Farmborough Compression Zone should not be expected (cf. note 3 of 'Mendips' model, Fig. 11.1). The contribution of leakage to flow in the aquifer depends on the ratio of hydraulic conductivities K/K' (Fig. 11.4) but becomes small (about 5% of flow from the outcrop) for $\log(K/K')$ greater than 3.

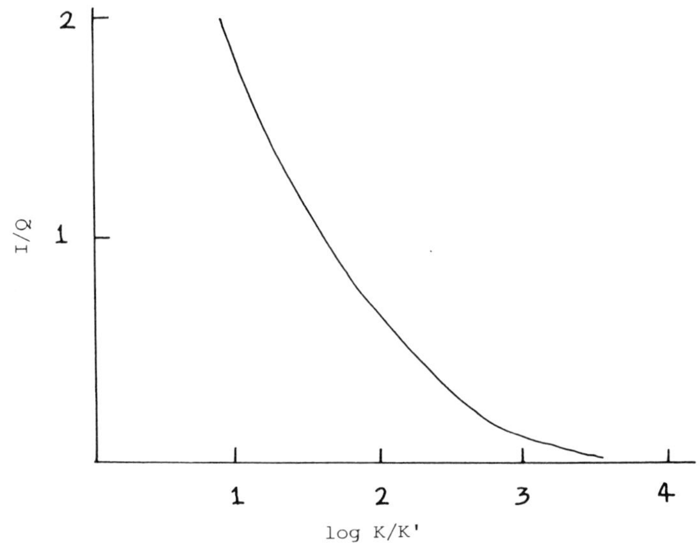

Fig. 11.4. The contribution of leakage from the aquitard to flow in the aquifer, as influenced by the relative permeabilities K/K', where K is the aquifer permeability, K' is the aquitard permeability, I is the flow derived from leakage and Q is the flow derived from outcrop.

Implications for the Mendips model

The 'Mendips' model for the Bath thermal springs is justified under a range of conditions which reflect uncertainty in the magnitude of the controlling hydraulic parameters, and there are therefore no hydraulic grounds for challenging its plausibility.

However, the analysis highlights one feature that was not considered previously; the importance of downwards leakage to the aquifer via overlying strata. It is suggested here that this water might subsequently re-equilibrate chemically within the Carboniferous Limestone, and account for the small proportion of very ancient water that has been identified in the thermal springs. The transit time for flow across the aquitard, for $\log(K/K')$ of 3 and a kinematic porosity of 5%, is calculated as 200,000 years. If the bulk permeability of the Carboniferous Limestone is 10^{-7} m/s, and the porosity is 1% (karstic), then a similar calculation for transit time for the aquifer route is 10,000 years.

While the 'Mendips' model is broadly supported by this hydraulic analysis, it is stressed that uncertainty in the hydraulic characteristics remain. Strictly, data from the central region of the basin are required before the flow system can begin to be quantified. However, the difficulty of investigating a system such as this which is likely to be at least incipiently karstic, by drilling, would demand enormous resources.

References

Andrews, J. N., Burgess, W. G., Edmunds, W. M., Kay, R. L. F. & Lee, D. J. 1982. *Nature*, **298**, 339–343.

Black, J. H., Alexander, J. & Noy, D. J. 1984. Proceedings BNES Symposium, 161–166.

Black, J. H. & Barker, J. A. 1981. Report Inst. Geol. Sci. ENPU 81–83.

Burgess, W. G., Edmunds, W. M., Andrews, J. N., Kay, R. L. F. & Lee, D. J. 1979. Unpublished report, Inst. Geol. Sci.

Donovan, D. T. 1957. Unpublished notes to Bath Natural History Society.

Freeze, R. A. & Witherspoon, P. A. 1967. *Water Resources Research*, **3**, 623–634.

Kellaway, G. A. & Hancock, P. L. 1983. *The Variscan fold belt in the British Isles*, 88–107.

Rastall, R. H. 1926. *Geological Magazine*, **63**, 98–104.

Toth, J. 1963. *Jour. Geophys. Res.*, **68**, 4795–4812.

Williams, G. D. & Chapman, T. J. 1986. *Journal of the Geological Society, London*, **143**, 63–73.

From Kellaway, G. A. (ed.) *Hot Springs of Bath*, pp 179–204

12. The hot springs of the Avon Gorge, Bristol, England

A. B. Hawkins & G. A. Kellaway

Although the Clifton Gorge and Suspension Bridge attract many tourists to Bristol, few visitors are aware that the gorge had a thriving spa as recently as the 18th Century, with the thermal waters greatly in demand for medicinal purposes.

Hot springs were developed at several sites within the city of Bristol, although only the one in Hotwells at the southern end of the Clifton Gorge was used for any length of time. Here the water issues from fissured Hotwells Limestone (Plate 15), which crops out in the river bank about two hundred metres south of the Suspension Bridge (Fig. 12.1). The total yield from the spring is reported as 26 gals/min by Burgess *et al.* (1979) but as much as 60 gals/min by Pascoe (1981). The temperature is variously recorded as 73 to 76°F (22.8 to

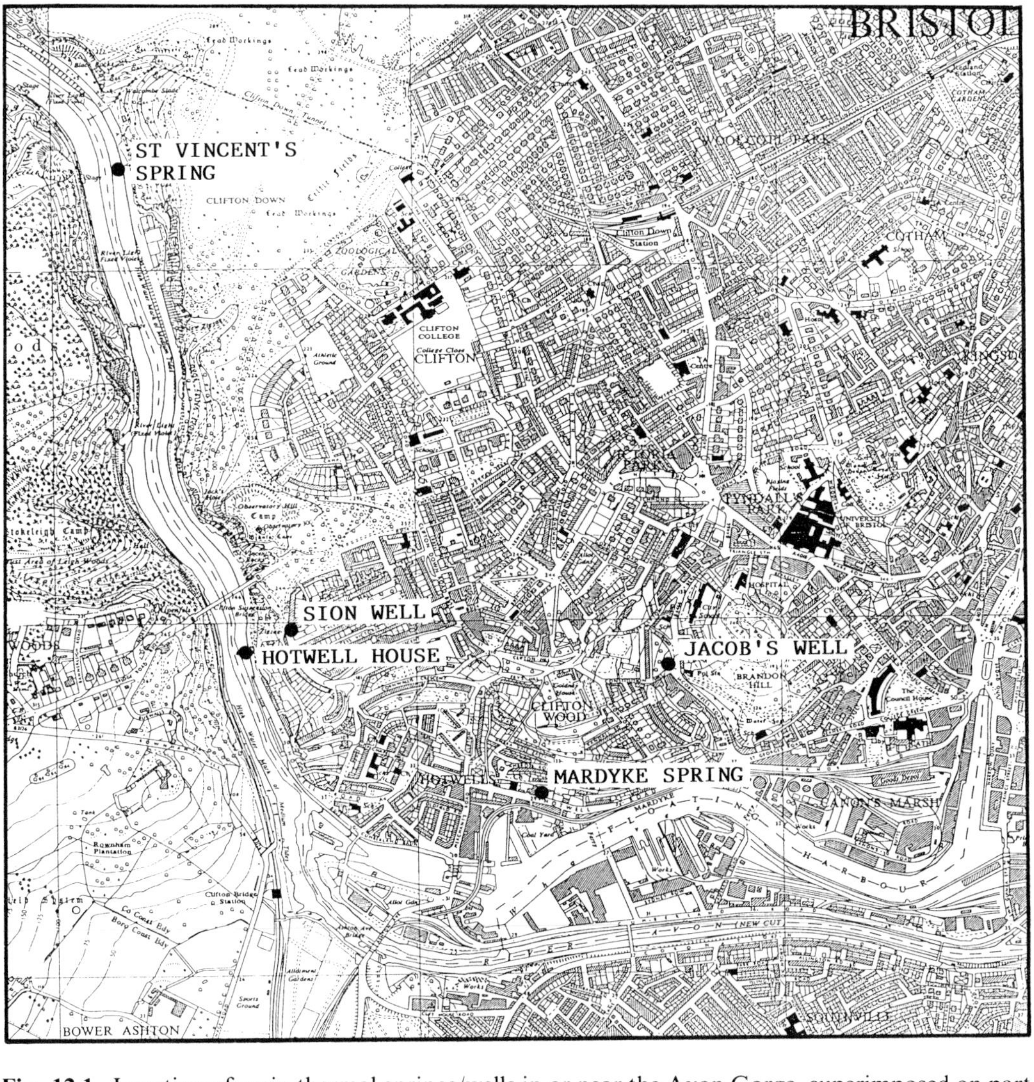

Fig. 12.1. Location of main thermal springs/wells in or near the Avon Gorge, superimposed on part of the 1:10,560 Ordnance Survey Sheet, ST 57 SE.

24.4°C); Burgess *et al.* (1979) quote a measured 24.3°C and 24°C was confirmed in January 1990. The chemical composition is said to indicate that the thermal water is similar to that of Bath (Andrews *et al.* 1982), although clearly there is a much greater dilution by near-surface waters. A variety of important ailments, including scurvy and consumption, are said to have benefitted by taking the waters.

In the 17th and 18th Centuries the area was much less accessible than today. This gave rise to problems, both in moving sick people to and from the spring and also in building accommodation to house those who travelled to Bristol to take the waters. There is no record of competition between Bristol and Bath, its more famous neighbour, since the waters at Bristol were used primarily for drinking. On the contrary, research suggests that many people would spend the winter in Bath where they could immerse themselves in the thermal water pools and then in the summer would come to Bristol to drink the waters, walk or horse ride on the Downs and eat strawberries in the surroundings of Long Ashton, two miles to the southeast of Hotwells.

Although it was thought that maximum benefit was derived when the waters were taken warm, they were nevertheless bottled and sold not only throughout Britain but also exported by sea from the Port of Bristol. This had an important secondary effect in stimulating the Bristol glass industry, which in turn was dependent on local coal to fire the numerous glass cones that existed at the time.

History and development of the hot springs

There is no single authoritative study of the Bristol hot springs. The account which follows has drawn upon a number of sources; some of these are given in the reference list rather than attributed to each and every comment made in the text.

The first mention of the Bristol hot springs is found in 1480 when William of Wyrecestre, a well known Bristol topographer and scholar, referred to the Clifton Gorge and St Vincent's Rock area, which in those days lay beyond the city boundary. At that time the water was bottled and taken by sailors as a cure for scurvy, a common ailment of seafarers.

In 1630 John Bruckshaw obtained a 40-year lease from the Crown to mine rocks and minerals in the Gorge (notably the crystalline quartz-filled geodes known as 'Bristol diamonds') and to make baths for those who frequented the springs. At that time Clifton was very busy and a steep winding path connected high Clifton to the banks of the Avon. Bruckshaw's petition to the King referred to a fountain or spring of water in the western part of the kingdom 'which cures many diseases far beyond any known bath in the kingdom'. He noted its position between high and low water marks and requested a licence to construct a wall between the spring and the River Avon. The level of the spring was later established as being 10 ft (3 m) above mean low water and 26 ft (8 m) below mean high water.

Three cavaliers are reported to have visited many of the southern counties in 1634 and their report refers to a strong, hot well which gushed out of the rock on the eastern side of the river between high and low tide. They are alone in recording the interesting fact that on the western side of the River Avon a cold spring (later called Scarletter Well) exists at a similar elevation, but with a flow of normal temperature water (measured at 49°F, 10°C in January 1990). The location of the springs as given by Buckland and Conybeare (1824) can be seen in Fig. 12.2. The cavaliers referred to the heavy medicinal demand on the thermal springs and drew attention to the water being sent to many parts of the country. Thomas Johnson also visited Hotwells in 1634 and wrote:

"Here from the clefts of the rocks issue forth a spring of warm water pleasant to the taste. It is a water of some repute and much commended for . . . affectations of the kidneys taken inwardly and for the old sores applied outwardly."

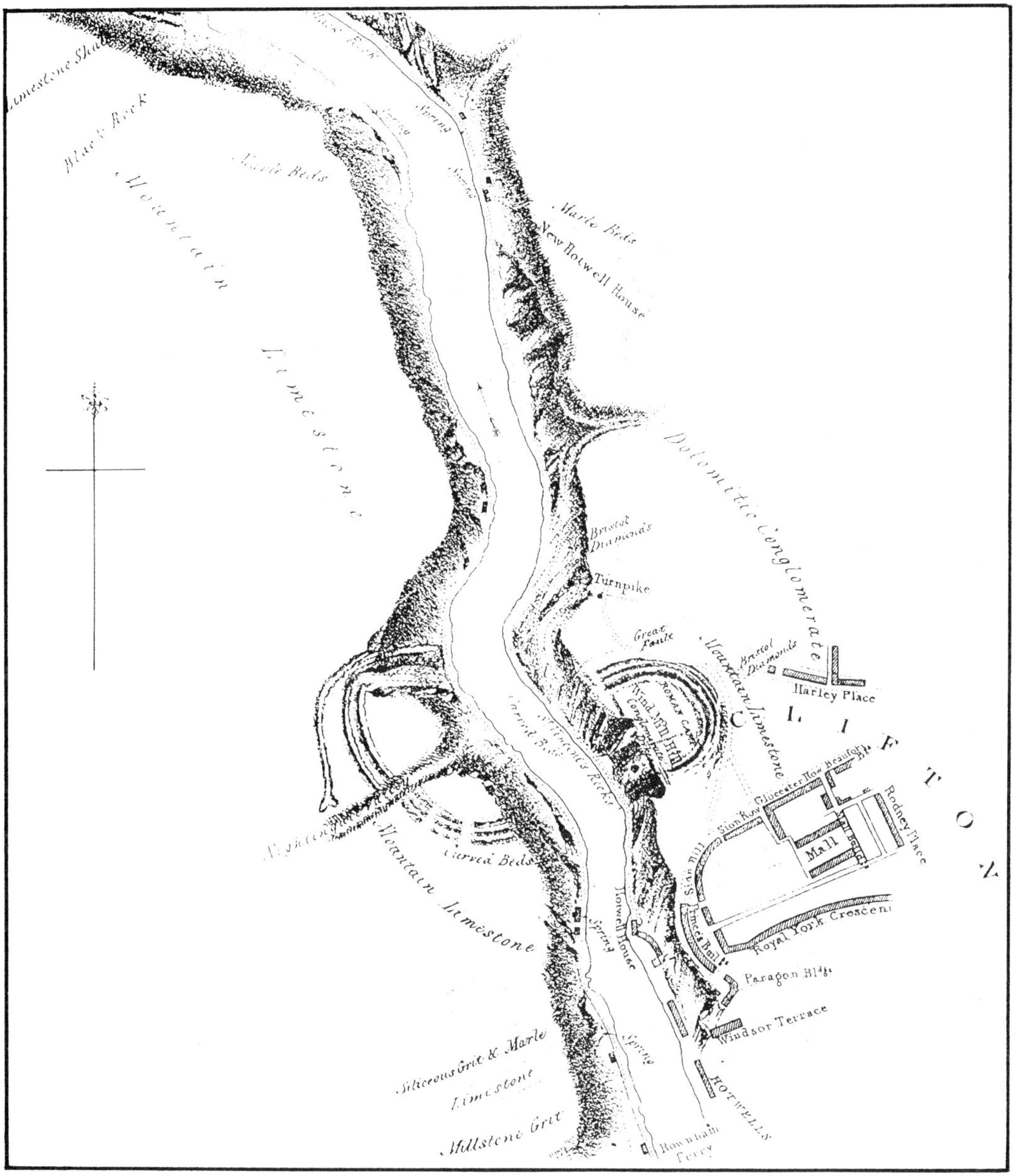

Fig. 12.2. Map from Buckland & Conybeare (1824) showing the springs as recorded in the 1820s and the location of the Hotwell House in the south and New Hotwell House in the north.

Queen Catherine of Braganza, the wife of Charles II, gave the spa greater respectability and standing when she visited the hot well in 1677. Following the recovery of a Mr Gagg, a baker from Castle Street, diabetes was included as one of the important ailments considered to be curable by drinking the water, and indeed the omens were fulfilled! Dr

Randolph's inclusion of this story when writing of Bristol did much to stimulate trade in the spa water; although some of the exaggerated claims made for the medicinal value of the water suggest quackery.

In 1676 the Bristol Society of Merchant Venturers purchased the Manor of Clifton, including Hotwells, and in 1678 they leased the wells to two tenants for £2 per annum. For this it was agreed they would construct a wall in an attempt to stop contamination of the water by the tide. There was concern, however, as the increased head of water began to divert the course of the spring, and it was feared the established exit point in the river bank might be lost if the water found it easier to disperse through a number of minor connecting fissures.

(a)

(b)

Fig. 12.3(a). Samuel Jackson's lithograph of the Old Hotwell House. (**b**) Pump Room and Meeting Rooms at Hotwells (1695–1822). Note the tree-lined entrance. The Suspension Bridge had not been started; the foundation stone was laid in August 1836.

In 1695 the well was leased for ninety years to Sir Thomas Day, Charles Jones, Thomas Callowhill and others for a rent of £5 per annum. The lease stipulated they should make proper foundations for pumps to raise the water to the level of the present Portway at about 30 ft (9 m) OD and to lay pipes through which the waste water could be diverted back into the River Avon via tide closure valves. They also agreed to construct Hotwell House and lay paths, thus producing better access to the area.

The Hotwell House was constructed on a small rocky ledge protruding slightly into the River Avon (Figs 12.1 & 12.3). Although there were properties on the edge of the tidal flats, at this time only tracks existed through the gorge and the area was still unaffected by quarrying. Some conditions, however, were similar to those of today, such that in 1696 Gerry Melford reported "My uncle was complaining of the stench occasioned by the vast quantities of mud and slime which the river leaves at low ebb under the windows of the pump room."

However, there is no doubt that the water of Hotwells had considerable attractions. In her great journey around England in 1698, Celia Fiennes visited "the hot spring of water which lookes exceeding cleer and is as warm as new milk and much of that sweetness; this is just by St Vincent's Rocks that are great clifts which seeme as bounds to the River Avon. This channel was hewn out of these rocks".

Robert Boyle introduced the Florentine thermometer to England in the 17th Century and the Fahrenheit scale came into use from 1712 onwards. The poet Alexander Pope may have been the first to note the uniform temperature of 76°F at the exit of the spring and the sparkling appearance of the water, frequently with air bubbles which settled on cooling. He also referred to the pleasant taste of the water and recorded it as being free of any odour and making excellent tea!

Dr Randall, in his writings of about 1745, stressed the use of the waters as a cure for tuberculosis. Naturally it was claimed that to be successful the water must be taken at an early stage of the disease. However, many practitioners did not send their patients until they wanted something or someone (apart from themselves) to blame for the inevitable demise of the patient. At this time, with so many people coming to Bristol at an advanced stage of their illness, one set of lodging houses became known colloquially as "Death Row". Despite this, many of the medical journals of this period contain papers and reports of people said to have been cured by the waters.

The waters were taken to all parts of England, the Continent, and as far as the West Indies, requiring a substantial output of the Bristol glass industry for water containers. In the 18th Century, for instance, Patrick McGarth quotes the London newspapers displaying the following advertisement. "Famous Bristol hot water, fresh from the well, will be delivered to any part of the town, for 6 s per dozen bottles. These bottles are of the largest size and by the extraordinary favour of the wind, arrived last week, eight days from Bristol, the common passage being a month to six weeks."[1]

So great was the popularity of the hot well that by 1717 work had started on Dowry Square (Figs 12.1 & 12.4) to house the wealthy visitors. Soon Dowry Parade and Albemarle Row (1763) were constructed and a second Assembly Room was added. This survived as a school until the 1950s when it was demolished to make way for the Cumberland Road complex.

The season began in April and lasted until September. A typical programme for taking the waters included a morning ride to the pump room by carriage, drinking the prescribed number of glasses of water (up to 12 glasses daily) and then sitting with the company talking scandal, playing cards and listening to the small orchestra. For the benefit of those who preferred the fresh air and exercise, there was a tree-lined promenade along the river bank. In about 1786 a small colonnade was built for people to browse in the wet weather,

[1] In the 18th and 19th Centuries many public water supplies were of poor quality, tainted or polluted by water draining from farms, industrial sites and insanitary towns. The use of clean bottled spring water may therefore have been conducive to good health as well as being more pleasant to drink.

part of this structure still remains (Fig. 12.5). This colonnade replaced the small piazza seen on some of the earlier drawings of Hotwells; the commemorative plaque can be seen in Fig. 12.6. Part of the 1828 New plan of Bristol, Clifton and Hotwells is reproduced here as Fig. 12.4. This shows the area at the heyday of the Hot Well, indicating the names of many of the streets/dwellings, e.g. Paradise Row and Hope Square, but not Hopechapel Hill which joined Love Street to Granby Hill. Many of the slopes in this area are excessively steep and were hazardous to horsedrawn vehicles.

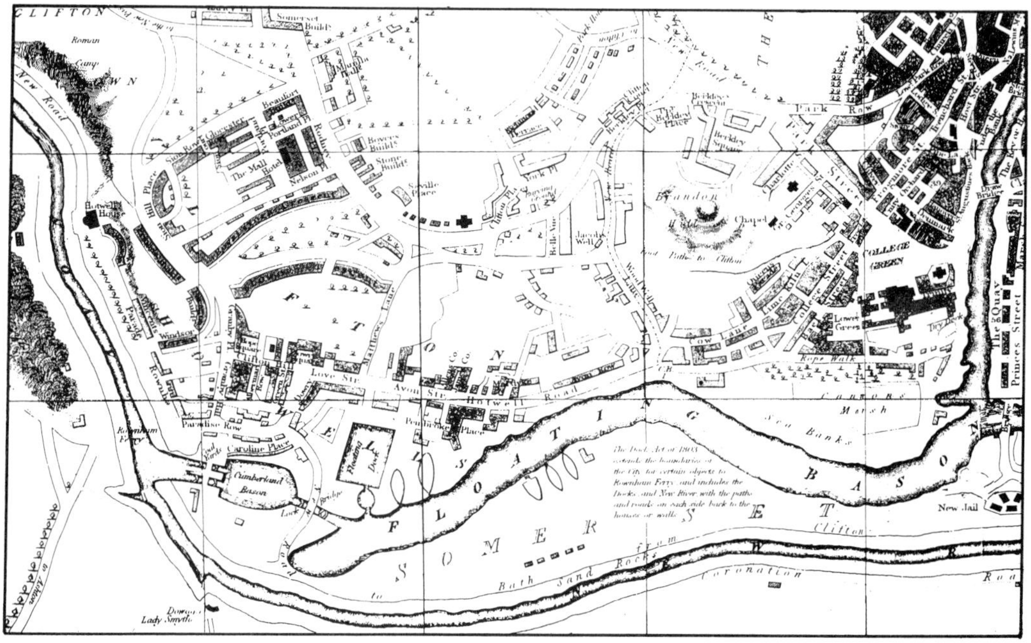

Fig. 12.4. Part of the 1828 New Plan of Bristol, Clifton and the Hotwells, engraved by Sidney Hall, showing the area of Clifton at the heyday of the Spa and Jacob's Well west of Brandon Hill.

Fig. 12.5. The Colonnade and the New Hotwell House in about 1860. Note the western pier of the Suspension Bridge has been constructed but work was in abeyance.

Fig. 12.6. Plaque seen at the river end of the Colonnade.

It is clear from the Diary of John Wesley that he had been seriously ill with 'consumption' in late 1753. He writes that he arrived in Bristol on 2nd January 1754 and "I began drinking the water at the Hot Well, having a lodging at a small distance from it". For a man of his great energy he records little activity during his stay, but notes that on Sunday 10th March, 1754 "I took my leave of the Hot Well, and removed to Bristol".

Although there was no hospital at Hotwells, the arrival of Thomas Beddoes MD in 1793 led to the establishment of the Bristol Pneumatic Medical Institution in 1797–1798 at nearby Dowry Square. Beddoes was drawn to Bristol by the large number of invalids at Hotwells and among other methods of treatment which he employed was the inhalation of gases. As a direct result of these activities the young Humphrey Davy came to be appointed as first Superintendent of the Institution. Here he discovered the properties of laughing gas or nitrous oxide, the forerunner of modern anaesthetics, which he tested on Coleridge and Southey, among others. The history of the valuable scientific work carried out at the Institution has been described by Eyles (1955). In spite of some eccentricities, Beddoes' work was important and he created a second medical institution for the treatment of poor and needy people at Broad Quay.

By 1784 it was clear that a considerable amount of remedial work was necessary in the area of Hotwell House, including building a quay wall in front of the pump house which cost about £1,000, and protecting the spring from the tide, £500. With such arduous conditions stipulated in the lease, it was not surprising that no new tenant was forthcoming and the owners, the Society of Merchant Venturers, had therefore to finance the work themselves. As a result of this expenditure, when Samuel Powell took the tenancy in 1790 he had to increase the charges considerably, raising them above the level many of the clients could afford. It is interesting to compare the descriptions of the area given by Dr Carrick, a Clifton physician, writing about Hotwells in 1816 as he remembered it in the mid-1780s:

"It was then during the summer one of the best frequented and most crowded watering-places in the kingdom. Scores of the nobility were to be found there every season, and

> such a crowd of invalids of all ranks resorted to the waters that it was often difficult for them to provide themselves with any sort of lodgings. About that period a considerable number of lodging-house keepers had in the course of a few years realised very handsome fortunes without any complaints of extortionate exactions. Three extensive taverns were constantly full and two spacious ballrooms were profitably kept open. There was a well-attended ball, a public breakfast, and a promenade every week, and often twice a week. The Pump Room was all day long the resort of invalids who left with the keeper of the Well many hundreds a year in voluntary donations, and from 12 to 2 o'clock it was generally so crowded that there was often some difficulty in getting up to drink the water. The adjoining walk was filled with fashionable company; the sublime scenery of the cliffs was enlivened by the sounds of music. The Downs and all the avenues to the Hot Well were filled with strings of carriages and with parties on horseback and on foot."

By 1861 however:

> "It has the silence of the grave, to which it seems the inlet. Not a carriage to be seen once an hour, and scarcely more frequently does a solitary invalid approach the neglected spring. One of the ballrooms and taverns has been long ago shut up, and the other with great difficulty kept open. The lodging-houses, or such of them as still remain open, are almost entirely empty in summer, and not very profitably filled even in winter."

Dr Carrick also recalls the very steep path from Clifton was too difficult to negotiate, another reason why the Hot Well became less used. To quote:

> "To many the hire of a carriage twice or three times a day at the increased charge, and the fatigue and terror of riding up and down Granby Hill, is an insuperable obstacle."

In 1822 Hotwells House was demolished to make way for a road which was to pass along the Hotwells end of the Gorge and up the incline now known as Bridge Valley Road. The 1916 1:2,500 Ordnance Sheet, reproduced here as Fig. 12.7, shows the tram and railway system to Avonmouth built in the 1860s. Although a new pump house was erected (Fig. 12.8), it failed to revitalize the Hotwells Spa and was demolished to make way for the new railway line and to cut back the rocks around the spring in order to improve the approaches to the Floating Harbour. In 1877, however, under the direction of Mr Howard of the Docks Committee staff, the spring was again enclosed and the waters were piped to a small grotto hollowed out next to the Clifton Rocks Railway (Fig. 12.9). Although a Dr Griffin wrote in 1880 that the pump waters were not from the original spring, the pump was kept in operation until 1913, at which time approximately 350 people per day were supplied with water. Now only the Colonnade and the small grotto opening in the rock wall at the side of Hotwells Road testify to the past glory of the thermal resort.

Analyses of the two samples of Hotwells water undertaken by the Bristol Public Analyst in 1909 are given in Table 12.1 (Richardson 1930).

Further work was undertaken as part of the investigation of the geothermal potential of the UK undertaken by the British Geological Survey (Burgess *et al.* 1979). It is noted that Ca^{2+}, HCO^{3-} and $SO_4{}^{2-}$ are the dominant ions and when the chemistry is plotted on a trilinear diagram its position is intermediate between most shallow groundwater samples and the thermal water of King's Spring, Bath. The levels of tritium and nitrate signify considerable dilution of the thermal component by recent near-surface groundwater.

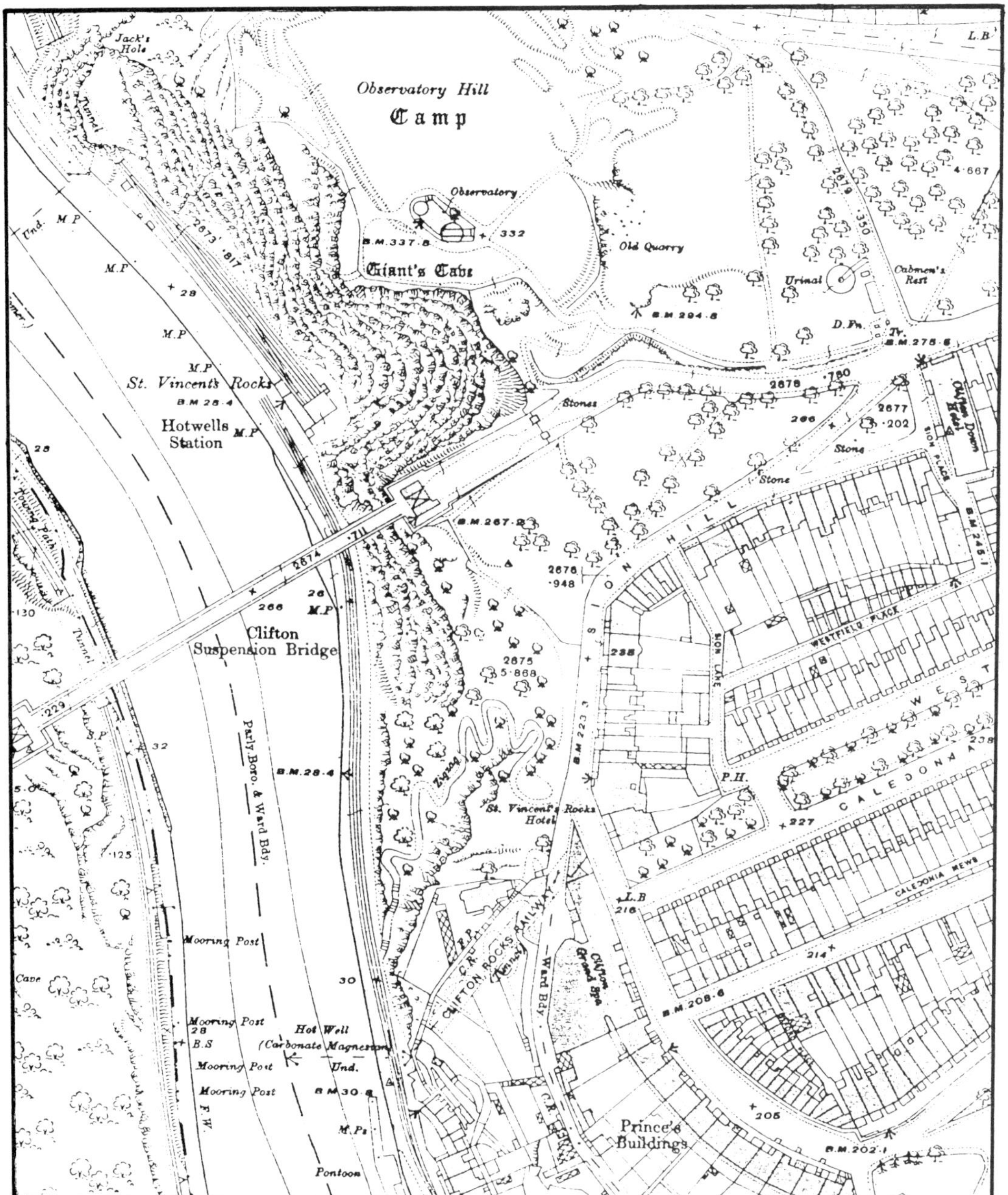

Fig. 12.7. The 1:2,500 scale Ordnance Survey map of the Hotwell area published in 1916 showing the line of the Clifton Rocks Railway, the line of the tramway from Bristol and Hotwells Station, constructed in the 1860s.

(a)

(b)

Fig. 12.8(a). The new Pump House, 1822–1867, from across the river showing the tree-lined pathway. The Hotwells Limestone from which the thermal water issues is seen behind the spa building. (**b**) The new Pump House, 1822–1867, looking downstream to show it was built partly over the mud flats, hindering navigation at high water.

Jacob's well

Evidence of a recently exposed lintel of Jacob's Well (Fig. 12.1) indicates that it was originally used to supply a Jewish *mikveh*—a natural purification bath involving total immersion. The exact date is not known, but must be between the arrival of Jews in Bristol *c.* 1087 and their expulsion from England in 1290. It is believed the *mikveh* supplied by the spring/well is the oldest in Europe, pre-dating that in Cologne which has been dated *c.* 1170 (Vaughan & Martlett 1987). These authors state that the inscription in the lintel over the spring (Fig. 12.10) has been translated by Ralph Emmanuel to mean "flowing".

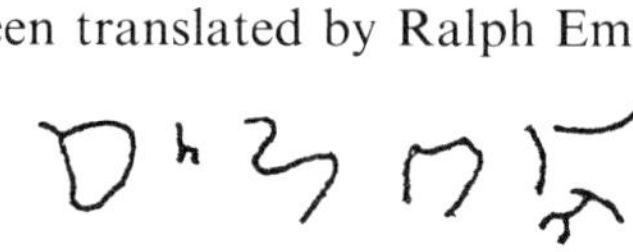

Fig. 12.9. The entrance to the Clifton Rocks Railway with the grotto immediately adjacent to it. (Fisher *et al.* 1983).

Table 12.1. Analysis of two samples of Hotwells water.

	parts per million	
Total solids	1092.5	1112.5
Mineral matter	875.0	847.5
Organic matter	217.5	265.0
Chlorine as chlorides	80.0	79.0
Hardness	—	—
Free ammonia-traces	nil	nil
Albuminoid ammonia	0.02	0.01
Total oxidized nitrogen	8.0	12.1
Nitrites	nil	nil
Permanent hardness	—	—
Lead	nil	nil
	grammes per litre	
SiO_2		0.0174
Al_2O_3 and Fe_2P_3		0.0048
CaO		0.2900
MgO		0.0472
SO_3		0.3100

Fig. 12.10. Jacob's Well, as seen in 1989, showing the lintel with the inscription translated as "flowing".

It is of note that although there are a number of springs in the vicinity of the Jewish settlement near Brandon Hill, the *mikveh* was constructed at what is known as Jacob's Well. It can only be assumed that this spring was chosen because the heat of the water was somewhat above that of normal groundwater and retained its temperature during the winter. When the well was rediscovered during construction works in December 1986 the temperature was recorded in the Temple Local History Group's Proceedings as 53°F (11.7°C) but in January 1990 57°F (13.8°C) was measured, considerably below the 76°F (24.4°C) measured in the hot well in the gorge. However, the level at which the water emerges in the gorge is only about 2 m OD, so the temperature of the rising thermal water under hydrostatic pressure is less likely to have been reduced significantly by dilution with near surface groundwater. In the vicinity of Jacob's Well and Park Street, the springs emerge on the southern slopes of the ridge on which the University and the Victoria Rooms are situated, between 30 and 40 m OD. It is likely, therefore, that the percolating groundwaters make a significant contribution to these springs and that the thermal water is diluted with cold groundwater. The flows from Jacob's Well (42 gals/min), Dean and Chapter Spring adjacent to Gorse Lane and St John's Spring at the top of Park Street are described by the Temple Local History Group (1984, 1987).

For a short time following the rediscovery of the thermal spring in 1987, a small bottling industry was set up on the site of Jacob's Well and visitors were welcome to taste the waters. Although the old Bristol glass industry has long since died, the spring water was again made available as "Genuine Superior Aerated Waters" as advertised from 1840 to the 1880s by Rough, Sedge & Summers and bottled in the traditional blue glass. The chemistry of the waters was determined by Waterford & O'Brien (1987). Part of the chemical analysis is given in Table 12.2 by permission of Terry Gardiner and Terry Noble who acquired the site in October 1988. It has since been purchased by a Mr Jacob who intends to re-open the bottling plant in the future.

During a wet period in 1986 a collapse occurred in the north retaining wall of Constitution Hill, adjacent to the building which now houses Jacob's Well; another failure took place where Lower Clifton Hill joins Gorse Lane; both are situated a short distance west of the springs. Boreholes drilled on behalf of Avon County as part of an investigation into the collapse of the retaining wall at Constitution Hill proved the presence of a fault within approximately 10 m of the well.

Jacob's Wells Road (referred to as *Woodwell Lane* on plans dated 1796) is situated in a valley reminiscent of a fault controlled feature, particularly as it dissects the very strong siliceous sandstone feature which forms the ridge extending through Cliftonwood to the Brandon Hill area. The thin veneer of Triassic deposits overlying the Namurian Quartzitic Sandstone Group, proved in the borehole investigation adjacent to Jacob's Well, is now visible in the side of the well. This shows that the original valley is of pre-Triassic origin and the initial fault is therefore likely to be of Variscan age.

Table 12.2. Characteristics of water at Jacob's Well.

pH	7.1
	parts per million
ammonia nitrogen (N)	0.04
albuminoid nitrogen	0.06
nitrate nitrogen (as N)	8.8
(as NO_3)	39
calcium	145
potassium	13
sodium	33
carbonate (CO_3)	162
sulphate (SO_4)	78
chloride (Cl)	55

There is no evidence of water from Jacob's Well being used for immersion after the departure of the Jews. In 1888, however, waters were diverted from the Dean and Chapter Conduit to the large red brick swimming bath constructed by Bristol City Council at the bottom of Jacob's Wells Road. The fresh, spring water was heated and used in the baths until their closure in the 1970s.

Other wells and thermal springs

St Vincent's Spring

About a mile north of Hotwells House another natural source of thermal water was known to the quarrymen as St Vincent's Spring (Fig. 12.1). This was first mentioned at the very beginning of the 18th Century as being about 200 yds beyond Black Rock Quarry, the southern of the two being shown by Buckland and Conybeare (Fig. 12.2) named New Hotwell House. At that time the main access was by a stony path along the banks of the river or the spring could be reached by travelling down the steep horsetrack known today as the Gully. A small pump room and lodging house were constructed, but as the area was lonely and the access difficult, its success was shortlived and by 1792 the building was taken over by the quarrying industry. Little evidence of this venture survives today; the photograph of the Black Rocks Pumping Station in 1845 (Fig. 12.11) shows the engine house designed by Brunel which took all the available fresh water from "a large spring to supply water for Clifton". In 1894 Bristol Corporation constructed a drinking fountain between the river and the railway which then ran through the gorge (see Plate 5a in Vaughan 1905). Although no longer in its original position, this disused fountainhead can still be seen on the eastern side of the Portway (Fig. 12.12).

Fig. 12.11. The Black Rocks Pumping Station in 1845, designed by Brunel to pump the spring waters from the northern part of the Gorge to Clifton (reproduced from Buchanan & Williams 1982).

Mardyke Spring

In 1786 a spring was located near the Mardyke ferry off Hotwells Road. Little is known about the Mardyke Spring but it is referred to in the New History or Complete Guide to Bristol (Anon 1794). In the following extract the special cures of the mineral water are referred to but no specific mention is made of any thermal qualities:

> "In this land of wonders our readers will not be surprised to have a lately discovered mineral water announced to them. A few years ago was discovered by sinking a well, at the house of Mr Davis, at the Two Nag's-heads, in the street leading from Mardyke to the floating-dock, and Hotwells, an inestimable spring of water, of the Chalybeate, and other useful and excellent qualities; which good judges have esteemed to be far superior to any other mineral water of its kind in the Kingdom. It has been known, since its discovery, to have performed many wonderful cures on persons who were deemed incurable by the faculty, in scrophulous, rheumatic and cutaneous disorders. We hear that the resort of the public to participate of the benefits of this lately discovered and valuable mineral water, is continually increasing. It sharpens the appetite, raises the spirits and strengthens the body. Its efficacy is great and real, and is ready to be attested by those who have drank it, and found its salutary effects by happy experience."

Again the proprietor built a small pump room, but by 1810 the development had ceased and the site was used for industry. Interestingly, this was referred to as Poole's Mineral Spa Coal Wharf.

Fig. 12.12. The old St Vincent's Spring drinking fountain, re-positioned by Bristol Corporation when the Portway was constructed in the 1920s.

Sion Spring

In 1793 a local lawyer, drilling on Sion Hill at a height of about 220 ft (67 m) to obtain water for his house, intercepted thermal waters at a depth of 245 ft (75 m), the level being slightly below the hot spring in the gorge. Having decided to take advantage of this prime location in the fast-growing area of Clifton, he built a small pump room and set up in competition with the established hot well in the Gorge. Unfortunately its exact location is not shown on the Matthews Plan of Bristol of 1794, which simply records "New Hot Well" in the area of the present St Vincent's Rocks Hotel. The Ashmead map of 1846 shows the hotel as a public building and Buckland and Conybeare indicate it in a similar manner (Fig. 12.13). Clearly the Sion Spring was pumped as de la Beche (1845) recorded that a steam engine was used to obtain the waters and that "The temperature is the same as that of the Hotwell water". De la Beche quotes a flow of 24 gals/min which was pumped to supply over three hundred houses in Clifton.

John Nott (1793) notes that

> "About an hundred yards from the Well house up the river a bubbling spring of warm water is seen at times, nearly midway between the high and the low water mark; it is further singular that, whenever this is visible, should the fire-engine of the upper hot spring at Clifton chance to work, it immediately disappears."

Nott records temperature as being 66°F (19°C), although the proprietor of the pump room argued it was 70°F. De la Beche (1845) also notes the existence of Richmond Spring near

Richmond Terrace, although he does not indicate it to be thermal. No spring is shown on Ashmead's (1846) map in Richmond Terrace although one is indicated where Richmond Park Road joins Kensington Place.

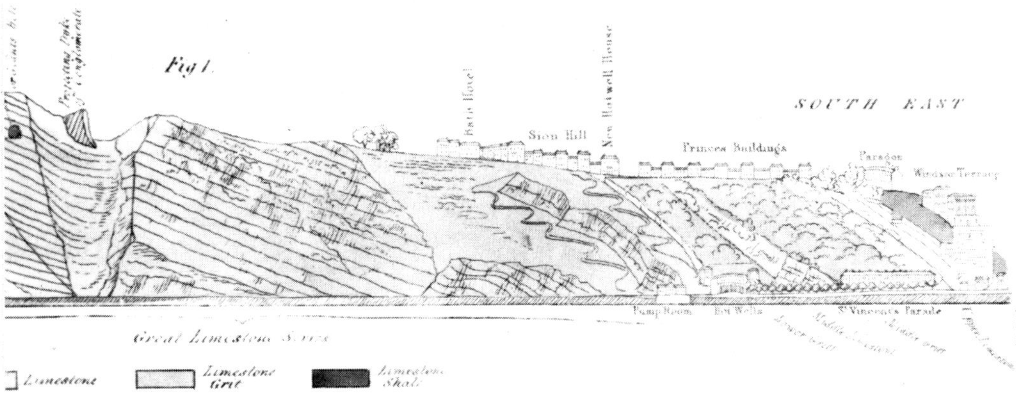

Fig. 12.13. Section from Buckland & Conybeare (1824) showing the Pump Room and Hot Well and the New Hotwell House at the end of the block forming Sion Hill. The Grand Spa (Avon Gorge) Hotel had not been constructed.

Grand Spa Hotel Spring

In 1793 George Newnes constructed a hydraulic cliff railway from Hotwells Road to Sion Hill to carry people from the river valley to the village of Clifton. It is understood that he encountered some hot water during this venture and consequently the Grand Spa Hotel (now the Avon Gorge Hotel) was constructed with various rooms set aside for the enjoyment of steam baths. It is difficult now to determine how much the large boilers at the hotel re-heated the water, but it is clear from the construction of the steam rooms that they were popular at least for a short period. The attempt to develop a thriving spa industry, however, was not successful, probably due to the general decline in spa activities. With limited usage and the improvement in road transport the Cliff Rocks Railway (see Fig. 12.9) ceased to operate in 1934.

The effect of the Lisbon earthquake of 1755

Although there is no record of any effect produced by the Lisbon earthquake on the Bath hot springs, Arrowsmith's *Dictionary of Bristol* (Anon, 1794, 2nd edn, 1906) states that on November 1st 1755 "the Hotwell water became very red, and so extremely turbid that it could not be drank" and that it "ran foul a long time before it recovered its former purity." Not until news was received in Bristol of the great Lisbon earthquake on that date was it realized that the discoloration was probably due to this disastrous event. It is also noted (Anon 1794) that "On the very same day, the water of a well in a field belonging to Mr John Harrison near St George's-church in Kingswood, which had been remarkably clear, became suddenly as black as ink, and continued unfit for use, nearly a fortnight."

It is difficult to assess these accounts but in the case of Hotwells the event would have been of such importance and such rarity, indeed as far as is known, unique in the life of the

Spa, that it is unlikely to have been due to any other cause. The distance from Lisbon to Bristol is about 950 miles (1500 km).

No records of any similar disturbances have been found at Bath, but it is possible they escaped notice. The most likely contaminant of the Bath hot spring would be Lias Clay which, in small quantities, produces only a slight turbidity without any marked change of colour. An increase in the volume of flow of the hot springs would only have been noticed if it was very substantial, as in 1755 much of the thermal water was leaking into debris overlying the Roman remains (Chapter 7).

Origin of the Avon Gorge

Although quarrying and the construction of road and rail communication have significantly modified the natural relief, the 80 m deep gorge at Clifton is a spectacular feature, enhanced by the presence of Brunel's Suspension Bridge (Buchanan & Williams 1982). Many writers have offered hypotheses to account for the presence of the gorge and some of these are of considerable historical interest. Almost the first was John Aubrey (1656–1691) who attributed the "great cracks and vallies" at Bristol and Bath to the work of an earthquake. The internal evidence in his *Natural History of Wiltshire* suggests that he wrote the passages shortly after the occurrence of three Italian earthquakes in 1688–1690. Aubrey concluded that the irregularities and changes on the earth's surface were due to earthquakes, and that the major valleys are following "a crack" (i.e. their courses are tectonically determined). He recognized that the limestones forming the plateaux at Claverdon (Claverton) and Lansdon (Lansdown) are "both downes of the same piece" and that their separation is due to the formation of a "crack". He continues "it may be at the same time the crack was thus made at St Vincents Rock near Bristow". In the light of recent studies at Bath, John Aubrey's words might truly be regarded as prophetic.

In Aubrey's time little consideration had been given to the effect of erosion in controlling the evolution of the land surface and almost nothing was known of the effects produced by glaciation or permafrost. However, by the beginning of the 19th Century erosion due to marine and fluviatile action and to frost and other natural weathering agencies was being studied almost to the exclusion of the other factors. This led to the rejection of Aubrey's tectonic concept by Jukes in 1867 in favour of what C. Lloyd Morgan (1885) called "differential action of general and special denudation". According to Morgan (1885) "The Clifton Gorge is no chasm rent in the rocks". Morgan gave a clear statement of the theory of superimposition of the Avon drainage. "The Lias then formerly reposed on the Carboniferous Limestone of Clifton Down, and the Oolite spread over that... The original form of the surface was such as to turn the Avon towards the Severn... The course it originally took it has ever since maintained."

The concept of superimposed drainage in some form or other has long had adherents (Gresswell 1954). It serves a useful purpose, focussing attention on the significance of the stages in the removal of an unconformable Mesozoic and Tertiary cover (Chapters 13 & 14). It fails, however, to deal with the reactivation of old faults and fissure belts and with the creation of new ones in response to changing stress. Trueman (1938) emphasized the fact that in the Bristol area it is possible to see streams such as the Avon and some of its tributaries, which are actively engaged in cutting through the unconformable Mesozoic cover onto the underlying Palaeozoic rocks, giving rise to a drainage pattern which is apparently unrelated to the underlying structure. In his discussion of the Bristol Avon, Trueman (1938) commented "Some believe that rivers may once have flowed across the position of the Bristol Channel in a direction opposite to that of the Bristol Avon". This is clearly an oblique reference to the work of Varney (1921), discussed at greater length elsewhere (see Chapter 13). Opposed to the views outlined above are those of Harmer (1907), who suggested that the Avon Gorge is a glacial overflow channel. For a long time, this was rejected as being without supporting evidence. Harmer's views have more recently been re-assessed in the light of the discovery of glacial deposits in the Kenn–Clevedon–Flax Bourton area where a pre-Devensian glaciation may have provided both a barrier and a source of glacial melt waters. When evidence of the glaciation of the area was first reported by Hawkins & Kellaway (1971) they suggested that the River Avon originally flowed westwards through the Flax Bourton Gap. However, the nature of the Drift deposits on the western and eastern sides of the gap indicates incursion of ice from the west, blocking the Gap and providing melt water which ponded against the Carboniferous Limestone ridge and which then escaped by eroding a channel through a zone of fissured limestone to join the River Trym at Sea Mills.

The location and alignment of the Clifton Gorge is therefore thought to be controlled by the tectonic fractures from which the thermal water rises between the Gully and Hotwells (Plate 16). Thus the downcutting of the gorge and the uprising of the thermal water are likely to be interrelated. The absence of thermal water and the presence of cold water (seasonal) springs in the areas immediately to the east and west of the fracture zone (Kellaway 1967) indicate that the discontinuities extend to great depths, possibly to the contact of the Silurian rocks and the 'soft' Tremadoc shales.

There is some evidence suggesting that this line of deep fracturing is still subject to seismic activity. Thus the earthquake of 3rd April 1852 which was felt in Bristol and is described by Davison (1924) is said to have also affected Wells and Cheddar (Fig. 13.7). However, an examination of Davison's record and some additional information gleaned from old records and notes suggests that Davison's account may refer to two separate epicentres, one south of the Mendips with a disturbed area similar to that of the Wells earthquakes of 1893, the other with its epicentre near Clifton and with an isoseismic axis roughly parallel to the fracture zones seen in the Avon Gorge and at Kingsweston.

There is no evidence indicating that the shock was felt in the Radstock basin where it might have been recognized in what was the most densely populated area of the Somerset Coalfield. This again points to the possibility that there was a twin earthquake in April 1852, with epicentres near Wells and Bristol respectively. It would have been interesting to know whether this earthquake affected the flow at Hotwells, but by this time the Spa was in its death throes and there may have been no-one with sufficient interest to record it.

The fracture belt

The orientation of the Avon Gorge at Clifton coincides with that of a strong belt of vertical or steeply inclined features with a general NNW–SSE trend intersecting both the Carboniferous and the Triassic rocks. At the northern end of the gorge the direction changes locally to a WNW–ESE direction similar to the highly mineralized fractures containing Mesozoic rocks which have been mapped on the surface of Durdham Down (Geological Survey 1:10,560 Sheet ST 57 SE). In addition to the mineralized fractures containing barite, calcite, red mudstones, iron oxides and fish remains, fractures which are unmineralized except for stalactitic calcite and very thin films of reddish brown mud, appear to be of a more recent origin.

The roadside section at the top of Bridge Valley Road (Plate 17) shows unmineralized fissures cutting the Triassic Dolomitic Conglomerate and passing without deflection through very strong boulders of limestone and very hard siliceous sandstone (Hawkins & Kellaway 1971). In the large cliff sections in Carboniferous Limestone of the Clifton Gorge some of the fractures, such as those in the old quarry, east of the Bridge Valley Buttress and the Suspension Bridge face, can be seen to extend the full height of the Gorge. These and other parallel fractures (Plate 16) show evidence of limited movement, in some cases amounting to recognizable small-scale wrench faulting. In the area north of the Bridge Valley Road/Portway junction the rock slope is strongly fractured (Plate 18), with thick, iron-rich calcite veneers forming parts of the rock face both west of Bridge Valley Road and in the steep cliff north of the New Zig Zag. At the western end of the Bridge Valley Buttress very steeply inclined faults with slickensides generally parallel to the bedding form much of the higher part of the northern face.

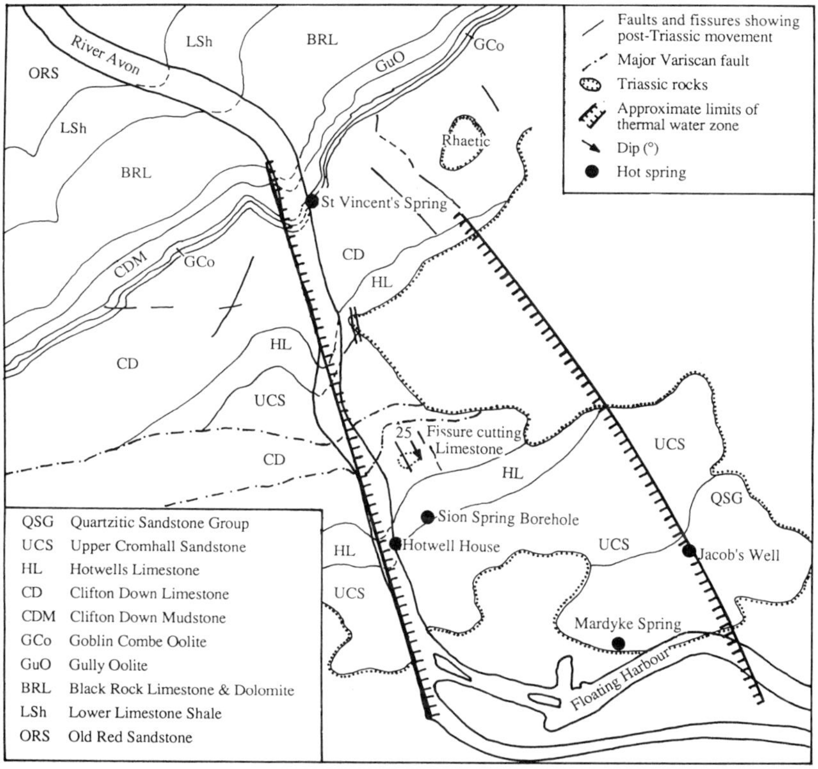

Fig. 12.14. Geological map of the Gorge area showing the main fracture belt and the hot springs. See also Plate A, p 274.

An attempt has been made to determine the position of the main fractures within a wider belt (Fig. 12.14). The main western fracture influencing the thermal waters (Fig. 12.15) runs through the Hot Well Spring and the St Vincent Spring. It is believed that the next one merges into this north of the junction of Bridge Valley Road and the Portway and explains the topographic features through the Windsor Terrace area where the Carboniferous rocks have been eroded to produce a "Triassic embayment". The iron ores formerly worked in this area were probably emplaced in the base of the Dolomitic Conglomerate. The Mardyke Spring probably rose via a fault zone, part of this disturbance being seen in excavations in 1989 for the installation of a petrol tank in the forecourt of the Esso Garage on Hotwell Road. Projected southwards the disturbed zone would pass through the fault in the Bedminster Great Vein where the geological map indicates the seam to have been vertically displaced by 30 m. This would have contributed to the mapped deflection of the Ashton and Bedminster Veins. The Mardyke Fault also forms the eastern boundary of the hollow in the Carboniferous strata in which the Triassic sandstone and breccia has been

Fig. 12.15. Part of the 1:10,560 Ordnance Survey Sheet ST 57 SE with postulated fault lines connecting locations of hot springs.

deposited. When the location of the fault in the Jacob's Wells valley is also taken into account it is possible to project a line through the site in Constitution Hill, the Richmond Park Road Spring and another in the Clifton Zoological Gardens (Fig. 12.15). Extended southwards this can again be identified as a disturbed zone noted in the mine plans of the Southville to Lower Knowle area.

Examples of these steeply inclined fractures can be seen in the eastern approach to the Suspension Bridge. The wall-like mass of Dolomitic Conglomerate seen in the quarried area immediately north of the Clifton approach to the Suspension Bridge (Fig. 12.14) has probably been isolated by quarrying. Initially it is likely it formed part of a Triassic fill, the greater part of which has been quarried away leaving a wall-like mass bounded by the fracture planes which cross the Triassic valley-fill almost at right angles. In an old quarry on Observatory Hill (ST 566 733) Variscan shear planes, picked out in white calcite (Hancock 1972), are cut by NNW–SSE-trending post-Triassic fissures; they provide an excellent example of the relationship between the two groups of structures.

In the steep cliffs immediately beneath the eastern abutment of the Suspension Bridge, the closely spaced fracture planes intersecting the inclined limestone strata (Clifton Down Limestone) present the appearance of the torn pages of a book (see photograph in Hawkins 1987). One of the fractures referred to by Hawkins & Kellaway (1971) in the eastern approach to the Suspension Bridge can still be detected today, partly concealed by ivy on the northern side of the buttress, the effects of this on the bedrock surface can be seen in Fig. 12.16. The preferential erosion over this fracture explains why the wall supporting the footpath around the eastern end of the Suspension Bridge pier varies so much in height and, indeed, why Brunel used an arched construction nearly 4 m high beneath the eastern end of the footpath parapet in this part of the Bridge, a fact not discovered until 1978.

Fig. 12.16. The wall necessary to support the eastern parapet of the Suspension Bridge due to the topographic hollow where one of the fractures has "opened out". Examination of the ivy-covered northern slope reveals the fracture continues below this feature.

In the initial stages of the formation of the Avon Gorge, possibly up to half a million years ago (Table 13.3), water would have passed directly over/through the lower part of the limestone ridge at the western edge of the fracture belt in which the thermal water was entrained at depth. Thus the formation of the Avon Gorge at Clifton marks the final stage in the process by which tectonic fracturing was exploited by water erosion, opening up the way for the confined thermal waters to rise to the floor of the Avon valley, possibly aided by seismic pumping (Sibson *et al.* 1975). It is interesting to note that at both the sites in the Clifton Gorge the springs rise to the surface just downstream of where the less permeable Clifton Down Mudstone and the mudstones of the Hotwells Group cross the lines of the fracture belts. These lithologies would not only be less permeable, they would also be less brittle and the muddy rocks more easily broken down by water; Quaternary reactivation would therefore be less effective.

Conclusions

The paper draws attention to the history and exploitation of the hot springs in the Avon Gorge area of Bristol and highlights the geological explanation for their occurrence. It is hoped that the text figures will allow those visiting the sites mentioned to identify what remains today of the once-thriving spa in the Hotwells/Clifton area and better understand their geological significance.

Acknowledgements. The authors would like to express their appreciation of the co-operation extended by Terry Gardiner and Terry Noble, the former owners of Jacob's Well, and Julian Lee-Jones of the Temple History Group. Simon Powell is thanked for photography of the modern as well as reproduction of the old plates; some of the photographs came from originals held in the Bristol City Museum. The assistance of Marian Trott throughout the preparation of the script is gratefully acknowledged.

References

Andrews, J. N., Burgess, W. G., Edmunds, W. M., Kay, R. L. F. & Lee, D. J. 1982. The Thermal Springs of Bath, *Nature*, **298**, 1–4.

Anon 1794. *The New History, Survey and Description of the City and Suburbs of Bristol, or Complete Guide to the Hotwells and Clifton.* W. Matthews, Bristol.

Arrowsmith's Dictionary of Bristol, 1906, 2nd edition. J. W. Arrowsmith, Bristol. See under 'Hotwells', Water Supply', 'Hydro and Spa'.

Atkyns, R. 1712. *Ancient and Present State of Gloucestershire.* W. Bowyer, London.

Aubrey, J. 1656–91. *The Natural History of Wiltshire.*

Buchanan, R. A. & Cossons, N. 1969. *The Industrial Archaeology of the Bristol Region*, David & Charles, Newton Abbott.

—— & Williams, M. 1982. *Brunel's Bristol.* Redcliffe Press, Bristol.

Buckland, W. & Conybeare, W. D. 1824. Observations on the South-western Coal District of England. *Transactions of the Geological Society*, **1**, 210–316.

Burgess, W. G., Edmunds, W. M., Andrews, J. N., Kay, R. L. F. & Lee, D. J. 1979. Investigation of the geothermal potential of the UK, *Report of the Natural Environment Research Council EEC Contract 084-76-EG, Institute of Geological Sciences, London.*

Davison, C. 1924. *A History of British Earthquakes.* Cambridge University Press.

De la Beche, H. T. 1845. *Report on the State of Bristol and other Large Towns*, Health of Towns Commission Report, HMSO, London.

Eyles, V. A. 1955. Scientific Activity in the Bristol Region in the Past *In* MacInnes, G. M. & Whittard, W. F. (eds) *Bristol and the Adjoining Counties*, Chapter 8, 123–143. Published for the British Association.

Fisher, J., Fisher, D., Ford, M. & Ford, F., 1983. *Bristol on Old Postcards*, Reflections of a Bygone Age, Keyworth.

Greenacre, F. & Stoddard, S. 1986. *The Bristol Landscape, the watercolours of Samuel Jackson, 1794–1869.* City of Bristol Museum & Art Gallery.

Gresswell, R. K. 1958. *The Physical Geography of Rivers and Valleys*, Hulton Educational Publications, London.

Hancock, P. L. 1972. Shear zones and veins in the Carboniferous Limestone near the Observatory, Clifton, Bristol. *Proceedings of the Bristol Naturalists' Society*, **32**, 297–306.

Harmer, F. W. 1907. The origin of certain canyon-like valleys associated with lake-like areas of depression. *Quarterly Journal of the Geological Society, London*, **63**, 470–514.

Hawkins, A. B. 1987. Rock Slope Stability in the Avon Gorge. *Proceedings of the Bristol Naturalists' Society*, **47**, 65–78.

—— & Kellaway, G. A. 1971. Field Meeting at Bristol and Bath. New Evidence for Glaciation. *Proceedings of the Geological Association*, **82**, 267–292.

Kellaway, G. A. 1967. The Geological Survey Ashton Park Borehole and its bearing on the geology of the Bristol District. *Bulletin of the Geological Survey of Great Britain*, **27**, 49–153.

Latimer, J. 1900. *Annals of Bristol in the Seventeenth Century.* Bristol.

Little, B. 1954. *The City and County of Bristol.* T. Werner Laurie, London.

Plate 15. The Hotwells Spring in the Avon Gorge: the limestone from which the water issues has been cut back to the line of the principal fissure. The rock face behind the Portway is Hotwells Limestone.

Plate 16. Quaternary fissuring crossing the Avon Thrust at the foot of Bridge Valley Road. In the distance the relationship of the major fissures in the original form of the valley side can be seen in the Great Quarry.

Plate 17. Massive Dolomitic Conglomerate (Triassic) cut by fresh immineralized fissures. The fracturing is nearly vertical and slices through boulders of tough limestone and very hard quartzitic sandstone. It can be traced vertically and laterally into Carboniferous Limestone and forms part of the fissure belt in which the thermal water rises.

Plate 18. Fissured rock loosened by quarrying and weathering has fallen on to the Portway at the foot of Bridge Valley Road and is being cleared away (May 1978). The fissures form part of the linear zone of fissuring marking the site of the Avon Gorge at Clifton.

—— 1976. The Gloucestershire Spas: an eighteenth-century parallel, *In* Essays in Bristol and Gloucestershire History, *Transactions of the Bristol and Gloucestershire Archaeological Society, Bristol.*

Loupekine, I. S. 1951. Fluorite from the Carboniferous Limestone of the Avon Gorge, Bristol. *Proceedings of the Bristol Naturalists' Society*, **28**, 208–219.

Matthews, W. 1794. *Bristol Directory*. Bristol.

Morgan, C. Lloyd, 1885. Sub-aerial denudation and the Avon Gorge. *Proceedings of the Bristol Naturalists' Society*, **4**, 171–197.

Nott, J. 1793. *Of the Hot Well Water near Bristol*. S. Bonner, Bristol.

Owen, 1754. *Observations on the..... Nature of the Hot-Well and the Virtues of its Water*, London.

Pascoe, M. 1981. *The Clifton Guide*, Redcliffe Press, Bristol.

Reynolds, S. H. & Vaughan, A. 1935. The Carboniferous Limestone Series (Avonian of the Avon Gorge) *Proceedings of the Bristol Naturalists' Society*, **8**, 1–15.

Richardson, L. 1930. Wells and Springs of Gloucestershire. *Memoir of the Geological Survey of Great Britain.*

Sibson, R. H., Moore, J. M. C. M. & Rankin, A. H. 1975. Seismic pumping—a hydrothermal fluid transport mechanism. *Journal of the Geological Society, London*, **131**, 553–575.

Trueman, A. E, 1938*a*. *The Scenery of England and Wales*. Gollancz, London.

—— 1938*b*. Erosion Levels in the Bristol District and their relation to the Development of Scenery. *Proceedings of the Bristol Naturalists' Society*, **8**, 402–428.

Varney, W. D. 1921. The Geological History of Pewsey Vale. *Proceedings of the Geologists' Association*, **32**, 189.

Vaughan, A. 1905. The Carboniferous Limestone Series (Avonian) of the Avon Gorge. *Proceedings of the Bristol Naturalists' Society*, **1**, 1–15 & 74–168.

Vaughan, R. & Martelett, J. 1987. Jacob's Well Rediscovered, *Temple Local History Group*, 3/87, 7–15.

Waite, V. 1960. *The Bristol Hotwell*, Bristol Branch of the Historical Association.

Wickham Legg, L. G. 1904. (ed.) *A Relation of a Short Survey of 26 Counties*. F. E. Robinson & Co. London.

Zoetl, J. G. 1989. Bibliography of the History of Karst Research. *International Contributions to Hydrogeology*, **10**, 1–58.

From Kellaway, G. A. (ed.) *Hot Springs of Bath*, pp 205–241

13. Structural and glacial control of thermal water emission in the Avon Basin at Bath

G. A. Kellaway

In the Bristol and Somerset Coalfield and the Mendips we see evidence of the major structural movements which took place in Devonian, Carboniferous and Permo-Carboniferous times (Fig. 13.1, Table 13.1). These are generally described as Variscan and are of great complexity (Kellaway & Hancock 1983). The central region of the Coalfield is marked by the Malvern Fault Zone and this shows evidence of lateral and vertical movement from Devonian times onwards. The resulting structures include the Coalpit Heath Basin in the north and the Radstock Basin in the south, the two basins being separated by a transverse structure known as the Kingswood Anticline, consisting of highly folded and thrust-faulted Coal Measures (Kellaway & Welch 1991).

Table 13.1 Timescale in millions of years B.P.

Period (duration)				Main episodes of earth movement
Quaternary (1.64)	Holocene	0.01		
	Pleistocene	1.64		
Tertiary (64)	Pliocene	5.2		Alpine (mid-Tertiary)
	Miocene	23.5 (n.r)		
	Oligocene	35.5		
	Eocene	56.5		
	Palaeocene	65.0 (n.r)		
Mesozoic (180)	Cretaceous	Upper	Chalk 97	
			Upper Greensand and Gault 112	
		Lower 146		
	Jurassic 208			
	Triassic 245			
	Permian	Zechstein 256 (n.r)		
		Rotliegendes 290 = 'Permo-Carboniferous' (n.r)		Variscan
	Carboniferous	Coal Measures 311		
		Millstone Grit 333		
		Carboniferous Limestone 363		
	Devonian	Upper Old Red Sandstone 377		Caledonian
		Lower Old Red Sandstone 409		
	Silurian	439		
	Ordovician	510		
	Cambrian	570	concealed	
	Pre-Cambrian c.	4000	concealed	

(n.r) not represented by strata in the regional succession.
Dates based on Harland *et al.* (1990).

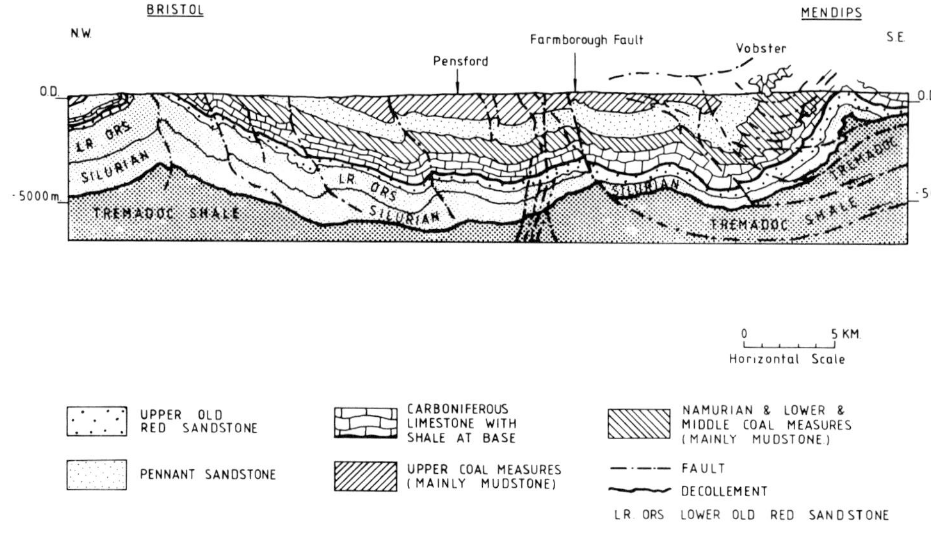

Fig. 13.1. Generalized section between Bristol and the Mendips (Mesozoic rocks omitted).

The Carboniferous Limestone which underlies the Millstone Grit and Coal Measures (Table 13.2) extends locally to depths which may be about 4000 m (13,000 feet) below sea level in the deepest parts of the basin. Whether the Carboniferous Limestone forms a hydrologically continuous mass at depth is questionable as most of the Variscan structures are highly mineralized and in this condition they may act as a barrier to water movement rather than as a channel. This is illustrated by some major Variscan structures including the Avon thrust at Clifton where the thermal water is making use of intersecting reactivated Mesozoic fissures to migrate across the thrust, without (as far as is known) travelling laterally along it. An analysis of the post-Variscan faulting is therefore essential if the thermal water movements are to be understood.

Tectonics and glaciation

Triassic to mid-Jurassic

Large scale warping, folding, faulting and fissuring of the Palaeozoic rocks of the Mendips and Bristol and Somerset coalfield took place in late Triassic to mid-Jurassic times producing substantial modification of many Variscan structures. The strongest phases of movement pre-date deposition of the transgressive limestones of the Upper Inferior Oolite (Middle Jurassic) which rests on an erosion surface incised in Palaeozoic and older Mesozoic rocks (Donovan & Kellaway, 1984, Fig. 8, p. 22).

Some of the fissures and the associated karstic features, such as caves and sink-holes produced by weathering of the Carboniferous Limestone in Triassic times, were wholly or partly filled with sediment and strongly mineralized soon after their formation. Cavities and channels were filled up and their walls were lined by the deposition of calcite, dolomite, barite, galena, pyrite and other minerals. Mineralization continued in the bedrock and the overlying sediments on the sea floor throughout Jurassic times, though on a diminishing scale while the Upper Jurassic strata were being laid down (Alabaster 1976). This process probably terminated with the emergence of the sea floor at the onset of the Lower Cretaceous earth movements.

Silicification of the mineralized Palaeozoic rocks and the thin Triassic and Jurassic strata resting on them took place in Albian and Cenomanian times in the shallow water areas of the central and eastern Mendips. Infillings of Quaternary sediment and re-cementing of the vein rock by calcareous tufa and stalactite may in some instances provide evidence of renewed movement or re-opening in Quaternary times; Mesozoic fissures and cavities reactivated by Quaternary movement exercised control over the migration of thermal water as exemplified by the warm springs issuing in the Avon Gorge at Clifton (Chapter 12).

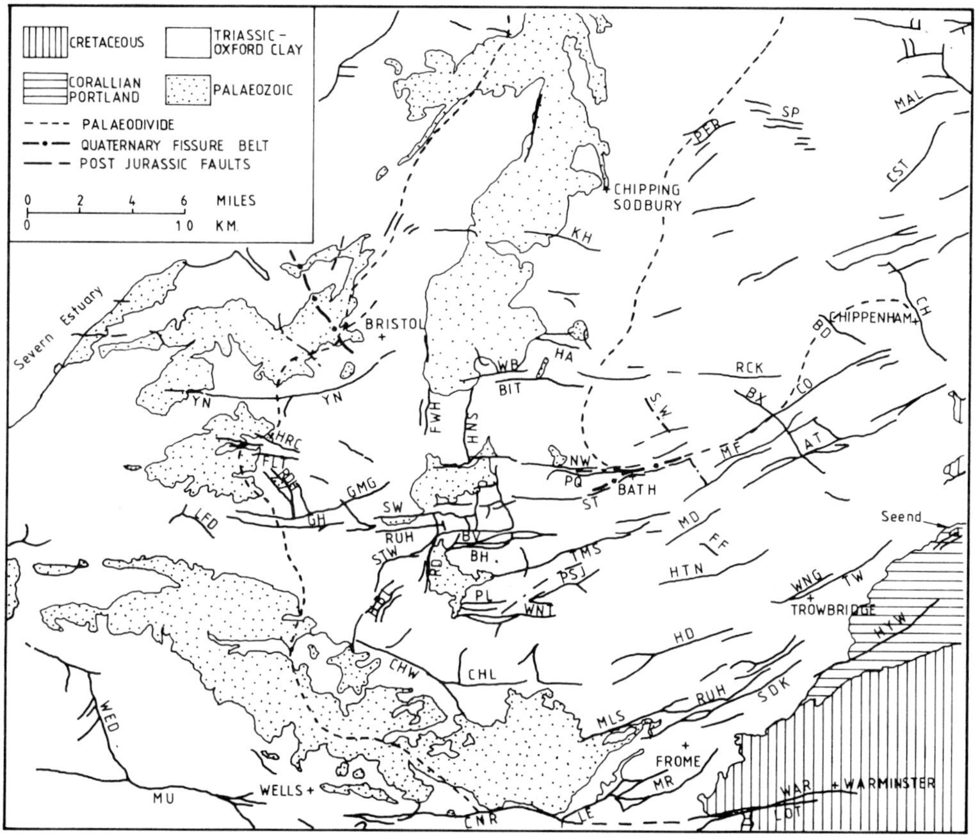

Fig. 13.2. Post-Variscan faults of the Bath district. Based on the maps of the British Geological Survey. Key to faults:

AT	Atworth Fault	FF	Freshford F.	LOT	Longleat F	RH	Roundhill F.
BA	Battlefields F.	FLT	Felton F.	MAL	Malmesbury F.	RUH	Rudge Hill F.
BD	Biddestone F.	FWH	Flowers Hill F.	MD	Midford F.	SDK	Standerwick F.
BH	Barrow Hill F.	GH	Gravel Hill F.	MF	Monkton Farleigh	SP	Sopworth F.
BIT	Bitton F.	HA	Homeapple Hill F.	MLS	Mells F.	ST	South Twerton F.
BV	Barrow Vale F.	HBL	Hinton Blewett Faults	MU	Mudgeley F.	STW	Stowey F.
BX	Box F.	HD	Hardington F.	NW	Newton F.	TMS	Timsbury F.
CH	Chippenham F.	HNS	Hunstrete F.	PFR	Petit-France F.	TW	Trowbridge F.
CHL	Chilcompton F.	HRC	Hartcliff Rocks F.	PL	Paulton F.	WAR	Warminster F.
CHW	Chewton F.	HTN	Hinton Charterhouse F.	PQ	Pennyquick F.	WB	Willsbridge F.
CMG	Chew Magna F.	HYW	Heywood F.	PSJ	Peasedown St. John F.	WED	Wedmore F.
CNR	Cranmore F.	KH	Kidney Hill F.	RCK	Rockhouse F.	WNG	Wingfield F.
CO	Corsham F.	LE	Leighton F.	RD	Red Hill F.	WNT	Winterfield F.
CST	Corston F.	LFD	Langford F.	RDH	Ridgehill F.	YN	Yanley F.

Table 13.2 Succession in the Palaeozoic rocks of the Bristol–Bath region

Permian rocks are absent; the main *Variscan* movements took place between 290 and 260 Ma B.P.

Carboniferous	(363–290 Ma)		
	Coal Measures (2500 m)		
	Millstone Grit = Quartzitic Sandstone Group (*c.* 300 m)		
	Carboniferous Limestone (900 m)	Hotwells Formation (300 m)	Upper Cromhall Sandstone Hotwells Limestone
		Clifton Down Formation	
		Black Rock Formation	
		Lower Limestone Shale	
Devonian	(409–363 Ma)		
	Upper Old Red Sandstone	Tintern Sandstone (125 m) Quartz Conglomerate (30 m)	

Major unconformity: Strong tectonic movement along the Malvern Fault Lines accompanied by erosion (*c.* 390–370 Ma).

	Lower Old Red Sandstone	Thornbury Beds (600 m) Downton Castle Sandstone (8 m)	
Silurian	(439–409 Ma)		
	Ludlow and Wenlock (*c.* 300 m)		
	Llandovery, basal beds not represented (*c.* 500 m)		
Ordovician	(510–439 Ma)		

A major unconformity exists beneath the Llandovery rocks, the higher Ordovician formations including Arenig–Ashgill being absent. The Tremadoc is represented by the Micklewood Beds and Breadstone Shales (1500–2000 m). The contact with the underlying *Cambrian* is not exposed.

Carboniferous Limestone of Bath. It is thought that the Hotwells Formation may be about 300 m thick at Bath the contact with the Clifton Down Limestone having been proved at −65.7 m O.D. in the Kingsmead Borehole. The Stall Street inclined Borehole proved that Rownham Hill Coral Bed at about −50 m O.D., this is approximately 400 ft (122 m) above the contact of the Hotwells Limestone and Clifton Down Limestone.

All thicknesses given in metres (m); ages in millions of years B.P.

Lower Cretaceous (pre-Albian)

A fairly dense concentration of ENE–WSW trending faults associated with some transverse faults and parallel or sub-parallel folds affects the Palaeozoic and Jurassic rocks of the central and eastern Mendips, the Bristol and Somerset coalfield and adjacent parts of the south Cotswolds and western Wiltshire (Fig. 13.2). Some of these structures appear at Seend near Devizes where they were truncated by pre-Aptian erosion before being buried by the pebbly ironstones of the Lower Greensand. Elsewhere the faults cannot be so precisely dated. Thus the Heywood Fault, formerly seen in the Corallian ironstone workings at Westbury, Wilts, cuts folded Corallian and Kimmeridge Clay, the folds being of pre-Albian age. Although the Heywood Fault is not seen in contact with Cretaceous rocks, its trend suggests that, like the Berkeley Marsh, Marston and Leighton faults which mark the southern limit of the eastern Mendips, it is of Lower Cretaceous age.

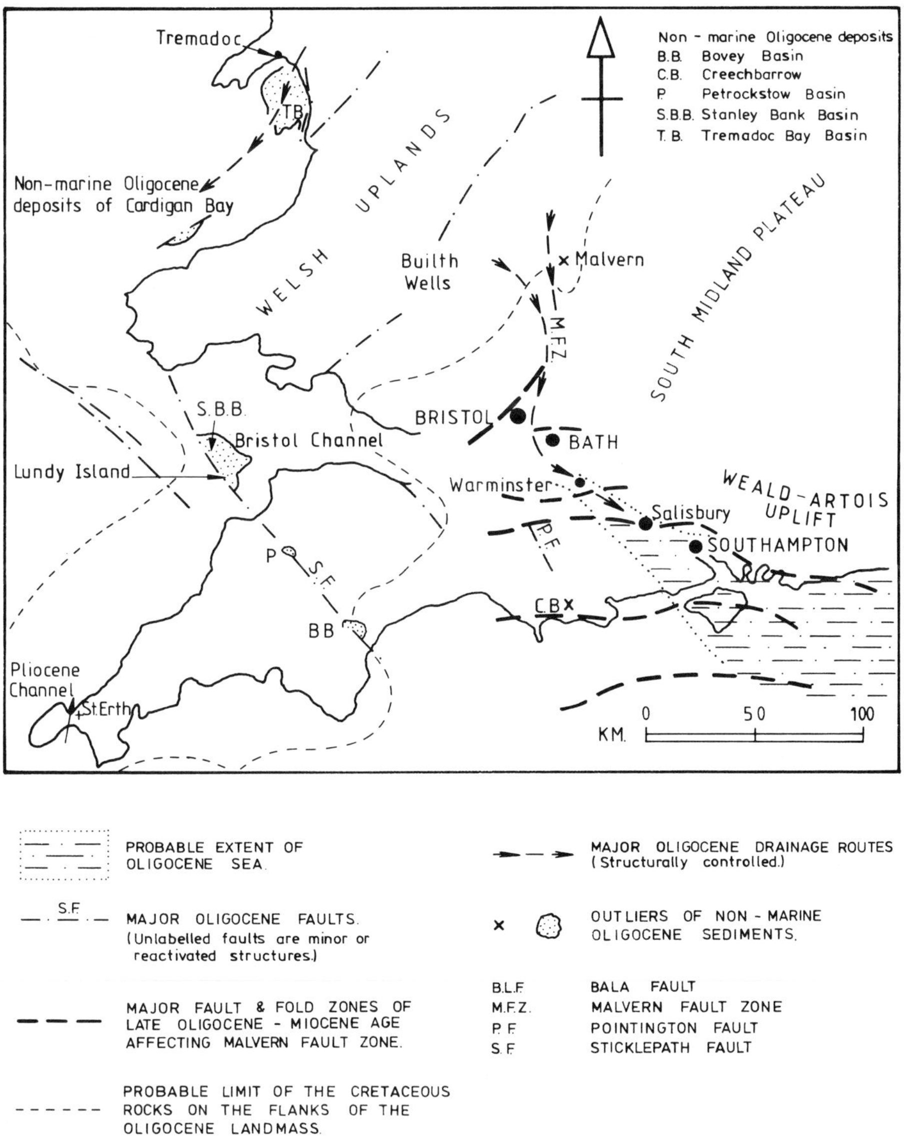

Fig. 13.3. Tectonic control of mid-Tertiary relief and sedimentation.

A point of some importance is that the ‘dead’ or quiescent Lower Cretaceous faults, formed during the structural evolution of the Wealden basin farther east, had little effect upon the development of the Oligocene–early Miocene NW–SE trending drainage system (Fig. 13.3) which was itself the forerunner of the Plio-Pleistocene palaeovalley of the

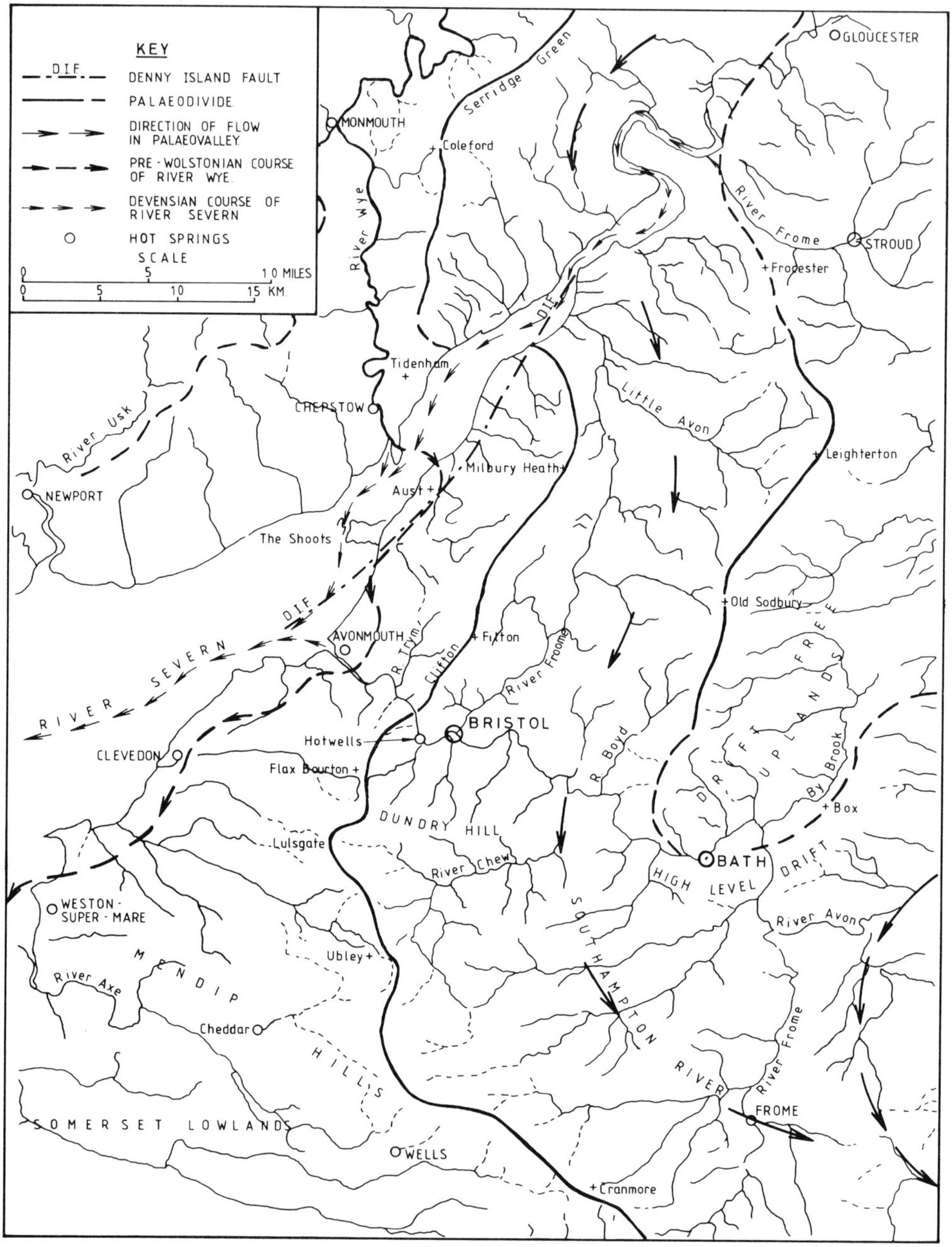

Fig. 13.4. Palaeodivides of the Severn estuary region. The Stroud Frome, Little Avon and Bristol Avon are highly transgressive (obsequent) streams. The Wye above Chepstow is the oldest river valley surviving in this region. Below Chepstow, the Wye has been dismembered by the Severn and Bristol Avon. A Tertiary–Upper Cretaceous cover existed in early Quaternary times and the strong Lower Cretaceous fault swarm (Fig. 13.11) would only have affected the drainage pattern in the Palaeovalley, where reactivation had taken place.

Southampton River (Fig. 13.4). In the early–mid Pleistocene stages of the physiographic evolution of the Bristol–Bath region, when the faulted and folded Palaeozoic, Triassic and Jurassic rocks were exposed by the removal of an unconformable cover of soft gently dipping Upper Cretaceous and Eocene deposits, the older formations would have exerted an indirect influence on the development of the local stream pattern. The latter reflects in some degree the hardness, permeability and other physical attributes of the Palaeozoic and older Mesozoic rocks which influenced the development of many of the smaller streams in Pleistocene and Recent times. Only where late Tertiary or Quaternary reactivation of the older faults and fissures has taken place is there evidence of *active* control of the drainage pattern by Mesozoic faults or of association of those faults with thermal water movements.

Reactivation of some Lower Cretaceous and Tertiary faults has taken place at Bath and reactivation of older Mesozoic faults and fissure belts at Bristol (Chapter 12). However, with the possible exception of a doubtful record of a thermal emission at Dunkerton (Jones 1572), none of the Mesozoic faults south of the Avon valley gives rise to thermal water springs. In the high ground of the Mendips the lack of hot springs would be expected owing to the altitude of the fissured Carboniferous Limestone.

Although tectonic movements took place on a limited scale in Upper Cretaceous–Eocene (i.e. Palaeocene) times, little is known about them. Nevertheless the original thickness and structure of the Chalk and Eocene strata are important in relation to the former extent of the cover rocks on which the mid-Tertiary drainage was established (Fig. 13.5). This aspect of the geological history has a bearing on the stages whereby the Upper Cretaceous and Tertiary cover rocks were eroded away in later Tertiary and early Pleistocene times as shown by the movement of boulders and detritus formed by erosion of the Palaeozoic rocks of Wales and the Welsh borders and the translation of this material eastwards towards Bath and Salisbury Plain in successive stages to which the denudation chronology is related.

Oligocene and Miocene

Marine Oligocene deposits are present in the Hampshire Basin, both in the Isle of Wight and the adjacent part of the New Forest. Farther west, at Creechbarrow in Dorset and Bovey and Petrockstow in Devon as well as on the floor of the Bristol Channel off Lundy Island (Davies 1987), non-marine strata of Oligocene or early Miocene age are found in isolated basins (Fig. 13.3). These are described here as being of Oligocene age. They appear to have been laid down in intermontane basins or depressions aligned along the Sticklepath Fault in Devon and other NW–SE trending fold and fault belts which may be situated above reactivated Variscan structural zones such as the Malvern Fault zone. The Hampshire deposits are of shallow freshwater, brackish and marine origin and exceed 600 feet (183 m) in thickness; they were laid down in a tectonic basin with a NW–SE or WNW–ESE orientation. Intermittent subsidence reflects the structural instability of the basin and may also be related to the evolution of the Weald–Artois ridge on its northeastern flank.

As long ago as 1892 in his 'Building of the British Isles', Jukes-Brown envisaged the deposition of the Hampshire basin sediments in a tectonic depression extending in a NW direction towards Bath and the Malverns. In 1902 he suggested that the growth of the Wealden uplift (or Weald–Artois ridge) was initiated in Eocene times and culminated in the separation of the Tertiary London and Hampshire basins, a concept which was further developed by Dudley Stamp (1921) and by several French geologists (Pomerol & Feuguer 1969, Fig. 8).

At the western end of the Bristol Channel the granite mass of Lundy Island (142 m A.O.D.) is situated west of the Sticklepath Fault (Fig. 13.3). East of the fault is an area of non-marine Oligocene strata (Fletcher 1975) constituting the Stanley Bank Basin (Davies 1987). Non-marine Neogene sediments are also found in the area north of the main basin. The Oligocene sediments rest on Devonian, Triassic and Jurassic rocks; no Cretaceous or Eocene strata have been found.

The Lundy granite is said to be of Eocene age (Owen, 1971). It has obviously undergone considerable uplift since its emplacement but this is due primarily to regional uplift in Tertiary times. Vertical movement along the Sticklepath Fault is not considered to have played much part as the fault is a dextral wrench fault which shows some evidence of sinistral reactivation in Tertiary times in South Devon (Holloway & Chadwick 1986). Regional uplift in Upper Cretaceous and Eocene times produced an Oligocene landmass which extended from South Devon to North Wales, marine sedimentation being eventually restricted to the New Forest and Isle of Wight. By mid-Miocene times the sea had also been excluded from the Hampshire Basin.

While the uplifted areas of southwest England were undergoing erosion in Eocene to Miocene times large areas of Cretaceous rocks were exposed to weathering and dissection (Plate E, p 282). By Oligocene times the western limit of the Upper Cretaceous cover lay near the eastern margin of the Bovey Basin in south Devon and west of the Blackdown Hills in southwest Somerset. In the Bristol Channel it was certainly confined to the east of Lundy Island and may not have extended west of a line from Nash Point to Minehead, with its northern limit at the foothills of the South Wales coalfield. At this time all but the highest points of the Mendips were buried, while the Somerset lowlands and the Bristol and Somerset coalfield as well as the hills around Bath probably retained a fairly continuous Upper Cretaceous cover on which the Oligocene drainage pattern was established (Fig 13.3).

The writer believes that the well known '430 ft platform' of Cornwall and Devon is of Palaeogene age, owing its altitude primarily to late Oligocene–Miocene tectonic uplift moderated by Plio-Pleistocene warping. Incised in hard rocks, it was probably exposed in late Pliocene or early Pleistocene times by the removal of a soft Palaeogene cover. The marine Pliocene rocks of St Erth have been emplaced in a depression incised in the older surface, possibly when part of the unconformable Palaeogene cover was still in place on the platform.

Comparable high-level platforms in western South Wales were considered by T. N. George (1970) who concluded that it is unlikely that they are of Cretaceous age. While accepting his arguments for a post-Cretaceous origin, the writer feels some uncertainty as to his conclusion that they are all of Neogene age, post-dating the main Alpine movements. It is true that these movements produced strongly folded structures in the Hampshire Basin and the English Channel, but when traced laterally or downward into the indurated Palaeozoic rocks, these folds are replaced by faults. Step-faulting, rift and horst-faulting and wrench faulting become dominant at depth. If these movements were active during and after the deposition of the Tertiary cover then subsequent removal of the soft sediments with their growth-folds and post-depositional contortions would reveal a series of stepped platforms bounded by erosional fault scarps and cliffs which would resemble those of southwest Wales and Devon and Cornwall. These platforms do not extend into the Severn estuary region where, if Palaeogene strata were formerly present, they were resting unconformably on Mesozoic rocks.

The possibility that some of the hard rock platforms may have undergone modification by Plio-Pleistocene warping or faulting and glacial action has, of course, to be accepted.

The regressive effect of the late Eocene and early Oligocene movements produced an expanded Oligocene landmass stretching from the high ground of the Cotentin to the mountains of North Wales (Fig. 13.3). On its western side the mountainous northern part

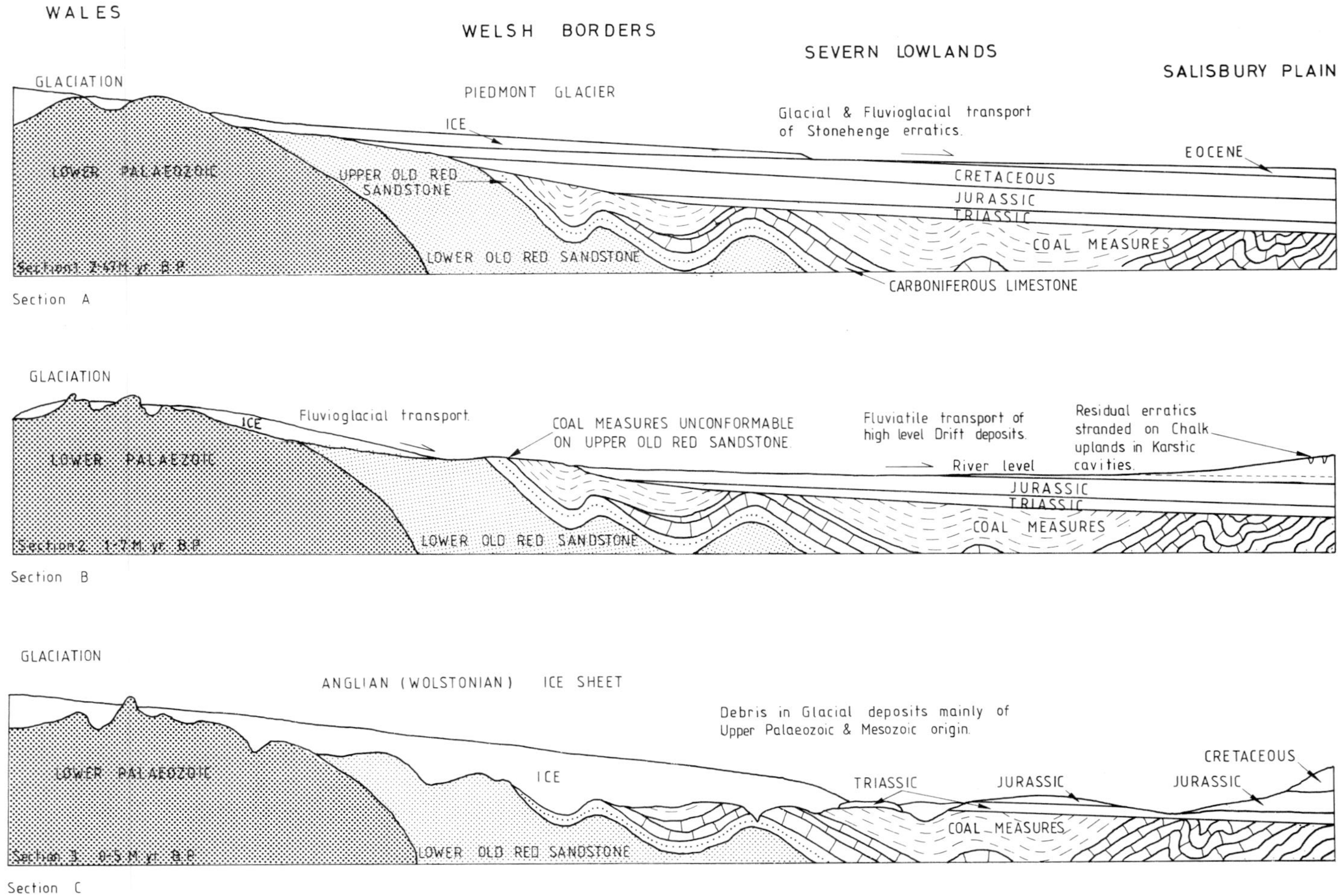

Fig. 13.5. Comparative sections showing the distribution of the Tertiary–Mesozoic cover in Plio–Pleistocene times in the Welsh borders and Severn valley. The representation is diagrammatic and not to scale. Length of section is about 100 miles (161 km).

of the landmass was drained towards Cardigan Bay and St George's Channel (Dobson & Whittington 1987); on its eastern flank the principal line of drainage was probably directed from the Welsh borders towards Southampton Water, much as indicated by Jukes-Brown (1892, fig. 14). At its southeastern end the estuary of the Oligocene forerunner of the early Quaternary Southampton River (Fig. 13.4) may well have become blocked in late Oligocene or early Miocene times by the growth of E–W striking folds of which the monoclinal folds seen in the Isle of Purbeck and the Isle of Wight are outstanding examples. Other mid-Tertiary structures with strong E–W trends, such as the Mere Fault and the post-Eocene folds at Salisbury, may have interrupted the Oligocene drainage pattern and are also likely to have affected the movement of surface and underground water in Somerset and Wiltshire.

In the western part of the Bristol Channel a submarine platform is incised at 46–52 m below O.D. across the surface of the non-marine Oligocene and Neogene sediments of the Stanley Bank Basin (Davies 1987). It is unlikely, however, that this surface extends far to the east where the relief and geology of the sea floor is known in some detail (Evans & Thompson 1979). The sea floor rises towards Croyde Bay and Bull Point and the erosion surface must be limited in this direction. The submerged drainage channel or palaeovalley leading from the Severn estuary to the Celtic Sea (Collins 1987, fig. 1) lies roughly midway between Gower and north Devon and passes north of the Stanley Bank Basin. It is probably of mid-Pleistocene age, the last period of substantial downcutting being Devensian, at the time of very low sea level. The evidence from the raised beaches of the Weston-Super-Mare and Clevedon area indicates that marine incursion in this region dates from before the Ipswichian interglacial and may be 'Anglian' or even pre-'Anglian' in age.

Pliocene to mid-Pleistocene

The boundary problem and onset of glaciation[1] Before proceeding with an analysis of the evidence for Plio-Pleistocene and neotectonic movement it is necessary to decide on the Plio-Pleistocene boundary and, if possible, to define its absolute age. Many geologists regard the base of the Red Crag of East Anglia as the bottom of the Pleistocene and in the case of the British succession this has some practical advantages. Recently, however, a strong case has been made for selecting an Italian stratotype and on this basis Harland *et al.* (1990, p 68) have proposed that a date of 1.64 Ma BP should be adopted for the boundary. This has the effect of excluding from the Pleistocene the major glaciation recorded at 2.47 Ma in the O^{18} isotope calibration of the onset of ice-rafting in the North Atlantic based on analysis of material from the Rockall Bank Core 552A (Shackleton *et al.* 1984, pp 620–623). This core which showed alternations of white carbonate ooze and dark glacigenic layers with glacial debris has been correlated with the standard magnetic stratigraphy and provides evidence of a major glaciation which affected the North Atlantic but which does not appear to be represented by till in the east Midlands, East Anglia, the southern North Sea or eastern English Channel.

If the age of the Plio-Pleistocene boundary is taken at 1.64 Ma as indicated above, then the major glaciation recorded at 2.47 Ma in Core 552A lies within the Pliocene. The first evidence of strong temperature lowering is at 2.5 Ma but the presence of ice reaching tide-water in force dates from 2.47 Ma. The log and O^{18} record shows a number of minor cycles

[1] The best general map covering the regions referred to below is the 1:250,000 map of the 'Sub-Pleistocene Geology of the British Isles and the adjacent Continental Shelf', Institute of Geological Sciences, 1970.

within a total duration of about 0.5 Ma during which the ice masses on land must have spread at least some way on to the western continental shelf of the British Isles at times of lowered sea level. At the first glacial maximum at 2.47 Ma the volume of ice is thought to have been similar to that developed in mid-Pleistocene times on the British mainland (Shackleton *et al.* 1984, p 620). At its maximum extent this might have caused a substantial lowering of sea level, though it is difficult to estimate its lateral extent in the absence of definable limits and relief of the sea floor.

If, as seems likely, this ice developed primarily over Brittany, Ireland, Iceland and the western uplands of England, Wales and Scotland, then this would be the most likely time for movement of glacial erratics from Wales to Salisbury Plain to have taken place (Chapter 14). As will be seen, this timescale also provides for the very large amount of erosion which would have been required to remove the Mesozoic cover in the Welsh border area and to explain the retreat of the Cotswold escarpment to its present position.

Table 13.3 Principal events, Miocene to Holocene

Holocene (or Recent)		0– 10,000
Late Pleistocene	Devensian (or last) Glaciation	*c.* 50,000
	Ipswichian (Eemian) Interglacial	*c.* 150,000
	Wolstonian–Anglian Glaciations	
	(Saale-Elster Gl. of N. Germany)	
Mid-Pleistocene	including Hoxnian–Holsteinian	
	Interglacials	*c.* 500,000
	Cromerian Interglacial	*c.* 600,000
	Beestonian Glaciations	*c.* 750,000
Early Pleistocene	Northern Drift Glaciation (Cotswolds)	*c.* 1,100,000
——	Plio–Pleistocene boundary at 1,640,000	——
	Stonehenge Glaciation	*c.* 2,470,000
Pliocene		
	Iceland; Onset of glaciation	*c.* 3,000,000
——	Mio-Pliocene boundary at 5,200,000	——
——	Oligocene-Miocene boundary at 23,300,000	——

Dating of the Oligocene, Miocene and Pliocene boundaries is that given by Harland *et al.* (1990). All dates are given in years before present (BP).

Pliocene sea level and tectonic movements The thickness of Neogene sediments in the southern Irish Sea and the North Sea basin indicates that coastal deformation, both warping and the movement of blocks, was active throughout the period of basinal deposition (King 1977, pp 142–143). The change in the direction of flow of the Wolstonian–Anglian ice sheets (p 252) may be regarded as indicative of a similar process of warping in mid-Pleistocene times. The broad depression in Bideford Bay and the western part of the Bristol Channel could have been submerged in Pliocene or early Pleistocene periods of high sea level, but few marine deposits of this age have yet been found on land. Such marine Pliocene deposits as have been found in Britain are limited to SW Cornwall (St Erth), SE England and East Anglia, a distribution which suggests that down-warping towards the Atlantic in the west and the North Sea in the east was an important factor in controlling marine deposition.

Marine Pliocene deposits are also found in Brittany and Normandy where they vary greatly in altitude and character. They are generally thin and consist of Redonian marine deposits at the base, overlain by unfossiliferous sands with pebbly beds known as the Sables Rouges. The latter may be unrepresented in Britain, the Coralline Crag of East Anglia being roughly equivalent to the Redonian. Redonian deposits have been found at Valmont near Fécamp in a big karstic depression. They are overlain by pink and yellow sandy deposits described as Sables de Lozère which in turn are overlain by the Argile à Silex or Clay-with-flints (Cavelier & Kunz 1974).

The Valmont section is important for several reasons. Rocks derived from the Massif Central are found at the base of the Sables de Lozère[1] indicating that in post-Pliocene times (using the term Pliocene in the sense that it has been employed in France) water was still moving from the Massif Central to the English Channel.

Subsequently in Pleistocene times this movement ceased, and the pre-Seine Loire which formerly drained northwards from Orléans was deflected to the west and to the Bay of Biscay at Nantes. This drastic change in the drainage, formerly supposed to be Miocene (Pomerol 1974; Rat 1974), is therefore of younger date. The Chalk plateau, in which the karstic depression at Valmont is emplaced, stands at an altitude of about 100 m (330 feet) A.O.D. and is dissected by the westward flowing Seine which now crosses the old drainage line of the Sables de Lozère. The latter is structurally determined by movement along the major basement structure which passes beneath St Valéry in the direction of Orléans.

Although the dating of these movements may not be identical with those responsible for the break-up of the 'Southampton River' and the creation of the Bristol Avon, the effect of Pliocene and post-Pliocene warping between the Redonian Gulf on the Atlantic coast and the Belgian Diestian basin in NW France is comparable. The St Valéry line marks the tectonic hinge and the presence of Redonian deposits at Valmont on the west side and possible Pliocene marine sediments at Beachy Head on the east suggests that substantial movement may have taken place in the hinge area.

At the broad eastern end of the Kennett valley which merges with the low ground of the London basin, the northern Pliocene sea penetrated inland before falling sea levels about 3–2.5 Ma BP caused its regression. It has been quite commonly supposed that there is a maximum altitude of about 600 feet (182 m) A.O.D. to which sea level rose in Pliocene times in southern England, but while some marine Pliocene deposits achieve this altitude in southeast England, the known effects of Pliocene and post-Pliocene earth movements, cyclic climatic change and accentuated erosion in late Pliocene times make it impossible to rely solely on extrapolated 'base levels' even if these can be dated or identified with sufficient accuracy.

Depression of the western coastal region of NW France and the western parts of the British Isles due to late Pliocene ice-loading would have been combined with a lowered sea level. The relationship between this and tectonic subsidence in the North Sea basin cannot yet be accurately assessed, but warping of the intermediate region would be expected.

Mio-Pliocene developments north of the Mendips North of the Mendips the Miocene faulting and folding reactivated some large Variscan structures including the Bitton Fault and Kingswood Anticline on the north side of the Avon Valley. These movements may have created a physical barrier which forced surface water draining off the Welsh uplands to escape over the eastern divide of the palaeovalley. However, any water draining from

[1] In terms of the dating used here the Sables de Lozère would be classified as Pliocene.

the west into the synclinal Kennet valley would inevitably have been cut off at a later stage by the development of the Bristol Avon and Salisbury Avon whose headwaters now compete with the Kennett in the western part of the Marlborough Downs and the Vale of Pewsey.

The Kennett syncline shows a modified structural relief at its western end though it retains synclinally disposed Eocene strata in its floor east of Newbury. NW of Avebury the Chalk has been strongly eroded and to the SW the big dry valley leading from the Bishops Cannings gap near Roundway Hill has been severed from its original source area by the development of the streams at the head of the Salisbury Avon. These cross the Vale of Pewsey obliquely before combining to flow southward by way of Upavon to Salisbury (Fig 14.3). Varney (1921) suggested that some of the gravels in the Vale of Pewsey may correlate with those of Bathampton Down and regarded their position and altitude as indicating the direction and inclination of the valley floor of the ancient river. Even if the gravels do contain some material similar to that seen on Bathampton Down the effect of recycling has to be considered. This problem is particularly difficult in the Chalk downland areas and the Vale of Pewsey because of the mass effects produced by freeze–thaw processes in Devensian times.

The breaching of the West Mendip barrier Dateable deposits occur on high ground in the Mendips and are of non-marine origin and Cromerian age (Bishop 1974). There is no evidence to show that the Severn estuary existed in early Pleistocene times, though it has to be borne in mind that the Wye valley is of great antiquity and its course below the Severn Bridge (Fig. 13.6) was later adapted by the Severn estuary.

In the Somerset levels the presence of Devensian Head deposits masking the lower slopes of the Mendips and the foothills of the Middle Jurassic escarpments between Castle Cary and Ilminster, shows that a large physical depression existed between the Mendips and the Quantocks at that time. The regional Devensian river pattern closely resembles that of the present day and any major changes in the drainage pattern must date from Ipswichian times or earlier. However, there are a number of indications of neotectonic movement affecting the evolution of the west Somerset and north Devon coast line. These are a continuation of earlier Plio-Pleistocene movements along the neotectonic belt marked by the Taunton (1868, 1885) and Barnstaple (1894, 1920) earthquakes recorded by Davison (1924).

Of greater significance in the investigation of the history of the Bristol Avon is the western Mendip ridge, now represented by Brean Down, Steep Holm, Flat Holm and reefs known as The Wolves (Plate 28). These are masses of Carboniferous Limestone emerging from a Triassic cover of variable hardness and thickness. The geology of the coast and off shore area at Barry (Banner *et al.* 1971) shows that this area forms a continuation of the Mendips which is bounded on the south by an extensive tract of soft Lower Jurassic strata. The position of the SW–NE trending Denny Island Fault zone (Fig. 13.6) is uncertain south of Denny Island but the evidence for Variscan folding and the presence of unmineralized wrench faults seen cutting ripple-marked Flat Holm Limestone on the SW side of Flat Holm (Plate 27) suggests that the main disturbance may be close to Lighthouse Point. The bathymetry of the Channel floor and the structure of Denny Island indicate that the Denny Island Fault zone passes between Flat Holm and Steep Holm and that it extends in a northeasterly direction toward the eastern end of the Severn Tunnel and thence to a silted-up channel of the lower Severn between Aust and Littleton-on Severn. This is thought to be an important line of wrench-faulting responsible for the deformation of the palaeovalley (Fig. 13.6).

Raised beaches developed between Weston-Super-Mare and Clevedon and the dating of the glacial and interglacial deposits at Kenn and Clevedon shows that marine incursions into this area may have been made as early as the period immediately preceding the Anglian glaciation i.e. 0.4 to 0.5 Ma BP. At this time it would have been possible for water to cross the Mendip barrier and to enter the Bristol Channel. The presence of possible Wolstonian (or Anglian) till on the floor of the Bristol Channel (Evans & Thompson 1979, p 10) and the general distribution of the glacial deposits in South Wales and the Clevedon area tend to confirm the concept of Welsh ice penetrating an area which was laid bare at a time of low sea level when the Mendip barrier had already been substantially breached by the Wye. The final deepening of the channel leading westward to the Atlantic and the destruction of large areas of land in the Severn estuary was achieved during the melting of the Anglian–Wolstonian and Devensian ice sheets.

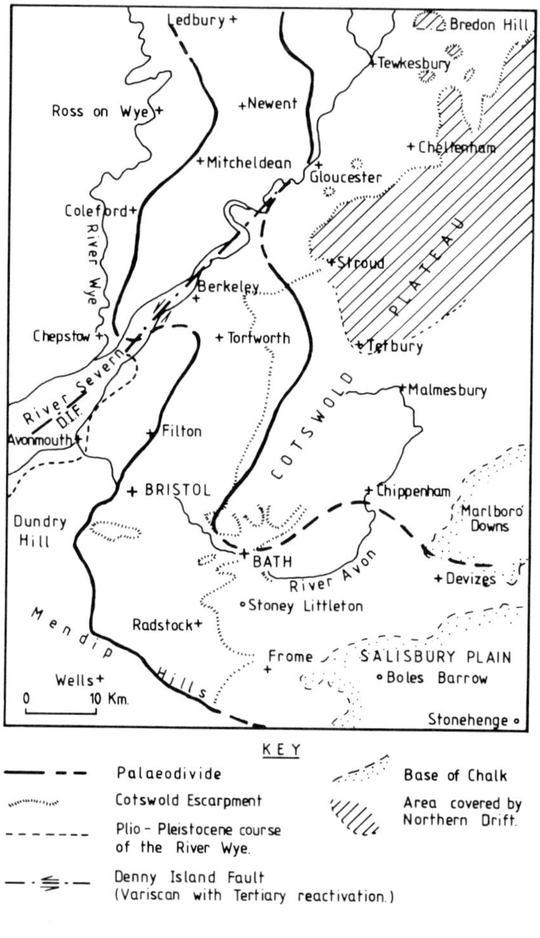

Fig. 13.6. Palaeodivides seen in relation to the Avon valley, the Cotswolds escarpment and Denny Island Fault (D.I.F.).

Pleistocene structures

The large Quaternary structures such as valley and fault bulges, trough-gulls and cambers date for the most part from pre-Devensian or pre-Ipswichian times. One of the problems in dealing with these structures is the lack of reliable stratigraphical data on which to base a chronological assessment. Archaeological evidence relating to the occurrence of Acheulian

Plate 19. Quarried inlier of Carboniferous limestone at Wick looking SE towards Freezing Hill and Tog Hill which form part of the south Cotswolds.

Plate 20. Swainswick valley seen from the landslip on the west side of Solsbury Hill. In the background is Freezing Hill beyond which lies the Wick inlier (Plate 19). The village of Swainswick is situated on a strongly cambered slope. The high ground on the right is formed by the Great Oolite plateau of Charmy Down.

Plate 21. The Avon Valley at Bath seen from Beechen Cliff.

Plate 22. Cambered Great Oolite limestone showing dip-and-fault structure. Approach road to Bath University, Bathampton Down.

Plate 23. Cambered Great Oolite showing dip-and-fault structure, approach road to Bath University, Bathampton Down.

Plate 24. Road widening at Bathampton (NE slopes of Bathampton Down) on Bath–Warminster Road showing isolated masses of Inferior Oolite in foundered strata.

Plate 25. St Catherine's valley, Charmy Down and Solsbury Hill looking south from near Ashwick. The valley joins the Avon at Batheaston. The deep valleys of the south Cotswolds near Bath are of mid-Quaternary age.

Plate 26. The chalk escarpment at Roundway Down at the western end of the Marlborough Downs near Devizes. View from the eastern slopes of the Seend outlier, the foreground being composed of Upper Jurassic and Lower Cretaceous rocks from which the Upper Cretaceous cover has been removed in (geologically) recent times.

Plate 27. Wrench faults cutting ripple-marked Carboniferous Limestone near the lighthouse on the island of Flat Holm.

Plate 28. Brean Down, the most westerly of the Mendip Hills, seen from the island of Steep Holm. This gap in the Mendip barrier existed before the Bristol Avon cut the gorge at Clifton.

hand axes in Drift deposits can be helpful (Wymer & Straw, 1977; Wymer 1985) and the general distribution of Acheulian-type implements in the Bristol and Bath area (Davies & Fry 1929) has a bearing on the evolution of the Bristol Avon and its gorges.

Recent investigations in East Anglia, the Midlands and southeast England have revealed much new evidence pointing to the importance of tectonic movements in Quaternary times. In particular it has become apparent that some of the larger Quaternary structures, formerly regarded as being of non-diastrophic origin, are continuous with belts of trough and ridge-faulted rocks (including large trough gulls) which penetrate hard bedrocks to such depths as to be independent of cryoturbation or glacial action. The presence or absence of ice or snow capable of producing large quantities of melt water is clearly a matter of importance in interpreting the superficial effects of Quaternary structural deformation, while the absence of valley bulges and cambers on the coast of Normandy, where the bedrock geology and strong fissuring of valley floors is similar to that of southern England, but where severe and rather arid periglacial conditions appear to have prevailed during the Quaternary cold periods, indicates that deep seasonal freezing or permafrost are not the primary causes of the formation of this group of structures.

The close association of superficial bulging and trough faulting with deep-seated tectonic features such as the fault-bounded Stamford Block (Kellaway in Hains & Horton 1969, pp 118–119) and the development of faults and big trough gulls which cut glacial deposits and extend at depth into the underlying rocks as seen at Stretton-on-Fosse in south Warwickshire (Kellaway *et al.* 1971, pp 18–22), point to tectonism as being the primary factor involved in their formation. The existence of an impervious zone of frozen ground may have affected the behaviour of rocks at the surface, but a high water content (mainly from melting ice and snow) and the development of tectonically generated stresses appear to be the two most essential elements. The effect of loading and unloading by glacial ice is difficult to assess when ice limits and thicknesses are uncertain.

At Bath the belt of faulted and crumpled Lower Lias Clay in the valley floor and also the strongly cambered and trough-faulted Lower Lias limestones and clays recorded by Winwood (1897) at Newbridge, west of the City, appear to be related to the deep fissures and reactivated faults emitting thermal water. The formation of spring-pipes above the fissures may pre-date the formation of the Twerton Terrace; the main phase of cambering and trough-faulting certainly does.

However, if both the hydrological and structural manifestations of Quaternary tectonic movement pre-date the deposition of the Devensian and Ipswichian river gravels, then they show comparability with east Midland Quaternary structures. These pre-date the formation of the post-Chalky Boulder Clay terrace gravels of rivers such as the Nene and Welland. All the east Midland river terrace gravels contain rocks derived from the Chalky Boulder Clay. The northern Chalky Boulder Clay ice sheets extended from East Anglia to the northeast Cotswolds but they did not penetrate the Bath area (Fig. 13.8). We cannot, therefore, date the Quaternary tectonic movements at Bath precisely in terms of the Chalky Boulder Clay glaciation of the Vale of Moreton and the Midlands. Nor is there any evidence to show that the ice of the Penultimate Glaciation that reached Clevedon and Kenn Moor (Gilbertson & Hawkins 1978) extended east of Stanton Drew. Ice may have extended over all or part of the Bath region in Plio–Pleistocene times, but most of the effects produced by its presence would have disappeared by the time that the cambers and valley bulges were formed. Thus direct intervention by continental scale glaciation cannot be demonstrated to be a causative factor in the formation of the large Quaternary structures at Bath.

Big rifts and fissures extend to depths of more than 70 m below the valley floor at Bath (Fig. 13.7) and cut hard Carboniferous limestones and quartzitic sandstones beneath the

Mesozoic cover. These rifts can only be of tectonic origin, though the accompanying near-surface movements in the overlying Jurassic rocks exposed in the Avon valley will have been at least partly controlled by loading and water content and by gravitational movements related to the physical relief. Under these conditions the presence of a thin ice cover which is capable of valleyward creep, but not of mass movement across interfluves and valleys, might have promoted creep-cambering at shallow depths and also the generation of terminal landslips by rotational shear.

In the steep-sided valley of the Avon at Bath it is usually difficult to see the structure of the formations cropping out on the valley sides, but in some of the tributary valleys such as those at Swainswick and Lyncombe there is evidence of strong cambering, particularly of the Inferior Oolite and Midford Sands. The Great Oolite Limestone is seldom strongly cambered, possibly because the underlying Fullers Earth Clay is greatly weakened by the passage of water in the basal part of the overlying limestones and is sufficiently thick to generate large landslips and mudflows. An exception is the western slope of Bathampton Hill (Plates 22 & 23) where a section on the road leading to Bath University (769647) shows cambered Great Oolite limestone with relict masses of Greensand chert and Chalk flints filling gulls in dip-and-fault structure (Hawkins & Kellaway 1973). This provides conclusive proof that the cambering post-dates the deposition of the high level Drift deposits described by Varney (1921) but the infrequency with which such slopes of cambered Great Oolite are found and their relatively high altitude, indicates that they may be pre-Anglian and much older than the Twerton Terrace. It is also of interest that the gull infillings seen in the Bath University section show that the high level deposits were once more widely distributed and that their destruction accompanied the early stages of cambering at Bath.

Many of the large landslips at Bath such as the one seen at Swineford on the west side of Lansdown and the huge slip cutting the eastern margin of Bathampton Down (Kellaway & Taylor 1968) were formed in Devensian times when the buried channel of the Avon was in course of formation (Plate D, p 280). There is no reason to suppose that these are causally related to the emission of the thermal water but the magnitude of the later Quaternary landslipping at Bath reflects the intensity of the mid-Quaternary tectonic movements which distorted and shattered the floor and sides of the valley. These movements resulted in the emission of the hot springs and were accompanied by the tilting and warping which caused drastic changes in the drainage pattern of southern England. The movements were not confined to Bath; similar evidence can be found elsewhere in Great Britain and northern France and relates to roughly the same period.

A mid-Quaternary age for the formation of the hot springs at Bath may, at first sight, appear to contradict the conclusions which have been drawn from isotope analysis and the study of the radioactivity of the water and its contained gases (Chapters 9 & 10). However, the age of the thermal water now reaching the surface and the date of the first emission of the thermal water are unlikely to be identical and the difference in age could have an important bearing on the structural and hydrothermal evolution.

The high level gravels and Varney's 'Solent River'

The Drift deposits of Bathampton Down

In 1921 W. D. Varney published an account of the high level Drift deposits resting on Great Oolite and Forest Marble at Bathampton Down, recording Upper Carboniferous and Upper Old Red Sandstone rocks which he considered to have come from South

Wales.[1] The residual gravels on the top of Bathampton Down and Farleigh Down are decalcified and emplaced locally in solution hollows and fissures in the Middle Jurassic limestones. On Claverton Down at a slightly lower level, there are patches of recycled pebbly or stony material mixed with clay derived from the Forest Marble and Great Oolite limestone. These are classified as Head on the published Geological Survey One-Inch to One Mile Bath (265) Sheet and have probably been derived by solifluction of the country rock and the residual pebble deposits that are present on the highest part of Bathampton Down (204 m). On the opposite side of the River Avon scattered residual deposits occur up to 205 m A.O.D. on Farleigh Down (Plate C, p 278).

Elsewhere, as on Kingsdown and in the valley at Limpley Stoke, the original base level of the gravel is indeterminate, though the relict and recycled deposits give some idea of its general distribution before dissection and degradation took place.

As the gravels have been derived from sources west of the Severn valley it follows that they have been transported from northwest to southeast, passing south of the Drift-free Lansdown plateau (230 m) at an altitude of over 670 feet (204 m). Since their deposition, downcutting of the River Avon has amounted to about 180 m and the greater part of this dissection is of mid-Quaternary date.

Varney's hypothesis led him to reject the theory of the overflow of glacial Lake Trowbridge which Harmer (1907) proposed as an explanation of the origin of the gorges of the Bristol Avon. In doing so Varney accepted the only viable alternative open to him, namely that the Severn valley did not exist at the time of the deposition of the high level Drift deposits at Bath. It will be seen however that the ponding of water to produce Harmer's Lake Trowbridge would have been an Anglian event (Fig. 13.8) while Varney's river is likely to have been much older, possibly Beestonian or Plio–Pleistocene in age (Fig. 14.1).

The dating of these ancient gravels is very difficult and they have little in common with the glacial deposits of Clevedon or Kenn (Gilbertson & Hawkins 1978). They are obviously of extraneous origin though deeply weathered and disturbed, have yielded no fossils or artifacts and are seldom exposed. However, if they have not been glacially transported and deposited (and there is no evidence for this), then they are almost certainly of fluvioglacial or fluviatile origin. Most of the clasts are of 6 inches (152 mm) or less in diameter, a few are rounded, others angular or subangular. No ice-scratched rocks have been found. The possibility that the deposits could be Eocene has been considered, but well rounded flint pebbles are rare and no large Greywether sandstone masses have been seen. Varney (1921) concluded from the condition of some of the flints that fresh Chalk may have been undergoing erosion in the vicinity at the time of their deposition. Greensand chert is one of the commonest rocks present and the most likely explanation is that the river was flowing through a valley cut in transgressive Upper Greensand and Chalk, the bottom of the river channel being incised in Forest Marble.[2]

Welsh border rocks including Coal Measures (with coal) and Upper Old Red Sandstone conglomerate, Chalk flints and Upper Greensand chert are associated with Middle Jurassic clay and limestone, the latter having been introduced by solifluction (Fig. 14.4). The upper part of the valley may well have been ice-fed in eastern Wales but the stones may have been conveyed by solifluction and fluvioglacial transport for up to 40–50 km (Fig. 13.5). Gulls in the cambered Great Oolite on the hillside are filled with chert and flint clasts. The Middle Jurassic limestones on the top of Bathampton Down are cavernous and

[1] The name Southampton River is used here in preference to that of Solent River as being more appropriate to the general trend which approximates to that of Southampton Water (Fig. 13.3).

[2] The resemblance of the high level deposits to the Clay-with-flints is discussed in Chapter 14.

large openings have been found in excavations for high-level reservoirs. Some of these cavities had been formed by selective solution of individual limestones and many were empty despite the fact that they were separated by less than a metre from the base of the overlying pebbly Drift deposits. They may have been ice-filled at the time that the overlying deposits were being moved along the valley floor. In this sense they differ from the gulls in the cambered limestone which are filled with flint and chert debris, quartz pebbles and other rocks derived from the high level Drift deposits. The deposition of the high level Drift preceded both the downcutting of the Avon valley and the cambering of the Jurassic rocks at Bath.

Palaeodivides and ancient water partings

In an attempt to test Varney's hypothesis an analysis of the total drainage pattern (including dry valleys) of the Lower Severn basin, the Mendips and south Cotswolds has been carried out. This method provides a valuable approach to neotectonic problems if used in combination with independent stratigraphical and structural evidence (Shearman 1967). It has only limited application in terrain which is heavily mantled by glacial moraine but in the case of the palaeodivides which pass through Clifton and Bath respectively this is not a problem and the boundaries are surprisingly well defined with a substantial amount of independent supporting evidence (Figs 13.4 & 13.7). At Bath, however, difficulties in identification are caused by Devensian and recent landslipping.

In assessing the reason for the excellent preservation of the palaeodivides or ancient water partings in the Bristol and Bath region it is worth observing that these can be traced continuously into the Chalk terrain farther east. Chalk has special properties affecting fluid flow (Price 1987) and these may have a direct bearing on the establishment of the Tertiary–Quaternary drainage pattern. Tectonic fissuring is one of the most important controls and the coincidence of fissure-dominated surface and sub-surface drainage is important where a substantial thickness of Chalk is present. The former presence of a Chalk cover may have contributed significantly to the stabilization of a tectonically controlled Tertiary drainage pattern which tectonic reactivation enhanced and modified in early Quaternary times.

In addition to the major palaeodivides shown in Figs 13.4 & 13.7 it will be observed that there are circular and elliptical drainage units centred on Tog Hill and Pitchcombe respectively. These are cycloliths, comparable with and similar in age to one centred on Chipping Norton (Kellaway 1964; Kellaway & Durrance 1978).

They are only known in plan and are circular or elliptical joint and fissure patterns up to 30–40 km in diameter which have a radial, concentric or spiral structure. They have been attributed by some to impact by extra-terrestrial bodies and referred to by others as being of magmatic or crypto-volcanic origin. In the writer's view the examples now being considered are of tectonic or magmatic origin, but in the absence of satisfactory evidence as to their form or persistence at depth it is difficult to account for them. It is possible that a group of these structures extends from east Devon to Northamptonshire and that they are of Cretaceous age.[1]

The Pitchcombe cyclolith is situated in an area of thick Triassic and Lower Jurassic strata and is not thought to have had much effect on thermal water emission. However, the Tog Hill cyclolith may have exercised some control on the form of the palaeodivide. It may also be observed that palaeodivides or ancient water partings are not necessarily aligned along existing high ground. The area enclosed by the palaeodivides is a palaeovalley and

[1] Cycloliths are known from many parts of the world and one has recently been described from Lake Huron in Canada by Forsythe *et al.* (1990). Though comparable in size with some British examples, the Lake Huron structure is of early Palaeozoic age.

formerly constituted a major line of surface drainage which, in this case, is of post-Upper Cretaceous (Oligocene) age and has since been dismembered or modified. The dislocation produced in the Severn valley by sinistral movement along the Denny Island Fault is particularly clear (Fig. 13.6) and may be important as indicating a tectonic origin for the formation of the Lower Severn valley.

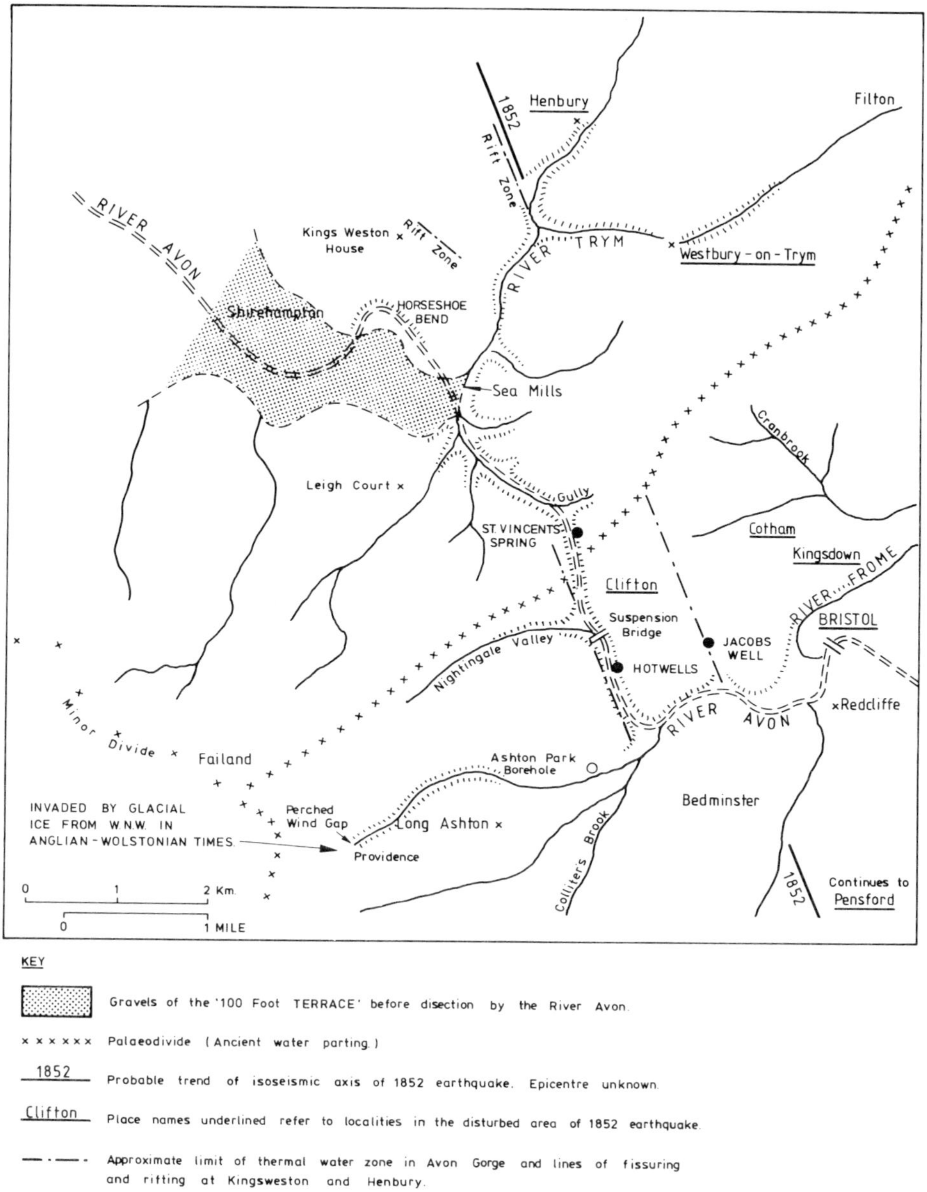

Fig. 13.7. Relationship between the palaeodivide and the thermal water zone in the Avon Gorge at Clifton, Bristol.

In the case of the palaeovalley of the Plio–Pleistocene Southampton River (Fig. 13.4) the principal rivers which now intersect it are the Severn between Chepstow and Gloucester, the Bristol Avon between Chippenham and Clifton and the Salisbury Avon at Wilton. Of the ancient Southampton River, little is visible though the general line of the Oligocene depression can still be recognized locally, as seen for example from Tog Hill north of Lansdown. The Anglian development of the Severn and its tributaries, such as the Little Avon in the Tortworth–Berkeley area, has greatly modified the old relief, and the Cotswold escarpment which developed subsequently as a result of erosion by the Bristol Avon, the Severn and their tributaries draining to the Bristol Channel now cuts across the palaeovalley between the Mendips and Cheltenham. The major outlier of cambered Middle Jurassic limestone situated at Dundry Hill South of Bristol shows that the height of the western palaeodivide south of Bristol was more than 718 feet (218 m) A.O.D. and that the crest at the eastern divide stood at approximately the same level or higher at Lansdown near Bath.

West of the western palaeodivide the course of the Wye above Chepstow is essentially pre-Anglian, probably late Tertiary in origin and its sinuous course is subparallel with the palaeodivide between Clifton and Coleford. The incised meanders of this beautiful river (Plate F, p 283) have long excited interest (Miller 1935) and it has a number of gravel terraces, including the '100 foot terrace' (Welch & Trotter 1961) which closely resembles the '100 foot terrace' of the Bristol Avon at Shirehampton (Fig. 13.7). There is reason to suspect that the sinuous Wye once flowed past Shirehampton and thence in a south westerly direction through the Clapton valley and the east Clevedon gap, passing north of Worle Hill and Weston-Super-Mare. Its intersection by a structurally controlled Lower Severn may be the key to later changes, but an adequate account of the development of the Lower Severn would require fuller treatment than is possible here.

Temporarily blocked and ponded by ice of the Penultimate glaciation at Clevedon, the Wye–Severn may eventually have broken through along the line of the reactivated Denny Island Fault, leaving the deep east Clevedon gap as a memorial of its former course. The Bristol Avon with its angular fissure-controlled course did not exist, or was in a an immature form, when the gravels of the '100 foot terrace' of Shirehampton were deposited and at that time the River Trym was a tributary of the Wye, the confluence being at Shirehampton (Plate F, p 283). The relationship of the palaeodivide at Clifton to the fissure belt in which the hot springs rise is shown in Fig. 13.7.

Two very large gaps or cols, one at Flax Bourton and the other at Ubley, are developed on the Clifton palaeodivide (Fig. 13.4). These are of glacial origin or were enlarged during the Penultimate glaciation of the Clevedon area. In spite of the development of these very broad cols the Clifton palaeodivide can be traced across them, climbing thence onto the Mendip Hills where the old water-parting is indicated by the heads of dry valleys, continuing thence in a southeasterly direction into the Cretaceous and Tertiary terrain of Wiltshire.

The trend of the original Oligocene valley draining towards Southampton was tectonically controlled and its form and function were drastically changed by the transverse Miocene and Pliocene movements (Fig. 13.3), but the NW–SE and WNW–ESE trending basement structures in the central and southeastern part of the Bristol and Somerset coalfield, like the NE–SW Denny Island Fault in the Severn estuary, were not excluded from the effects of renewed reactivation. As a consequence of this, the course of the early Pleistocene palaeovalley in which the Southampton River was re-established was probably more sinuous than its Oligocene predecessor and incised at a lower level. We do not know to what extent the Pliocene glaciation and the penetration of ice carrying the bluestones of Stonehenge affected its development, but it may well have contributed much melt water during deglaciation. Much depends on the interpretation of the Stonehenge erratics in the sense that though morainic in origin, they may have been carried for some considerable distance downstream in floating ice. In reconstructing these events however, it should be borne in mind that the Pliocene land surface on which this took place and which may have stood much higher than the Beestonian surface, has been almost completely destroyed by post-Pliocene erosion.

Several important aspects of the palaeovalley and its relationship to the area to the east of the Bath palaeodivide may be noted. South of Tetbury the Middle Jurassic limestone plateau of the South Cotswolds is essentially Drift-free, apart from deposits of Devensian Head of local origin and of alluvial deposits in valley floors. The Great Oolite limestone plateau of Lansdown attains an altitude of 238 m A.O.D. and the adjacent mass of cambered Inferior Oolite limestone at Tog Hill attains 219 m A.O.D. This elevated tract, of which the highest point coincides with the Tog Hill cyclolith, continues eastwards to Marshfield and Colerne and northwards towards Wotton-under-Edge and Tetbury. The Bath palaeodivide, which extends northwards to Frocester Hill does not follow the Cotswold scarp (Fig. 13.6) and where it is finally intersected by the edge at Frocester Hill its course is set almost at right angles to that of the escarpment. This relationship is consistent with their age, the Cotswold edge and deeply incised valleys of the Stroud Frome having been produced by erosion during and since the Penultimate glaciation.

Stratigraphical and geomorphological evidence for the origin of the hot springs

There are two distinct groups of Pleistocene deposits at Bath, the older one being the high level gravels of Bathampton Down, Farleigh Down and Kingsdown. At the highest points on Bathampton Down and Farleigh Down the gravels may be *in situ* at about 205 m A.O.D. Elsewhere, at lower levels they have probably been degraded. They are missing altogether on the higher plateau areas of Lansdown and Charmy Down north of the Avon valley, possibly because they were not deposited there.

The younger group comprises a series of low-level river gravels forming a series of terraces and filling the buried channel of the Avon with its floor at about 12 m A.O.D. The highest gravels of the Twerton Terrace (the uppermost member of the group) lies at or just below 50 m A.O.D. The relations of these deposits are shown by Chandler *et al.* (1976, fig. 2). The difference in level between the original floor on which the high level gravels were deposited and the floor of the late Pleistocene buried channel is 180–190 m and this gives a measure of the rate of downcutting (Table 13.3) in the Avon valley following reversal of drainage.

Ipswichian and Hoxnian interglacials: the 'Anglian' glaciation

Recent work on the dating by amino-acid racemization studies of the raised beaches of the Severn estuary (Andrews *et al.* 1984) indicates that the development of the Swallow Cliff raised beach may have taken place in two stages separated by a gap of about 270,000 years.

This implies an overlap of Ipswichian by Hoxnian sea level which would be expected from the evidence available on the south coast of England, particularly in Sussex, Here the Ipswichian Brighton raised beach lies at a lower level than the older beach deposits of Aldingbourne which may perhaps be broadly described as 'Hoxnian' and rest on a platform incised in Chalk at altitude of 27 m or more. Both in West and East Sussex there is evidence of substantial Quaternary tectonic movement. East of Brighton the Ipswichian beach levels are obscure but the non-appearance of the raised beach in East Anglia is due to the effect of subsidence on the flanks of the North Sea Basin. The altimetric classification of Quaternary beach and sea levels has therefore given much difficulty and attention has been directed on an ever-increasing scale to the subject of Quaternary tectonism (Preese *et al.* 1990).

In dealing with the tectonics and geomorphology of areas such as Bath, the lack of clearly defined limits for the 'Hoxnian', Holsteinian and Cromerian interglacial episodes inhibits accuracy particularly when changes in structure and drainage are concerned.

Definition of the cold phases of the Penultimate glaciation (Anglian and Wolstonian) is no less difficult than that of the interglacial episodes. In general the glacial deposits of the Clevedon area (Gilbertson & Hawkins 1978) exhibit a complex series of deposits which fall broadly between the Cromerian and Ipswichian warm periods (Table 13.3).

Low-level river gravels of the Avon above the Hanham Gorge

Between Bath and the Hanham Gorge there are three main low-level deposits of terraced river gravels. Some subsidiary divisions can be recognized but a simple classification into Flood Plain (No. 1), Saltford (No. 2) and Twerton (No. 3) terraces (Chandler *et al.* 1976, fig. 2, p 532) is sufficient for present purposes. The Twerton Terrace, though now in a somewhat fragmentary condition, stands at about 100 feet (30 m) above the flood plain of the River Avon at Twerton, while the Saltford Terrace gravels form a low to intermediate terrace (or series of terrace fragments) of somewhat variable altitude but seldom more than 40 feet (12 m) above the flood plain level. The No. 1 or Flood Plain terrace constitutes a valley fill mostly concealed by alluvium.

All three terraces include gravels composed of rocks of local and distant origin, but dominated by Middle Jurassic limestone. No reliable method of distinguishing between them on the basis of pebble content has yet been found. The critically important member is the Twerton Terrace. This is thought to be pre-Devensian, dating from the Ipswichian Interglacial. No sign of substantial tectonic deformation by cambering, the formation of trough gulls or bulging has been detected in the Twerton Terrace or any of the lower terrace deposits in the Avon valley below Bath. Many of the terrace gravels have been affected by landslipping and have been largely destroyed between Bath and Bradford-on-Avon.

At Twerton the terrace gravels rest on an erosion surface which truncates minor faults and folds formed during cambering of the Lower Lias (Chandler *et al.* 1976, fig. 2). If the Twerton Terrace is of Ipswichian age then its relationship to the main period of cambering and valley bulging is similar to that of the river gravels of the Severn valley and the Midlands. At least two periods of cambering were detected in the East Midlands, one before, or at an early stage of the Penultimate glaciation and the other during a later stage or at its termination (Hollingworth *et al.* 1944). The late stage movements may coincide in date with the trough-faulting which cuts the Chalky Boulder Clay and its associated sands and gravels at Stretton-on-the-Fosse in Warwickshire and possibly with the trough-faulting recorded by Shotton (1965) in the East Midlands.

These movements are likely to have preceded the deposition of the Twerton Terrace of the Avon Valley upstream of Hanham Mills. Twerton Terrace gravel deposits are fragmentary and the precise course of the Avon at Bath in Twerton Terrace times is not determinable, but in the later period marked by the deposition of the Saltford Terrace the course of the Avon at Bath may have lain through the col north of the hot springs. By late Devensian times when the buried channel was being cut, the river had shifted to its present situation at the foot of Beechen Cliff and has since retained this, with minor changes, to the present day. Thus the hot springs appear to have maintained their position despite changes in the course of the river over a period in excess of 100,000 years.

The character of the Lower Lias Clay is primarily responsible for this situation though the age and dimensions of the deep fissure system in the older rocks beneath the valley floor makes a major contribution to the maintenance of the stability of the springs.

When permeated by thermal water under hydrostatic pressure fissured Lower Lias Clay becomes greatly weakened and breaks down to form a porridge-like slurry. This characteristic property virtually excludes any possibility that the spring-pipes could have been drilled up from below and then filled with Drift deposits after a long pause. Once the thermal water had risen through newly formed fissures in the White Lias and Blue Lias and bored its way upwards through the Lower Lias Clay a continuous stream of water would be required to maintain the opening, though contemporaneous filling with coarse gravel would have a stabilizing effect.

On balance, therefore, the formation of the spring-pipes at Bath is likely to have taken place when dissection of the high level Drift deposits was far advanced and when tectonic movements were taking place preceding the aggradation of the Twerton Terrace.

Terrace gravels below Hanham Mills

Although the flood plain and low-level Devensian terrace deposits of Bath may be represented at Bristol, neither the Saltford nor the Twerton Terrace extend beyond the eastern entrance to the Hanham gorge at Hanham Mills (see Fig. Pf2). Their absence in the low ground at Bristol between the present exit of the Hanham gorge at Netham and the entrance to the Avon Gorge at Clifton has never been satisfactorily explained. If, however, the swiftly flowing Avon was cutting an almost continuous gorge or narrow trench-like valley in hard Pennant Sandstone at Hanham and Conham then it would certainly have continued to form a comparable trench in the calcareous Triassic sandstone (Redcliffe Sandstone) between Netham and the foot of Brandon Hill. Downstream of the site of Bristol Cathedral hard siliceous Carboniferous sandstones at Clifton Wood and Hotwells form the surviving steep north wall of the gorge. This coincides roughly with the contact of the Carboniferous rocks and the red Triassic sandstone which extends beneath the river to Redcliffe, Bedminster and Ashton. The northern wall of the gorge therefore survives as a steep cliff at Clifton Wood and Hotwells with the river at its foot, but the softer Triassic mudstones and sandstones of the south bank have since been deeply dissected by tributary streams including the Malago and Colliters Brook. These are now of insignificant size but were formerly torrents fed by melting ice which filled the Flax Bourton gap during the penultimate glaciation.[1] Even if some terraces had been formed on the soft red sandstone at Bedminster and Ashton they would have been unlikely to survive.

It therefore appears that in Twerton Terrace (and possibly in Saltford Terrace times) the Hanham and Clifton Gorges formed a continuous trench draining to the Trym valley. Below Sea Mills at the northern exit of the Avon Gorge and its confluence with the Trym, the Avon valley is incised in low plateau-like areas of Mercia mudstone at Abbots Leigh, Ham Green and Shirehampton (Fig. 13.7) where gravel deposits (Apsimon & Boon 1960) resting on this shelf contain fragments of slaty rock found by the late Mr J. A. Davies (pers. comm.), dark purplish grit and other non-local rocks. They also yield Acheulian-type implements (Davies & Fry 1929). These deposits may be fluvioglacial gravels and constitute the so-called '100-foot Terrace', resembling those of the same name in the Wye valley at Chepstow.[2] Palaeolith-bearing gravels of this kind exist neither in the low ground

[1] Dating of the Penultimate glaciation (or glaciations) in the Severn estuary is based partly on amino acid racemization (Andrews *et al.* 1984).

[2] A second terrace at a lower altitude is known as the 50 Foot Terrace. This is in ground long since built up and has not been seen for many years.

at Bristol nor above the Hanham Gorge at Hanham Mills. Palaeolithic implements have been found in small numbers, mainly in scattered Head deposits on high ground at Hanham and elsewhere on adjacent uplands but not in the low level terrace gravels.

The inference that in early Anglian times the downcutting of the Avon Gorge at Clifton had not been commenced, or was in a very early stage, seems inescapable. Thus the upper part of the steep slope at Shirehampton Park may represent the western bank of the Trym when it was a a tributary of the Wye and when the Avon had not yet penetrated the Trym valley, indicating that the intense fissuring of the Clifton–Failand ridge had not taken place, or was so recent that the Avon had not succeeded in developing a continuous gorge along it. This would point to a late Wolstonian–Anglian date as the minimum age for the emission of the hot springs at Hotwells and Clifton.

Both glaciation and tectonics have therefore played a part in the development of the Avon. It is safe to say that tectonic fissuring played an essential part in the formation of the Hanham and Clifton gorges and the valley at Bath, but without the presence of glacially ponded water it may be questioned whether the course of events would have been the same.

Changes in the English Channel including the breaching of the barrier at the Straits of Dover have recently been described by Alec Smith (1985). In this case it is suggested that a cataclysmic event took place and that glacial conditions caused the breakthrough. Nevertheless it is not impossible that neotectonic movement may also have occurred since changes in ice loading, the injection of melt water under pressure and other side effects of deep freezing and glaciation could trigger off seismic activity. The neotectonic movements now in progress in the Pays de Bray (Cavelier 1974) are not likely to be related to Devensian glaciation. Comparable movements may be in progress elsewhere in the English Channel region, and, as at Bath, these should be considered as an independent element in the evolution of the Quaternary land surface.

The Devensian cold period (periglacial)

During the last (or Devensian) glaciation (Fig. 13.8) the Bristol–Bath region and the Mendips were in a periglacial phase. Small ice fields may have come into being in upland areas during periods of high precipitation but there is no evidence of continental-scale glaciation. The climate was probably dry with long cold winters and a general lowering of sea level and decreased precipitation in the lee of the Welsh ice may have restricted the flow of the hot springs by reason of its effect on the recharge mound. It is therefore possible that groundwater levels at Bath were high in the closing stages of the Penultimate Glaciation and Ipswichian times, very low during the Devensian periglacial period and then recovered in the post-Glacial warm period. If any groundwater from Wales is reaching Bath at the present time it is probably doing so by way of the Malvern Fault Zone.

Another important effect produced by periglacial conditions in Devensian times was the increase in erosion rates due to the destruction of vegetation cover and to freeze–thaw and melt-water processes. Landslipping and scarp recession were both highly active. Downcutting of the buried channel of the Avon was accompanied by incision of the lateral thermal water streams issuing from the hot-springs at Bath which are situated at a higher level than the deepest part of the Devensian channel (Fig. 7.4).

The source of the thermal water of Bath and Hotwells

The uplands and mountains of Wales provide the only major region of high altitude catchment accessible from Bristol and Bath, Dartmoor and Exmoor being remote and structurally isolated, while the Mendip Hills, though well watered and composed largely of Carboniferous Limestone are small, of no great altitude and freely drained almost to sea level on their southern flank (Green & Welch 1965). The northern part of the central and eastern Mendips can be regarded as capable of affecting the thermal water balance at Bath at the present time. Other Carboniferous Limestone uplands such as Broadfield Down and those at the margins of the Coalpit Heath basin may also play a role (Fig. 13.9). The problem with the eastern Mendips is that it may be severed at depth from the Carboniferous Limestone farther north by major Variscan structures.

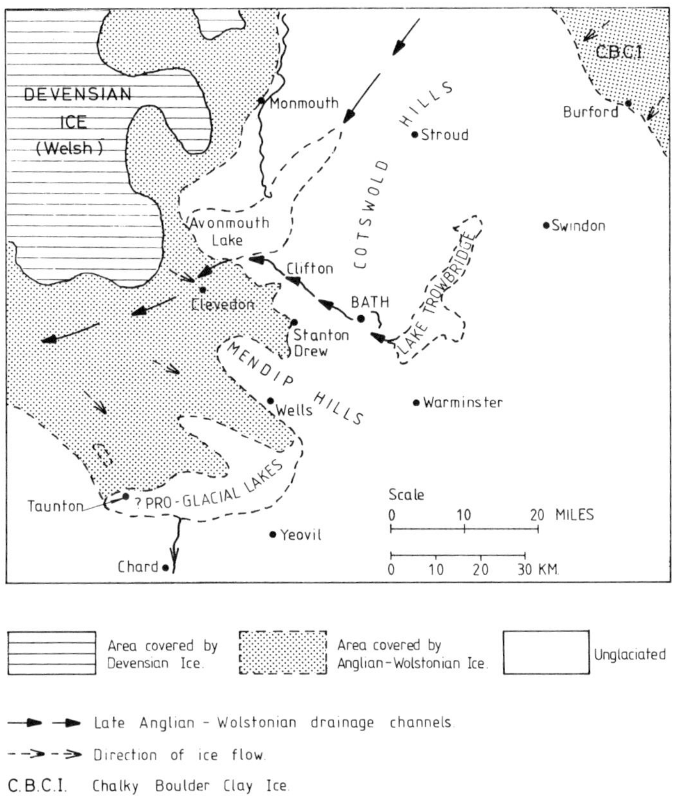

Fig. 13.8. Anglian–Wolstonian (Penultimate) glaciation.

Water levels recorded by Dr Stanton in Broadfield Down show a generally high level in the northern part of the Carboniferous Limestone uplands. Thus the overflow level of the big Cold Bath Spring at Barrow Gurney on the northern margin of Broadfield Down is +81 m O.D. while water levels of 134 m are found at Felton and 99 m O.D. at Winford, situated on the NE margin approximately 20 km west of Bath. Felton and Winford are located in a zone of Mesozoic and younger faults and folds aligned along a group of major Variscan structures including the northern boundary fault of Pensford Colliery (Fig. 13.2). These structures were reactivated in Mesozoic and Tertiary times and are continuous with the zone of fracturing which intersects the mid-coalfield zone of N–S faulting and then continues eastwards towards Newton St. Loe, Bath and Bathampton Down (Plate C, p 278). There is some evidence for Quaternary reactivation, particularly in the North Wick valley on Dundry Hill. This broad structural zone, which includes the Pennyquick Fault, is

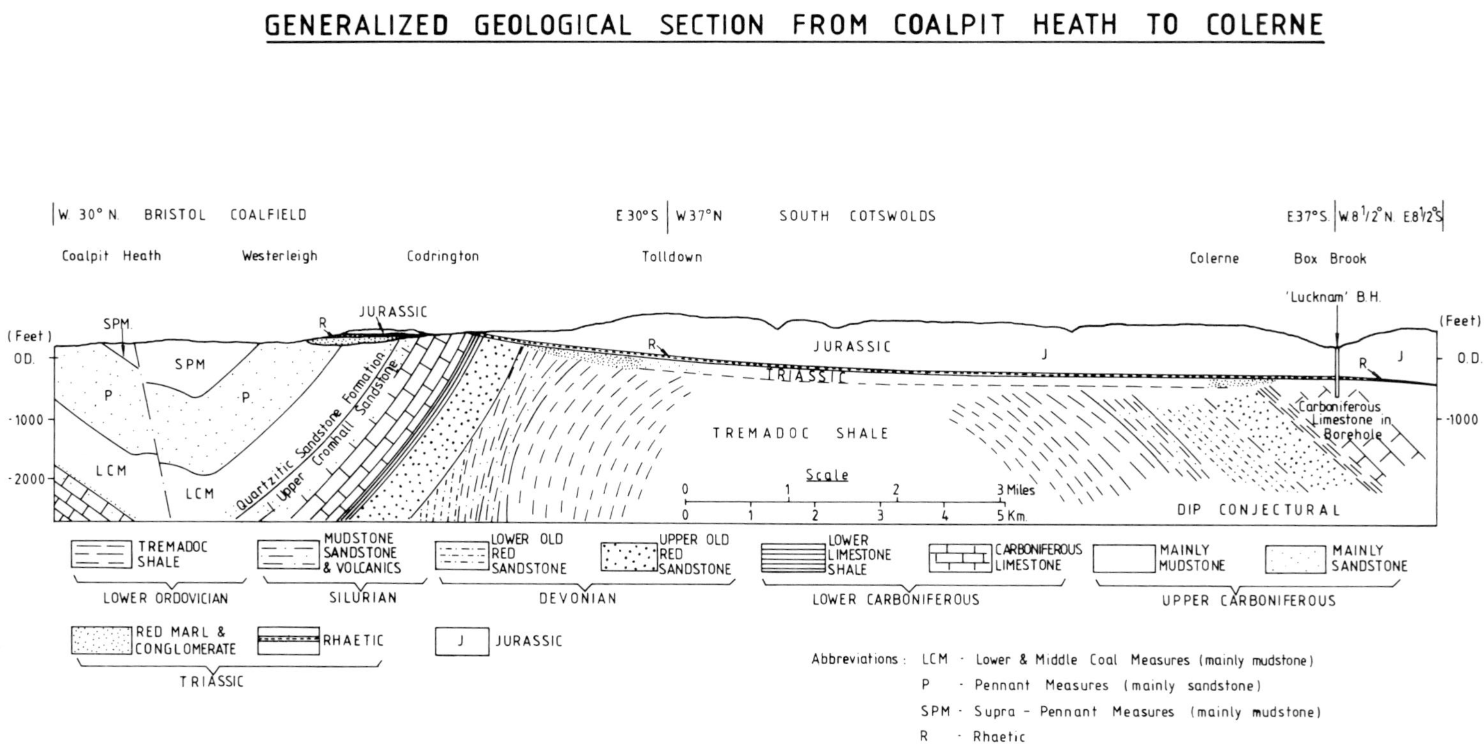

Fig. 13.9. Generalized section from Coalpit Heath to Colerne showing the inferred structure beneath the Mesozoic cover of the South Cotswolds. The anticlinal structure coincides with a gravity 'high' and is responsible for the pressure of the Tremadoc Barrier which impedes the early migration of groundwater from the Coalpit Heath basin. The thickness of the Triassic rocks in the South Cotswolds is uncertain.

clearly one which may be acting as a channel for the migration of ground water at depth. Even if it is not a channel for deep basinal migration of the thermal water, it may play an important part in controlling the water movement in the recharge mound at Bath.

It will be observed that though the water levels at Felton in Broadfield Down are lower than those in the central Mendips (Andrews *et al.* 1982, fig. 2) the structural relations of Felton and Bath are more favourable to thermal water migration than those of Bath and the central Mendips.

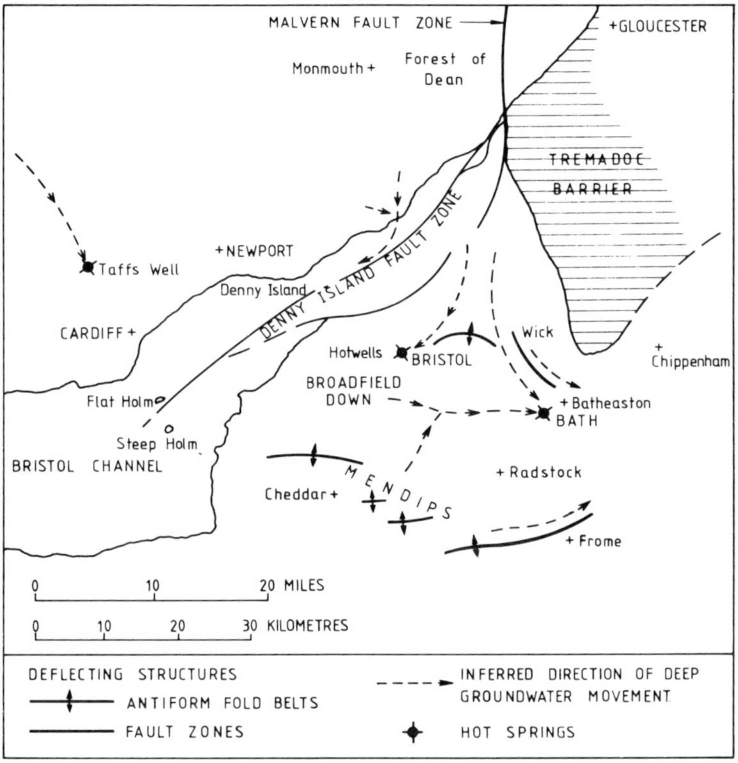

Fig. 13.10. Inferred deep groundwater movements in mid-Quaternary–Recent times in relation to deflecting structures and the Tremadoc barrier.

Resulting from its Tertiary–early Pleistocene history the Oligocene movement of water in the Palaeozoic rocks is likely to have become established at depth below the central area of the palaeovalley. One of the key points in the control of underground water movement is the line of the Berkeley Fault at Tites Point (Fig. 13.10) where the fault crosses the Severn. West of the Berkeley Fault folded and faulted Silurian and Lower Old Red Sandstone mudstones are succeeded by Upper Old Red Sandstone and Lower Carboniferous rocks at the margin of the Bristol Coalfield.[1] East of the fault, the Silurian rocks of Tortworth and the Tremadoc shales of Halmore and Breadstone are exposed at the surface. Here the Lower Old Red Sandstone is absent (Fig. 13.10) and Upper Old Red

[1] The Lower Old Red Sandstone is missing over much of the Tortworth inlier where Upper Old Red Sandstone rests unconformably on marine Silurian (Fig. 13.11). All the marine Silurian rocks (including the Llandovery) are calcareous. Destruction of any pre-existing Lower Old Red Sandstone beneath the South Cotswolds was advanced by Carboniferous times as Tremadoc achritarchs are found in the Millstone Grit at Yate where the Coal Measures contain pebbles of pre-Cambrian rocks (Kellaway & Welch in press).

Sandstone rests unconformably on Silurian. The presence of impervious Tremadoc shale, overlain as it is at Slimbridge by red Triassic mudstone and Lower Jurassic clays which thicken to the north and east and continue southwards towards Bath, suggests that easterly migration of groundwater is minimal at depth between Huntley in the north and Charfield and Chipping Sodbury in the south (Fig. 13.10). Any water migrating southwards in the fractured Silurian and Upper Palaeozoic rocks is likely to be confined to the region of the Berkeley fault and Tortworth, though migration in a SE direction towards Charfield might be possible at Woodford. South of Woodford the principal faults of the Malvern Fault Zone extend lengthwise down the Coalpit Heath basin. All the north–south and NW–SE structures in the basin show evidence of Mesozoic movement, the deposition of the Triassic Redcliffe Sandstone having been controlled by them between Frampton Cotterell and Dundry.

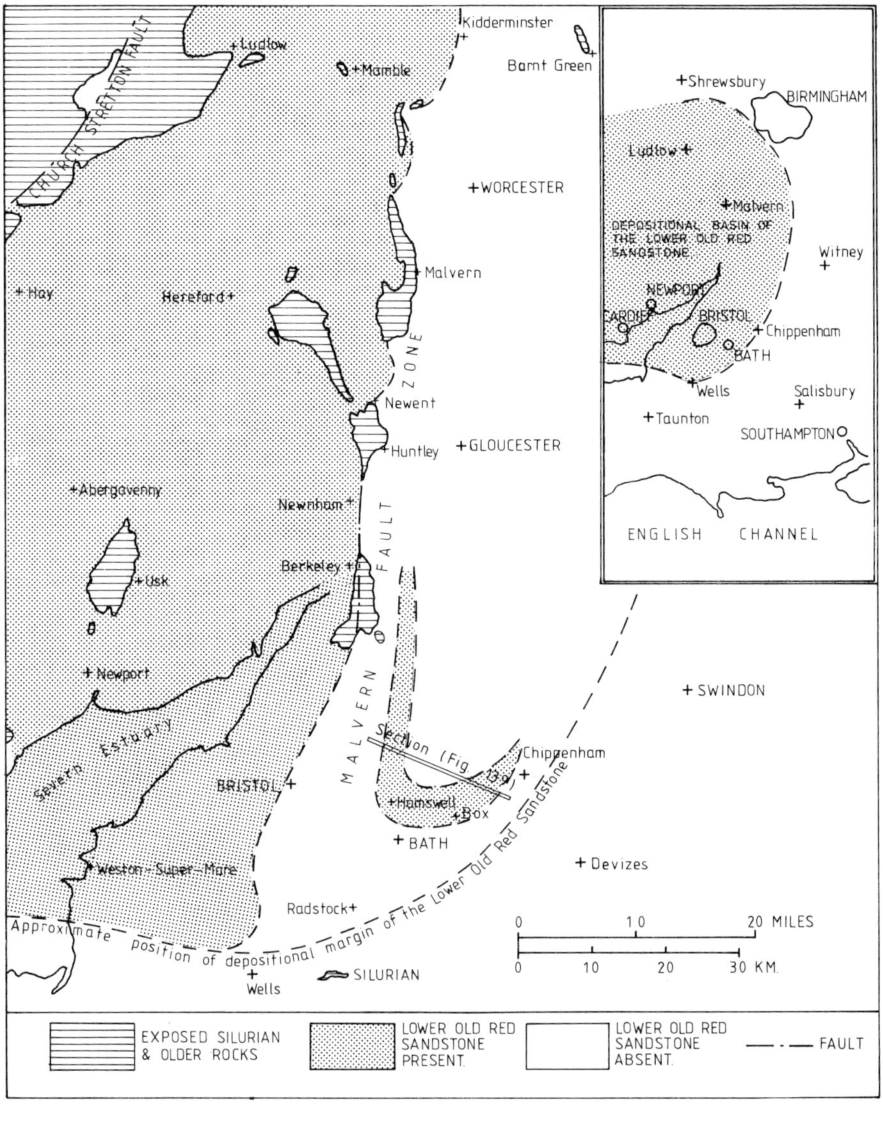

Fig. 13.11. Distribution of Lower Old Red Sandstone following episodes of faulting, folding and erosion in pre-Upper Old Red Sandstone, Carboniferous and Permo-Triassic times.

At the time of the Plio–Pleistocene palaeovalley when the Severn valley did not exist, the presence of thick Triassic and Jurassic rocks overlain by Upper Cretaceous strata would therefore have formed a continuous and relatively impermeable cover obliging groundwater migrating from the region north of the Forest of Dean to proceed in a southerly direction. On approaching the Kingswood Anticline, however, some of the thermal water trapped in the Carboniferous Limestone may have been forced to rise. Its emission in the Avon valley awaited the formation of the fracture system along which the river valley developed and which was probably dependent on movement along the Denny Island Fault Zone.

The present-day movements of thermal water at depth are very difficult to determine, not least because its emission is controlled by the physical relief as well as the age, direction and intensity of fissuring. The flow directions indicated on Fig. 13.10 are intended to suggest one possible pattern of deep groundwater movement which fits the geological structure, the physical relief and the age and intensity of fissuring. Like other systems it could be disrupted or unbalanced by the ruthless extraction of groundwater. Systematic monitoring of the thermal water yields and other features may throw light on these problems in the future.

It has not yet been possible to establish whether the thermal water issued initially in much larger quantities than it does at the present time at Bath or whether the springs were originally much hotter. On balance it seems likely that both are true and that very ancient thermal water forced its way upward through the ground under great pressure as a result of fissuring and seismic shock, tapping the deep channels carrying thermal water of which some may still be migrating slowly southeastwards towards the Hampshire Basin, or has been trapped in compartments between transverse Miocene folds and faults (Fig. 13.3).

If this scenario is true then tectonic warping, fissuring and thermal activity marked the rapid if not cataclysmic birth of the Avon in mid-Quaternary times. In any event it seems likely that the old very hot stored water, fed by deep fissure systems and under considerable hydrostatic pressure beneath the Mesozoic cover, would gradually be replaced at higher levels by younger cooler water moving in from sources in the recharge mound at the margins of the Bristol and Somerset Coalfield. This implies that the hot springs of Bath are slowly becoming cooler as young water replaces the older stock. The thought is disturbing, but need not cause apprehension if we remember that the temperature of the springs may already have been declining for more than 100,000 years.

Acknowledgements. To my former colleagues in the British Geological Survey and to Dr W. I. Stanton I am indebted for much useful discussion. The paper has been read by Dr A. B. Hawkins and Professor Alex Smith who have advised me on its presentation. For the opinions expressed in the paper I accept full responsibility.

References

Alabaster, C. J. 1976. Post-Inferior Oolite Mineralisation at Whatley Quarry, East Mendip. *Proceedings of the Bristol Naturalists Society*, **35**, 73–84.

Andrews, J. T., Gilbertson, D. D. & Hawkins, A. B. 1984. The Pleistocene succession of the Severn Estuary: a revised model based upon amino acid racemization studies. *Journal of the Geological Society, London*, **141**, 967–974.

Apsimon, A. M. & Boon, G. C. 1960. An Exposure of the Bristol Avon Gravels at Shirehampton, near Bristol, June 1959. *Proceedings of the University of Bristol Spelaeological Society*, **9**, 22–29.

Banner, F. T., Brooks, M. & Williams, E. 1971. The Geology of the approaches to Barry, Glamorgan. *Proceedings of the Geologists' Association*, **82**, 231–238.

Bishop, A. J. 1974. A Preliminary Report on the Middle Pleistocene Mammal-bearing deposits of Westbury-sub-Mendip, Somerset. *Proceedings of the Speleological Society, University of Bristol,* **13**(8), 301–318.
Cavelier, C. 1974. Etude Géologique de la Feuille Creuil a 1:50,000. Mémoire, Université de Paris.
—— & Kuntz, G. 1974. 'Découverte du Pliocène Marin (Redonian) à Valmont (Seine-Maritime) dans le Pays de Caux'. *Comptes Rendus de la Sociéte Géologique de France,* **16**(6), 160–162.
Chandler, R. J., Kellaway, G. A, Skempton, A. W. & Wyatt, R. J. 1976. Valley slope sections in Jurassic strata near Bath, Somerset. *Philosophical Transactions of the Royal Society of London*, A, **283**, 527–556.
Collins, M. 1987. Sediment transport in the Bristol Channel: a review. *Proceedings of the Geologists' Association,* **98**, 367–383.
Davies, C. M. 1987. Seismic stratigraphical sequences in the Lundy Tertiary Basin. Bristol Channel. *Proceedings of the Geological Association,* **98**, 355–366.
Davies, J. A. & Fry, T. R. 1929. Notes on the Gravel Terraces of the Bristol Avon. *Proceedings of the Speleological Society, University of Bristol* (1928), 162–172.
Davison, C. 1924. *A History of British Earthquakes*. Cambridge University Press.
Dobson, M. R. & Whittington, R. J. 1987. The Geology of Cardigan Bay. *Proceedings of the Geologists' Association,* **98**, 331–353.
Donovan, D. T. & Kellaway, G. A. 1984. *Geology of the Bristol District: The Lower Jurassic Rocks.* Memoirs of the British Geological Survey.
Evans, D. J. & Thompson, M. S. 1979. The geology of the central Bristol Channel and the Lundy areas, South Western Approaches, British Isles. *Proceedings of the Geologists' Association,* **90**, 1–14.
Fletcher, B. N. 1975. A new Tertiary Basin east of Lundy Island. *Journal of the Geological Society, London,* **131**, 223–225.
George, T. N., 1970. *British Regional Geology, South Wales*, 3rd Edition, HMSO, London.
Gilbertson, D. D. & Hawkins, A. B. 1978. The Pleistocene succession at Kenn, Somerset. *Bulletin of the Geological Survey of Great Britain.*
Green, G. W. & Welch, F. B. A. 1965. *Geology of the Country around Wells and Cheddar*. Memoirs of the Geological Survey of Great Britain.
Hains, B. A. & Horton, A. 1969. *Central England* (3rd Edition) British Regional Geology, Institute of Geological Sciences, HMSO, London.
Harland, W. B., Armstrong, R. L., Cox, A. V., Craig, L. E., Smith, A. G. & Smith, D. G. 1990. *A Geologic Time Scale, 1989*. Cambridge University Press.
Harmer, F. W. 1907. On the origin of certain canon-like valleys associated with lake-like areas of depression. *Quarterly Journal of the Geological Society, London,* **63**, 470–514.
Hawkins, A. B. & Kellaway, G. A. 1973. Field meeting at Bristol and Bath with special reference to new evidence of glaciation. *Proceedings of the Geologists' Association,* **82**, 267–292.
Hollingworth, S. E., Taylor, J. H. & Kellaway, G. A. 1944. Large-scale superficial structures in the Northampton Ironstone Field. *Quarterly Journal of the Geological Society, London,* **100**, 1–44.
Jones, J. 1572. Bathes of Bathe's Ayde, London.
Jukes-Brown, A. J. 1892. The building of the British Isles (2nd Edition) George Bell, London.
Kellaway, G. A. 1964. Structured rings and fissure systems in relation to river development. *Proceedings of the Geological Society, London,* **1617**, 100–104.
—— & Durrance, E. M. 1978. Circular structures of large scale and great age on the Earth's surface. *Nature,* **273**, 78.
—— & Hancock, P. L. 1983. The structure of the Bristol district, the Forest of Dean and the Malvern Fault Zone *In:* P. L. Hancock (ed.), *The Variscan Fold Belt in the British Isles*. Adam Hilger, Bristol, 88–108.
——, Horton, A. & Poole, E. G. 1971. Development of some Pleistocene structures in the Cotswolds and Upper Thames Basin. *Bulletin of the Geological Survey of Great Britain,* **37**, 1–28.
—— & Taylor, J. H. 1968. The influence of landslipping on the development of the City of Bath, England. 23rd International Geological Congress. Czechoslovakia (1968), **12**, 65–76.
—— & Welch, F. B. A. 1991. *The Geology of the Bristol and Somerset Coalfield*, Memoirs of the British Geological Survey.
King, C. A. M. 1977. The early Quaternary landscape with consideration of tectonic matters. *In:* F. W. Shotton (ed.), *British Quaternary Studies*, Clarendon, Oxford.
Miller, A. 1935. The Entrenched Meanders of the Herefordshire Wye. *Geographical Journal,* **85**, 160–178.

Plate 29. The copse marks the site of the big sink or karstic depression at Clay Pit Hill near Chitterne in which Prestwich (1989) recorded the presence of the Tertiary 'Druid Sandstone' (p. 249).

Plate 30. Sarsens (or Greywethers) in floor of dry valley at Fyfield, Marlborough Downs.

Plate 31. Mass of unhardened Greywether sandstone (6.4 m in length) found in M40 excavation in 1974 at Christmas Common, Aston Rowant, Oxfordshire (p. 263).

Plate 32. One of three large masses of unhardened Greywether sandstone enveloped in red Clay-with-flints filling a solution cavity in Chalk. M40 excavation (see Plate 31).

Plate 33. Boulder of dolerite recovered from Boles Barrow near Heytesbury and now in Salisbury Museum. Photo – H. C. Bowen.

Plate 35. Boulder of ignimbritic tuft-lava believed to be of Ordovician age from North Wales (p. 267) found at Stonehenge in 1924. Shows appearance before sectioning (Plate 36). Striation on flattened surfaces and smooth blunt-nosed shape indicate its glacial origin. The boulder has been subjected to prolonged weathering in cold acid water. Photo – British Geological Survey.

Plate 34. Bleaching caused by immersion in acid peaty water and revealed by wasting of peat. Bleached rock exposed to atmospheric weathering shows progressive loss of white outer surface which is replaced by normal weathering. Wrynose Pass, Lake District.

Plate 36. The boulder in Plate 35 has been sliced to show the depth of weathering. The broken surface at the lower end of the boulder records an attempt by Neolithic or Bronze Age man to trim the rock. Some of the striations on the surface of the boulder are unrelated to the structure and may be glacial striae. There are a few small scattered patches of tufa; the absence of post-Neolithic weathering reflects the protective influence of the chalky debris in which the stone was buried. Photo – British Geological Survey.

Owen, T. R., The structural evolution of the Bristol Channel. *Proceedings of the Geological Society of London,* **1664**, 289–294.
Pomerol, C. 1974. Le Bassin de Paris *In:* Geologie de la France, Vol. 1 (ed. J. Debelmas) Doin, Paris.
—— & Feuguer, L. 1968. *Bassin de Paris*, Guides Geologiques Region, Paris.
Preece, R. C., Scourse, J. D., Houghton, S. D, Knudson, K. L. & Penney, D. N. 1990. The Pleistocene sea-level and neo-tectonic history of the eastern Solent, Southern England. *Philosophical Transactions of the Royal Society,* **B328**, 425–477.
Price, M. 1987. Fluid flow in the Chalk of England, *In:* J. C. Goff & B. P. J. Williams (eds), *Fluid Flow in Sedimentary Basins and Aquifers*, Geological Society, London, Special Publication, **34**, 141–156.
Rat, P. 1974. Visages de la France entre l'orogenèse hercynnienne et l'orogenèse alpine. *Geologie de la France*, Vol. 1 (ed. J. Debelmas). Doin, Paris.
Rose, J. 1987. Status of the Wolstonian Glaciation in the British Quaternary, *Quaternary Newsletter,* **53**, 53–199.
Shackleton, N. J., Backman, J., Zimmerman, H., Kent, D. V., Hall, M. A., Roberts, D. G., Schnitker, D., Baldauf, J. G., Desprairies, A., Homrighausen, R., Huddlestun, P., Keene, J. B., Kaltenback, A. J., Krumsiek, K. A. O., Morton, A. C., Murray, J. W. & Westberg-Smith, J. 1984. Oxygen isotope calibration of the onset of glaciation in the North Atlantic region, *Nature,* **307**, 620–623.
Shearman, D. 1967. On Tertiary fault movements in North Devonshire. *Proceedings of the Geologists Association,* **78**, 555–566.
Stamp, L. D. 1921. 'On Cycles of Sedimentation in the Eocene Strata of the Anglo-French-Belgium Basin'. *Proceedings of the Geologists' Association,* **34**, 70–76.
Shotton, F. W. 1965. Normal faulting in British Pleistocene deposits. *Quarterly Journal of the Geological Society,* **121**, 419–434.
Smith, A. J., 1985. 'The English Channel—by geological design or catastrophic accident?'. *Proceedings of the Geologists' Association,* **100**, 325–327.
Welch, F. B. A. & Trotter, F. M., 1961. Geology of the country around Monmouth and Chepstow. Memoirs of the Geological Survey of Great Britain.
Winwood, H. H. 1987. On a Rhaetic exposure at Boyce Hill. *Proceedings of the Bath Natural History and Antiquarian Society,* **8**(4), 306–316.
Wymer, J. J. 1985. The palaeolithic sites of East Anglia. Geo Books.
—— & Straw, A. 1977. Hand-axes from beneath Glacial Till at Welton-le-Wold, Lincolnshire and the distribution of Palaeoliths in Britain. *Proceedings of the Prehistoric Society,* **43**, 355–360.
Varney, W. D., 1921. The Geological History of the Pewsey Vale. *Proceedings of the Geologists' Association,* **32**, 189–205.

From Kellaway, G. A. (ed.) *Hot Springs of Bath*, pp 243–273

14. The older Plio–Pleistocene glaciations of the region around Bath

G. A. Kellaway

In carrying out an analysis of the evidence relating to the origin of the hot springs at Bath and Hotwells it was inevitable that attention would have to be paid to the sequence of events in late Tertiary and early Pleistocene times, if only to ensure that all the relevant information regarding the age of the springs had been considered. From this it soon became apparent that the geomorphology of the region around Bristol and Bath had undergone more drastic changes in late Pliocene and early to mid-Pleistocene times than had previously been suspected.

The Stonehenge Glaciation

One of the most important aspects of the Plio-Pleistocene history of the region is the question of the origin of the so-called 'foreign stones' of Stonehenge which, like the 'bluestone' boulder from Bowls Barrow (a Neolithic long barrow on Salisbury Plain, 18.5 km WNW of Stonehenge) are of Welsh origin. The first geologist to point out the similarity of some of the bluestones of Stonehenge and the igneous rocks of North Pembrokeshire and North Wales was Sir A. C. Ramsey in 1858. They were described as being of glacial origin by Judd (1902). H. H. Thomas (1921 in Hawley 1923) attributed their movement to human transport. Their presence on Salisbury Plain and the incorporation of the bluestones in the Bronze Age monument at Stonehenge became a matter of controversy for nearly 70 years, and the absence of a continuous chain of bluestone boulders heading all the way back from Stonehenge to the Preseli Mountains (thought by some to be the sole source of the bluestones) has been seriously urged as a valid reason for questioning their glacial origin! As is indicated below, the remarkable feature of the Stonehenge and Bowls Barrow stones is that they have survived at all, and they might very well not have done so had they not travelled in ice across an upland area composed largely of Chalk, coming to rest on thin Tertiary clay and sand in which a series of karstic depressions and sink holes developed by solution of the underlying Chalk. In this way the erratics may have been offered some protection against weathering and solifluction, at least for part of their sojourn on Salisbury Plain.

It is now 20 years since the writer challenged the then widely accepted theory of the human transportation of the so-called foreign stones or bluestones (Appendix 1) of Stonehenge and Bowls Barrow in favour of a glacial origin (Kellaway 1971). When Judd (1902) proposed a glacial origin for them, hardly anything was known of the glaciations prior to the deposition of the great mass of boulder clay or till which was laid down by the Anglian–Wolstonian ice sheets that immediately preceded the Ipswichian Interglacial (Fig. 14.1). Investigations of ocean floor sediments by Shackleton and his colleagues have since established a succession of glacial episodes alternating with warmer periods. These can be related to the magnetic stratigraphy (Harland *et al.* 1990, fig. 3.13), though many of the older glacial episodes have not yet been identified on land.

It must be borne in mind, however, that the $\delta^{18}O$ record drawn from ocean cores, such as the one obtained on the Rockall Bank (Shackleton *et al.* 1984), indicates neither the exact position nor the shape of the ice sheets. The standard British Pleistocene succession is largely based on that of East Anglia, but valuable though this is, it does not include glacial deposits older than the Anglian, with the possible exception of the Beestonian Kesgrave sands and gravels of Essex. These may be outwash from an ice front situated at no great distance to the west, or fluvioglacial gravels carried by large melt water streams conveying rocks from the Welsh borders and the Midlands to southern East Anglia. The older glaciations, recorded in the sediments on the ocean floor, are not represented by tills in the East Anglian and Midland succession. Either periglacial conditions prevailed in eastern England or the relevant strata have been destroyed by erosion. In those parts of Britain where the Anglian and Devensian ice sheets were thickest, little or no evidence of the older glaciations remains. Destruction of the evidence for older glaciations by younger ones is a common accompaniment of a history of multiple glaciation.

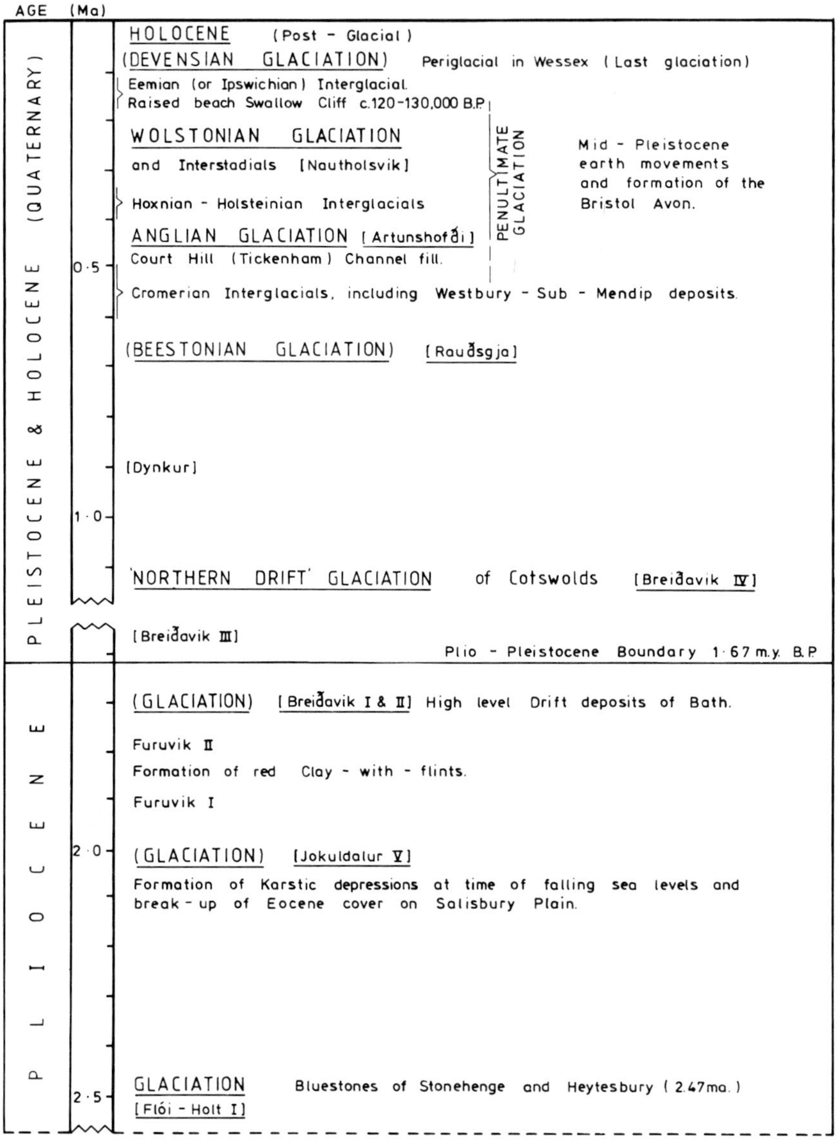

Fig. 14.1. Generalized glacial succession from 2.5 m yr. B.P. to present day. (GLACIATION)—brackets indicate glacial events not identifiable in Wessex. [Icelandic glaciations] based on Einarsson & Albertsson (1988). Plio-Pleistocene boundary based on Harland *et al.* (1990). The presence of glacial deposits of Wolstonian age in the Clevedon area is doubtful (Andrews *et al.* 1984).

It is in the sediments on the ocean floor and the geomorphology and residual sediments of the southwestern parts of the British Isles therefore that we may expect to find relics of the earliest Plio-Pleistocene glaciations and the largest of these is likely to have been responsible for the carriage of the Welsh rocks or 'bluestones' to Salisbury Plain. Another deposit which is probably of glacial origin is the 'Northern Drift' of the Cotswolds, while the high level Drift deposits of Bath (Chapter 13) constitute a third group of pre-Anglian deposits. These may be of approximately the same age as the red Clay-with-flints of the Chalk downlands to the east. Other ancient high level Drift deposits exist in the Thames valley but these are remote from Bath and are not considered here.

In the region around Bristol and Bath, the main problem is to correlate the various glaciations with the stages in which the Mesozoic rocks (which formerly covered all the Palaeozoic terrain in the lower Severn valley) were removed by erosion (Fig. 13.5). Any general synthesis must therefore take into account the development of the relief and drainage in the light of climatic and tectonic influences and the inferred sequence of events should accord with the development of the various geomorphological features as well as with the characteristics of the sediments transported by glacial, fluvioglacial and fluviatile means. Some of the geomorphological and tectonic aspects have already been discussed in relation to major changes in the drainage in mid-Quaternary times (Chapter 13).

Although the early stages of development of the drainage and the erosion of the Mesozoic cover have little immediate bearing on the mid-Quaternary emission of the thermal water at Bath, the later stages were of critical importance. In terms of general scientific interest, however, the identification of the glaciation responsible for the movement of the bluestones to Stonehenge is important both for understanding the geological history and the archaeology. The human transportation theory of the movement of the bluestones was based on an imperfect assessment of the petrological evidence and owed much to opinions strongly expressed by H. H. Thomas (in Hawley 1921, pp 39–40). His establishment of the similarity between the dolerites and rhyolites of the Preseli area and some of the 'bluestones' of Stonehenge was valuable, but its overstatement led to a widespread misapprehension (still surviving in some quarters) that all the stones came from a small compact area of the Preseli Mountains or from the supposed route along which the Preseli rocks were said to have been conveyed via Bath to Salisbury Plain by Neolithic man. This journey would have taken them up the Bristol Avon (Fig. Pf. 2) and thence by way of the Wylye and upper reaches of the Salisbury Avon to Amesbury and Stonehenge (Atkinson 1979).[1] An even longer sea route round Lands End was considered in an attempt to explain the presence of implements made of Preseli-type dolerite on the south coast of England. This however, was abandoned as impractical.

Some evidence regarding the bluestones and their sources is given in Appendices 2 and 3, showing that at least one of the stones found in Colonel Hawley's excavations at Stonehenge probably came from North Wales. Recent petrographical and geochemical investigations carried out at the Open University (Thorpe *et al.* 1991; Thorpe & Williams-Thorpe 1991) has thrown much new light on the distribution of the ultimate sources of the bluestones which are now known to have been derived from a much larger area than was formerly envisaged. Adherents of the human transport theory maintained that the presence of sandstones similar to the Altar Stone proved that the bluestones had been conveyed by sea from Milford Haven, but this interpretation has also been shown to be

[1] In putting forward the theory of human transportation archaeologists were influenced by authoritative geological statements regarding both the petrology of the bluestones and the number and extent of British Plio–Pleistocene glaciations. The extension of the human transport theory to the sarsens was an unnecessary complication. Without the aid of the theory however there would have been no explanation to offer for the presence of the bluestones—other than glaciation.

unsound. The concept of Neolithic (or Bronze Age) boatmen navigating the upper Avon and the tiny Frome and Wylye streams, conveying more than 300 tons of cobbles and boulders, many deeply weathered, of all shapes and sizes and up to 6 tons in weight, some so sheared or friable that they dropped to pieces when tested on arrival, has long appeared to the writer to lack any credibility (Kellaway 1971).

From the point of view of its general interest and of Pleistocene studies it is important that a new analysis has been made of the petrographical evidence relating to the inferred glaciation. This would have involved the movement in ice or melt water, of Welsh rocks across the site of Bath (or the country immediately south of it) at an altitude of about 230–245 m A.O.D. The deposition of the bluestones would have taken place at about 200 m at Bowls Barrow and possibly a little lower in the area south of Chitterne. The date of the glaciation is tentatively placed at 2.47 Ma BP, this being the date of the major glaciation identified by Shackleton *et al.* (1984) in sediments on the Rockall Bank. The relationship of this early glaciation to the dissection of the Mesozoic cover has already been mentioned in Chapter 13. Other aspects are discussed below.

The Upper Cretaceous–Eocene cover

Most of the older and high level Pleistocene deposits of the Mendips and the Bristol and Somerset coalfield including the highly disturbed Cromerian sands of Westbury-sub-Mendip (Bishop 1974) contain varying proportions of Upper Cretaceous material including foraminifera. Decalcified nodular flints as well as angular and subangular flint and Upper Greensand chert found in high level gravels at Bathampton Down and Farleigh Down near Bath indicate a local origin for this material. Farther south, in the eastern Mendips a small outlier of highly transgressive Upper Greensand and Gault resting unconformably on Lower Oxford Clay survives at Cloford Common south of Nunney. Upper Cretaceous chert and flint as well as Tertiary Greywether sandstones ('Sarsens') occur at Nunney and Whatley and reddish-brown loam with fragments of Middle Chalk resting in fissured Carboniferous Limestone has been seen nearby at Holwell. The surviving Upper Cretaceous rocks indicate a base level for the Upper Cretaceous of about 550 feet (167 m) A.O.D. at Cloford Common (Postlebury Wood) falling eastwards to about 350 feet (107 m) A.O.D. at East Woodlands (788442). From this it appears that much of the east and central Mendip surface may have had an Upper Cretaceous–Eocene cover in early to mid-Quaternary times. North of Frome the base level may have risen to over 180 m in the high ground at Faulkland and to levels about 205 m at Bathampton Down and over 238 m at Lansdown. South of Bristol the level at Dundry would probably have exceeded 245 m (Fig. 14.2).

The distribution of Albian–Cenomanian chert in southwest England has been described by Tresize (1961); there has been some controversy over its origin, but in the writer's view there is no question that it is diagenetic and that its distribution is controlled by conditions prevailing below the seafloor in Albian and Cenomanian times (Fig. 14.2). Consequently, the distribution of the silicified Jurassic rocks of the Central Mendips reflects the approximate position and altitude of the Upper Cretaceous sea floor and indicates the extent of the Albian sediments before their dissection in mid-Tertiary and later times. This shows that the Upper Cretaceous sediments rested on a surface of lower Mesozoic and Palaeozoic rocks at an altitude of about 250–260 m A.O.D. at Castle of Comfort in the central Mendips, figures which accord with those determined by other methods.

The concept of development of the Severn Valley and the Bristol Avon outlined in Chapter 13 carries with it implications for the drainage and erosion of the eastern flank of the Welsh uplands. Almost all geologists who have studied the Cretaceous rocks of Britain

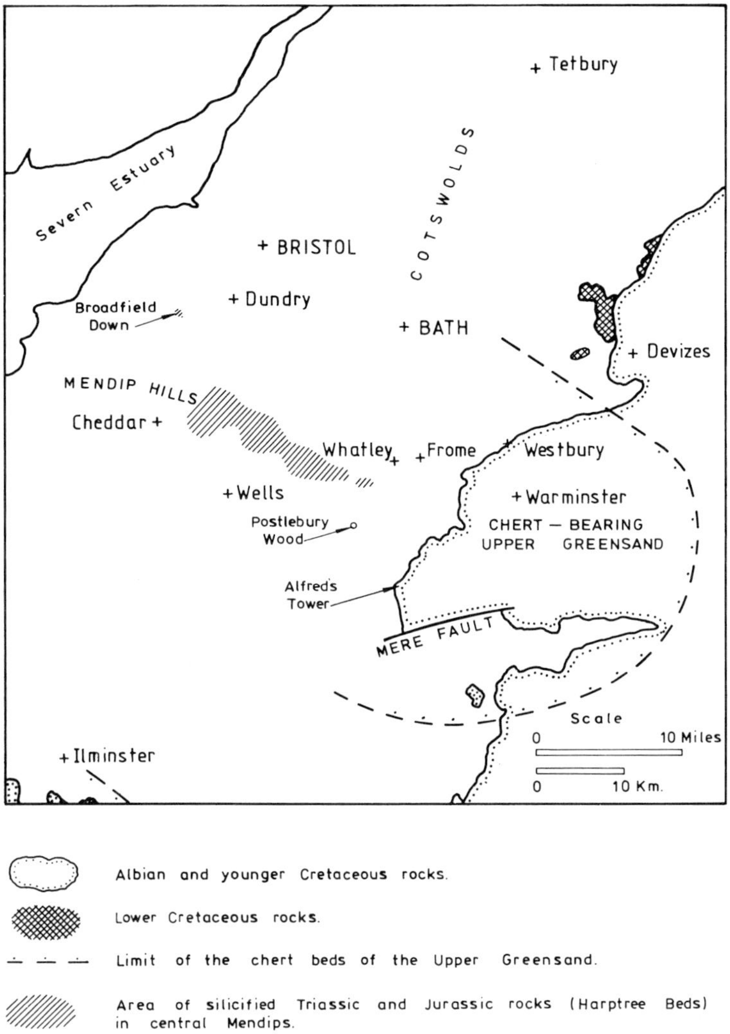

Fig. 14.2. Distribution of Upper Greensand chert in relation to the silicified Triassic and Jurassic rocks of the Mendips. The position of the chert-bearing sand is based on Tresise (1961).

have concluded that the greater part of the lowlands of central England were formerly covered by Cretaceous strata. Recent findings in SW Ireland, the English Channel and western approaches of the Bristol Channel have extended and modified the application of this hypothesis. With regard to the Bristol and Somerset coalfield and the area east of the South Wales coalfield and the Black Mountains, Jukes Browne (1982, p 280) calculated that the base of the Malmstone (Upper Greensand), if projected westward from Frome, "would cut the centre of the Mendip Hills at a height of 820 feet" (250 m A.O.D.) and that the height of the Warminster Greensand would be about 1600 feet (490 m) in South Wales. Some geologists have suggested that the whole of Central Wales may have been concealed beneath Cretaceous rocks, but the evidence for this is less satisfactory, being dependent on assumptions about high level erosion surfaces which have not yielded any evidence of Cretaceous deposits and could, conceivably be of Tertiary or early Quaternary age. The results of the Mochras Farm Borehole on the shores of Cardigan Bay (Woodland 1971) indicate that if Cretaceous rocks ever extended into central Wales they had been removed from the higher parts of the Welsh uplands by mid-Tertiary times. The same is true of most of Devon west of the Bovey basin and of the greater part, if not the whole of mainland Cornwall.

To some extent the distribution of the Greensand chert fragments and Chalk flints as recorded round Bath reflects the character of the Cretaceous rocks. Thus the absence or infrequent occurrence of Upper Cretaceous chert and flint in the south Cotswolds is likely to be due to its original distribution (Fig. 14.2) rather than to subsequent erosion.

In the western part of the Hampshire basin the Eocene Bagshot Beds show evidence of quite strong westerly overstep and rest directly on Chalk and Upper Greensand in West Dorset and East Devon. These Eocene gravels yield pebbles and fragments of Chalk flint, Greensand chert and Purbeck rocks. Palaeozoic rocks, some of which may have been derived from Permian breccias, are also present. Farther north there are indications that in late Pliocene and early Pleistocene times Eocene deposits may have been very thin or absent west of a line drawn from Exeter to the Central Mendips and the eastern margin of the south Cotswolds. This is due to a combination of circumstances including original attenuation and westerly overstep of the older Eocene strata by younger beds, and structural deformation and erosion in late Oligocene and Mio-Pliocene times.

About 15 km east of Bath however, Eocene strata may have rested on the Chalk in early Pleistocene times. Large masses of Greywether sandstone ('sarsens') are present on the eastern Mendips near Nunney and scattered on the lowlands of West Wiltshire at Rode and between Allington and Bishops Cannings at the western end of the Vale of Pewsey, while masses of Lower Tertiary sand and clay are found in deep solution pipes and cavities in the western part of Salisbury Plain. At least one of these, at Clay Pit Hill near Chitterne (Plate 29) contains pale grey and white sand with pipe clay and beds of well-rounded flint pebbles indicating that the Woolwich and Reading beds and London Clay are thin or have been overstepped locally by the Bagshot beds. Although Quaternary foraminifera were recorded from the Clay Pit Hill samples (Kellaway 1974) it is possible that they may have been contaminants.

One important characteristic of the uplands around Bath which distinguishes them from the eastern Mendips is the absence of big Tertiary Greywether sandstone masses ('sarsens') which mark the former extension of the basal Eocene strata in east Somerset and west Wiltshire. The beds of well rounded flint pebbles which are found near the base of the Eocene on the Marlborough Downs and in Wiltshire and Hampshire are also absent on the hills near Bath. Most of the flint and chert in the high-level Drift deposits on Bathampton Down and Farleigh Down consists of nodular or subangular fragments. In general character the high-level Drift deposits of Bathampton Down resemble the formless or poorly sorted Clay-with-flints of west Wiltshire, though they are grey, greenish or brown, not red in colour. They lack both the abundant well rounded flint pebbles and the red-brown colour of typical Clay-with-flints.

The red Clay-with-flints (Argile à Silex)

Compared with the Marlborough Downs, or with Great Ridge and Groveley Ridge on the south bank of the Wylye, Salisbury Plain is singularly deficient in Clay-with-flints. To the east extensive tracts of Clay-with-flints are seen on the Hampshire Downs but in some parts of Salisbury Plain there is little evidence of its presence.

Where it is well developed, red Clay-with-flints contains a high proportion of material derived from Eocene strata and locally, in the North Downs it may be so mingled with disturbed material of Red Crag (Pliocene) age that it is impossible to distinguish between the two. Both in northern France and in the south of England it can be seen that the formation of the largest karstic hollows or depressions on the surface of the Chalk took place before the final stage in the emplacement of the red Clay-with-flints. Several phases are recognizable in Normandy and south east England and similar conditions may apply on Salisbury Plain. Here the Clay Pit Hill karstic depression contains Tertiary (Eocene) sediments (Prestwich 1898) which subsided into the depression before the undisturbed Eocene had been removed from the surrounding areas where residual Clay-with-flints now rests on the eroded surface of the bare Upper Chalk.

Greywether sandstone ('sarsens')

Salisbury Plain also differs from the Marlborough Downs in the very small number of unused Greywether sandstone boulders which are visible on the surface of the Chalk at the present day. This prompted the suggestion, put forward by Atkinson (1979) that the big sandstone uprights at Stonehenge and their lintels are made of sandstone transported from the Marlborough Downs (Fig. 14.3). It is probable that the present-day distribution of sarsens is misleading in the sense that much of the stone originally present on Salisbury Plain has been used up by man and that the boulders have been cleared away if only to assist ploughing and other agricultural activities. There is, however, another factor which may have influenced their distribution, namely the destruction of the lower Tertiary rocks by the formation of solution holes and collapse structures in late Pliocene times. On Salisbury Plain this destructive period cannot be dated with any degree of accuracy, nor can the individual phases be identified as they can in Normandy, but the translation of the bluestones to the Heytesbury–Stonehenge area is likely to be pre-Red Crag and the lack of any appreciable deposits of typical red Clay-with-flints in the interior of Salisbury Plain may reflect the earlier destruction of the Eocene strata as well as subsequent erosion of evolved red Clay-with-flints.

Although there are some fairly large masses of weather-hardened Greywether sandstone which are almost all of local origin (the Heel Stone of Stonehenge being the most notable), the sarsens of Salisbury Plain are generally much smaller than those of the Marlborough Downs. An interesting example is provided by Bowls Barrow near Heytesbury, opened by W. Cunnington in 1801, where a dolerite boulder (Plate 33) was associated with at least 9 sarsens, forming part of 'a great heap of stones'. William Cunnington was obliged to discontinue excavation as 'the large stones came rolling down so fast that we were obliged to desist' (R. H. Cunnington 1975). The surrounding Downs are now bare of sarsens or bluestones, both having long since been gathered up.[1]

Weather-hardened sarsens of Greyweather sandstone are rare on some parts of the Chiltern Hills where large masses of Greywether sandstone are present in deep solution cavities, in some cases still embedded in the parent sand, in others in red Clay-with-flints. Prestwich's discovery (1898) of the 'Druid Sandstone' or its parent sand rock in situ in the Clay Pit Hill depression on Salisbury Plain has already been mentioned. Similar circumstances have led to the local preservation of masses of Tertiary sandstone in deep cavities and collapse structures in the Chalk downlands from Sussex to the Chilterns. It has been established that these masses of unweathered sandstone are in a soft workable condition,

[1] Bowen & Smith (1977) have published an interesting account of the general distribution of the sarsen stones of Wessex; this excludes eastern Somerset. In east Sussex silcreted sandstone of the Woolwich and Reading beds is found in collapsed sediments filling solution 'pipes' and large cavities in the Chalk. Some of these masses of silcreted sand are soft and workable. Elsewhere as at Falmer Pond near Brighton and in the adjacent valley at Stanmer Park the sarsens derived from the Woolwich and Reading Beds consist of intensely hard silcreted sandrock. A number of these weather-hardened sarsens have been preserved in the grounds of the University of Sussex at Falmer, where they were found in Devensian 'Combe Rock' in the valley floor.

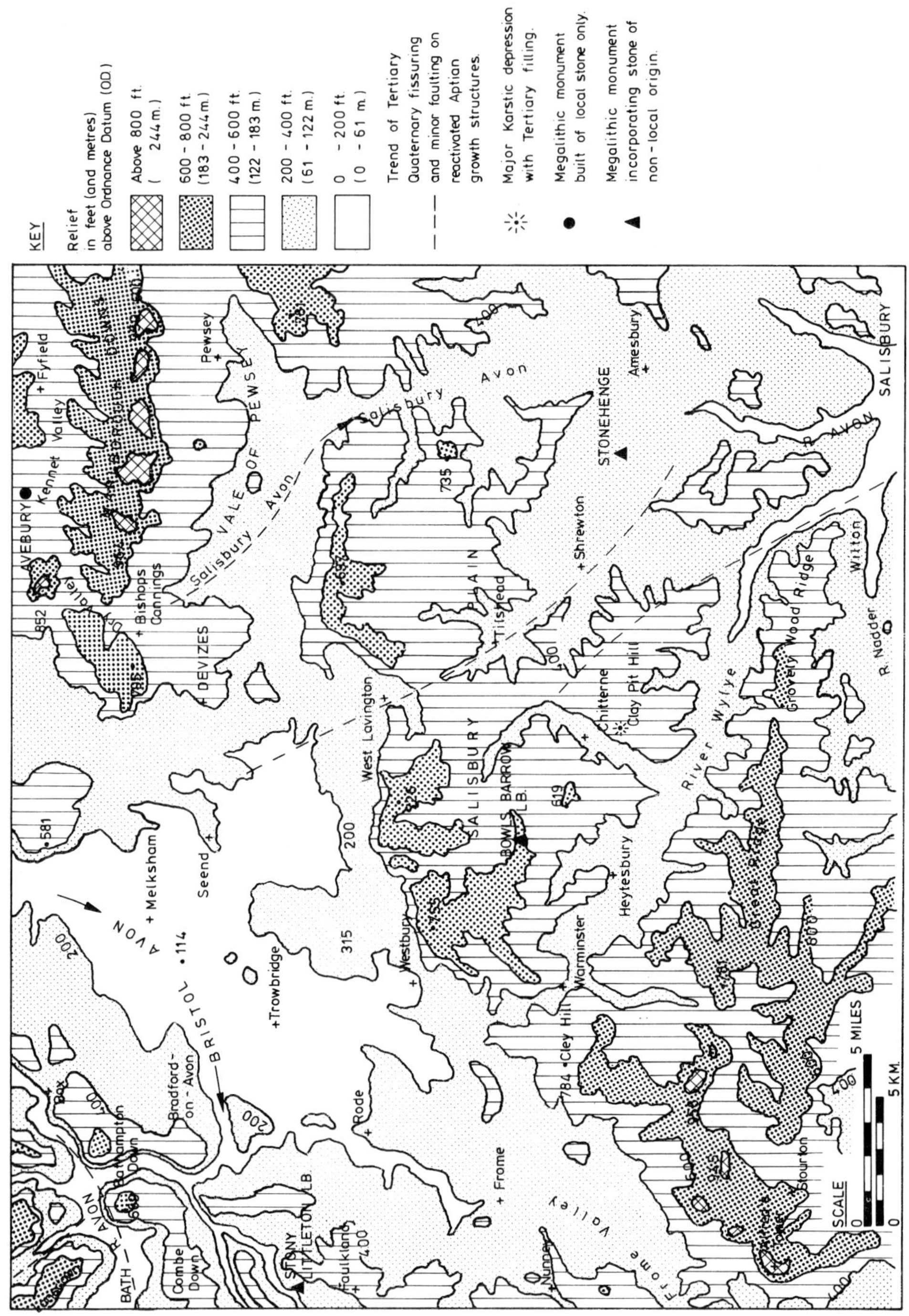

Fig. 14.3. Relief of the county SE of Bath.

unlike the hard massive weathered sandstones (sarsens) used in the construction of Avebury and Stonehenge I.

It is noticeable that in the Vale of Pewsey where substantial concentrations of sarsens rest on Upper Greensand and Lower Chalk, all of these are hardened bouldery masses. Only in those areas where large karstic depressions are developed in the Upper or Middle Chalk have masses of sandstone been found in workable condition. The builders of Stonehenge III must have discovered one or more of these subterranean storehouses and it is possible that some bluestones may have been concentrated in the upper part of shallow depressions marking the position of the cavities as well as lying scattered on the surface of the ground. The larger solution cavities containing Tertiary clay and sand would have given rise to acid humus soils where pockets of bluestones may have been recovered by Neolithic man during the clearance of wooded uplands. It would seem, however, that the extraction and use of unhardened silcrete was a Bronze Age achievement (Appendix 1), necessitating excavation of the cavities in the Chalk.

The Plio–Pleistocene glacial succession (Fig. 14.1)

In East Anglia there is a marked disconformity between the shelly marine Pliocene deposits known as the Coralline Crag and the overlying Red Crag. The latter is characterized by a boreal North Atlantic fauna which entered the North Sea and did not penetrate Normandy. The stratigraphical break between the Coralline Crag and the Red Crag marks a period of falling sea level and lowered temperature and for this reason the bottom of the Red Crag (pre-Ludhamian or Waltonian) was formerly adopted in Britain as the base of the Pleistocene. On stratigraphical grounds, however, an Italian section in Calabria takes precedence over the East Anglian sequence and if this boundary is adopted (Harland *et al.* 1990, p 69) then its age would be 1.67 Ma BP. This implies that the pre-Ludhamian Red Crag of East Anglia is of late Pliocene age.

In terms of the East Anglian succession (West 1977) the sequence below the Anglian Lowestoft Till is critical, since it bridges the gaps between the Red Crag and the Anglian glacial deposits. Where the Cromerian interglacial deposits are missing, the basal Anglian fluvioglacial sediments rest on the Kesgrave Sands and gravels. The latter are classified as Beestonian and it is in this cold period that pebbles of Midland and Welsh Border rocks reached southern East Anglia (Rose *et al.* 1976; Hey & Brenchley 1977). The date of the Beestonian glaciation is thought to be about 0.75 Ma BP.

Whether there is any record of a Beestonian glaciation in the west of England is uncertain. The only deposits in the Cotswold–Severn Estuary–Mendip region which could be related to it are the Northern Drift of the Cotswolds (Ramsay *et al.* 1858 pp 44–5; Lucy 1880; Richardson L. 1933; Cave 1977). The oldest known Pleistocene deposits in the Mendips which are datable on palaeontological grounds are the interglacial non-marine Cromerian sands and breccias which are preserved at a high level in fissured and cavernous Carboniferous Limestone at Westbury-sub-Mendip (Bishop 1974). The writer has assumed that these are younger than the high level Drift deposits of Bath which are unfossiliferous (? fluviatile or fluvioglacial) deposits, including rocks derived from west of the Severn. In their poor sorting these deposits bear a physical resemblance to the red Clay-with-flints of the NW part of Salisbury Plain and Cley Hill, Warminster.

The residual Northern Drift deposits of the Cotswold uplands also give the appearance of great age and are commonly found as clay-bound stony or gravelly deposits in fissures and solution cavities in the surface of the Middle Jurassic limestones of the Cotswolds. They are rich in Bunter pebbles of Midland origin and are found at altitudes of over 300 m A.O.D. on the central Cotswolds above Cheltenham and also on Bredon Hill where their deposition pre-dates the fissuring and foundering of the Inferior Oolite Limestone. The absence of any evidence showing that they came to rest on Cretaceous rocks may indicate that they are somewhat younger than the high-level Drift deposits at Bath. The high-level Drift deposits of Bath are confined to the palaeovalley (Chapter 13) while the Northern Drift deposits are all situated to the east of it (Fig. 13.6).

If, as is generally assumed, the Northern Drift deposits are of glacial origin and the high-level Drift deposits of Bath are fluviatile or fluvioglacial (perhaps ice-fed in Wales) then the difference in age may not be very great. For the high-level Drift deposits an age of about 1.7 Ma would be realistic and would allow for the presence of a Cretaceous cover for which there is evidence in the area around Bristol and Bath. Apart from the bluestones of Stonehenge, other indications of pre-Anglian glaciation found in the West of England and N.W. France include very large boulders of non-local origin scattered in the coastal regions of Brittany and the south and west of England. At Fremington near Barnstaple, deposits of till laid down by an intrusive ice sheet carrying rocks of Scottish origin represent an ice mass which is generally regarded as being of Anglian age, though it might be older. The associated post-glacial marine gravels contain erratics derived from the till and show no evidence of having been overridden by the ice. They are therefore classified as Ipswichian. However, the problem of dating some of the raised beaches and platforms are considerable and there is a possibility that some large erratic boulders, e.g. the Giants Rock at Porthleven, Cornwall may have been introduced in an earlier glaciation. Like many of the huge boulders stranded in the coastal regions of Brittany it may be residual, dating from much earlier times.

The Stonehenge Bluestones

An analysis of the etymology of some of the Stonehenge rocks including the name 'bluestones' is given in Appendix 1. From the analysis of the age of the Northern Drift and high-level Drift deposits of Bath it will be seen that an Anglian date for the Stonehenge erratics is out of the question. The date of the cold period corresponding to the deposition of the Beestonian Kesgrave gravels in Essex is about 0.65–0.75 Ma BP, possibly corresponding to O^{18} cycle 16. This is older than any of the glacial deposits discovered at Clevedon and Kenn but it is questionable whether this dating would allow sufficient time for the destruction of the Mesozoic rocks forming the surface over which Coal Measures and Upper Old Red Sandstone rocks were conveyed across the site of the lower Severn valley to Bathampton Down. To allow time for this, an age in excess of 1.7 Ma BP is suggested (Fig. 14.1).

Again it is possible that the abundance of Bunter pebbles in the Northern Drift deposits of the mid-Cotswolds may indicate a community of origin with the Beestonian Kesgrave gravels of Essex, but the Kesgrave gravels could equally well represent the recycling of older glacial deposits in the Midlands. The failure of the 'Northern Drift' ice to cross the eastern palaeodivide in the Cotswolds also suggests that the divide had not been seriously breached by erosion accompanying tectonic movement along the Denny Island Fault Zone (Chapter 13) when the glaciation took place. The balance of the evidence is therefore in favour of a pre-Beestonian age for the Northern Drift and a somewhat greater age for the high-level Drift deposits at Bath which were laid down while Upper Cretaceous rocks still existed in the locality (Fig. 14.4).

For the Stonehenge erratics, however, none of these Pleistocene dates is adequate to explain the composition of the bluestone assemblage, consisting primarily of Lower Palaeozoic rocks and lacking any sizeable fragments or boulders of Middle Jurassic, Lower Jurassic, Triassic, Lower Carboniferous or Upper Old Red Sandstone rocks. This implies that the Upper Cretaceous cover still extended over part of the Welsh borders at the time of glaciation and that an age of 2–3 Ma would be realistic for this event (Fig. 13.5).

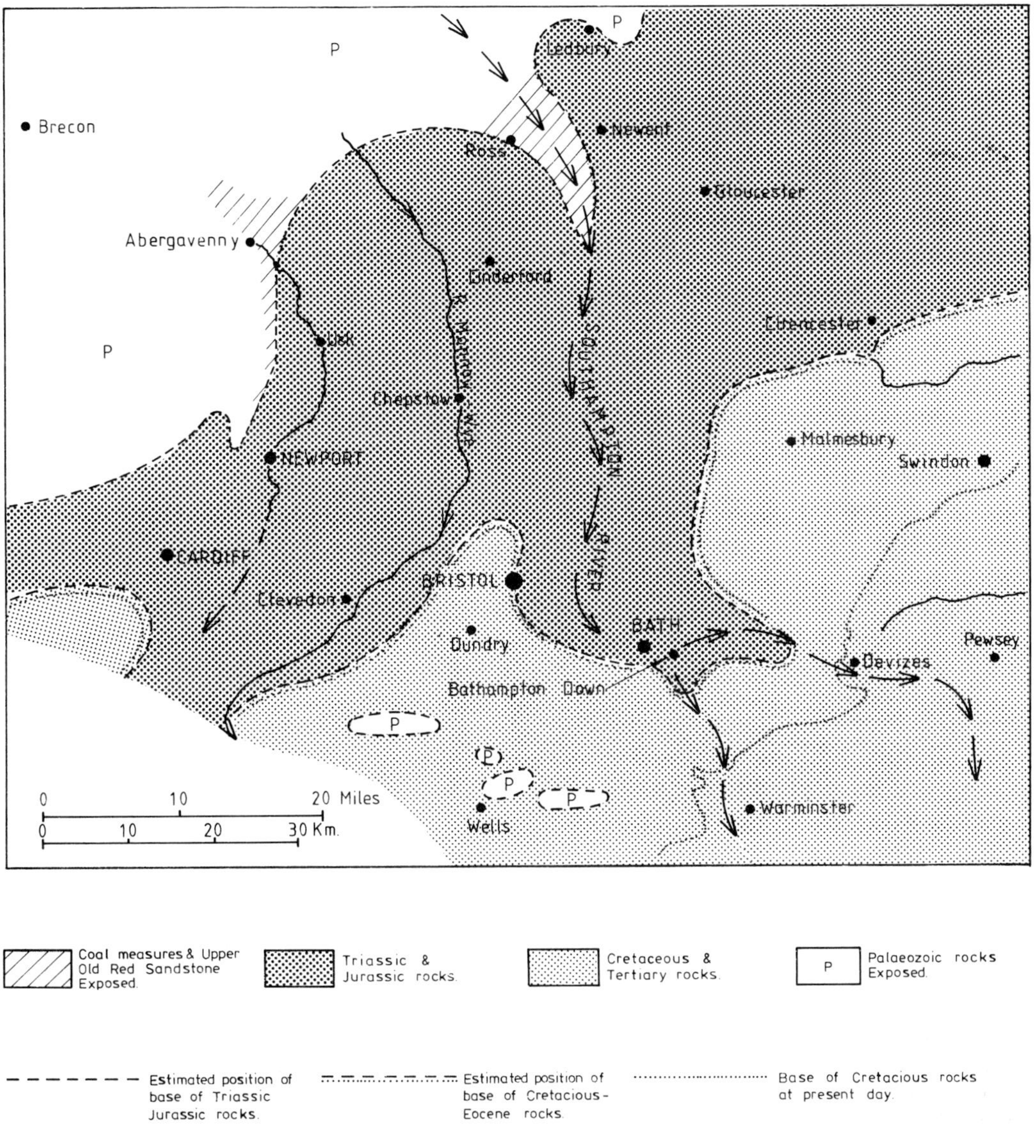

Fig. 14.4. The Southampton River: possible routes showing areas where Upper Old Red Sandstone and Coal Measures rocks found by Varney (1921) on Bathampton Down may have originated. Alternative courses are shown for the river east of Bath. Either or both of these may have been operative at sometime. The Mesozoic boundaries are conjectural. They fit the available evidence but are only intended as a general guide.

Among the rocks taken from the Stonehenge excavations carried out by Colonel W. Hawley in 1921, a number were seen by the late Mr R. S. Newall. At the time that these were recovered he drew attention to the fact that some appeared to carry glacial striae. The appearance of striations on some other bluestones also drew comment from observers at the time (Hawley 1921, pp 40–41); the Rev. G. H. Engleheart commented that the rocks which were exhibited were *boulders not stones*, that one was glacially striated and that he, like Mr Dale, regarded them as being of glacial origin. Commenting on a statement by H. H. Thomas who claimed the rocks had been transported by human agency, Engleheart pointed out that if the boulders had been brought by hand from Wales it was difficult to believe that they had been dressed only on arrival at Stonehenge since "transport of such an unnecessary weight argued lack of intelligence". Yet the dressing of weathered friable rocks did take place at Stonehenge (Appendix 3), and some of the rocks which were broken up by Bronze Age man were too soft or highly sheared to be of any use in building the monument. In terms of the human transport theory their presence is inexplicable.

The specimen preserved by Newall (Plates 35 & 36) is not thought to have been one of those exhibited at the 1921 meeting. It was passed to the writer for petrological examination and this was carried out by Mr R. Harrison, then Chief Petrographer of the Institute of Geological Sciences, at the Geological Museum in Exhibition Road. The result of this investigation is given in Appendix 2. Some other rocks supplied by Newall were also examined, including specimens of ophitic dolerite with 5 mm thick weathering crusts.

The principal specimen which is an ignimbritic rock probably of North Wales origin, has a typical white weathering zone which has been observed on a number of specimens from Stonehenge and which cannot possibly have formed since the rock was chipped or fractured by Stone Age man. The sheared Welsh ignimbrite (Plates 35 & 36) was found where it had been thrown away with other rock waste and chippings, having proved to be useless for the manufacture of implements. Newall regarded this rejection as indicating that the stone must have been found locally and, from the outset he considered it to be a deeply weathered glacial erratic.

The process of low temperature weathering which has affected the ignimbrite involves the conversion of rocks, mainly of basic or intermediate composition, into a white powdery substance without the formation of kaolin (Appendix 2). It is commonly found in stones and boulders which have been embedded in peat or humic matter for long periods of time at a relatively low temperature. These effects can be seen on rocks buried in upland peat bogs or coated in organic matter as in Wales, the Lake District or the Yorkshire moors (Plate 34). Very shallow bleaching can be produced by burial in a midden heap or soil; flint implements sometimes show these effects. Under these circumstances however the bleaching is superficial. Prolonged burial in an acid peaty soil is required to produce the depth of weathering (more than 5 mm) shown by some of the bluestones. At Stonehenge the effect of removing the weathered material by chipping it exposed fresh rock, but in the well drained chalky soil of Salisbury Plain the freshly exposed surfaces have remained unaltered since Bronze Age times. The same observation applies to the Bowls Barrow boulder which shows no evidence of prolonged weathering of its eroded surface. Prior to its inclusion in the long barrow it must have been embedded in a protective medium.

The weathered condition of some of the boulders and the unweathered condition of others indicates that some at least were completely enclosed in a fine-grained deposit of some chalky medium which protected them from weathering. The great karstic depression or sink at Clay Pit Hill near Chitterne, where Prestwich (1898) saw white Tertiary sand which he compared to the 'Druid Sandstone' of Stonehenge, is known to contain Tertiary sand and silt as well as deposits of pipe clay which were in work in the 17th Century. John Aubrey refers to these in his '*Natural History of Wiltshire*' and Prestwich's description of the pit in 1898 suggests that some working faces must still have been visible at that date. An attempt at investigation of the old workings has been made (Kellaway 1974) but without much success. White sand with flint pebbles was proved by boring alongside a vertical face of Chalk to a depth of 35 ft (10.6 m) on the north side of the collapse structure. The total thickness of the Tertiary deposits is unknown, while the central part of the old workings appears to be entirely occupied by made ground and further excavation and timber felling would be needed in order to explore it.

The general impression left by the dating of the large solution cavities and collapse structures on Salisbury Plain, the Chilterns and the South Downs is that they are of late Pliocene age (taking 1.67 Ma as the Plio–Pleistocene) boundary and that the process involved major changes in the water levels in the Chalk in which either tectonic movement or climatic changes, or both, were involved. It is most unlikely that the Clay-with-flints of Salisbury Plain was over-ridden by ice in Beestonian or Anglian times but a glaciation of Wales accompanied or followed by the formation of the big karstic depressions at a time of lowered sea level 2.47 Ma BP is quite a different matter. The intrusion of ice or melt water into the area of Salisbury Plain may well have been localized along a route roughly coinciding with the palaeovalley of the Southampton River and this would account for the restriction of the Welsh rocks to the Heytesbury–Stonehenge belt (Fig. 14.3). In any event the movement of the bluestones could have taken place in ice and melt water across the site of the Severn estuary at a time when an almost continuous Upper Cretaceous cover existed between Salisbury Plain and Gwent, with a thin Eocene cover east of Bath.

Climatic deterioration leading to the accumulation of ice on the Welsh mountains may have commenced soon after its appearance in Iceland about 3 Ma BP and the major glaciation at 2.47 Ma may therefore have recycled erratics carried previously in Welsh Alpine-type valley glaciers. Changes in the topographic relief are likely to have been produced by late Pliocene and early Pleistocene tectonic warping, with appreciable sinking of west Wales relative to central England. In the early stages of Pliocene glaciation (2.5–3.0 Ma B.P.) the Preseli Mountains may have been (relatively) much higher and more productive of snow and ice than they were after the Stonehenge glaciation of 2.47 Ma. A subsequent falling off in the size of the ice sheets and glaciers formed over the Welsh mountains is in keeping with the evidence of the $\delta^{18}O$ curve and the recorded glaciations. In terms of its depth of penetration into the central part of southern England the Stonehenge glaciation was probably the most extensive of all, but the distribution of the ice differed substantially from that of the Anglian–Wolstonian ice sheets which roughly equalled it in magnitude.

If the huge stones seen in Brittany at Locmariaquer, Dol and other places date from this early glaciation then the initial movement of some of the anomalous erratics in the regions adjacent to the Celtic Sea (Kellaway *et al.* 1975) could have been contemporaneous with the older glaciations, though later recycling of smaller boulders has undoubtedly taken place.

Physiographic evolution

The probable age of the Stonehenge glaciation as shown in Fig. 14.1 implies that it may have contributed substantially to the erosion of the Cretaceous rocks covering much of Gwent and the Bristol and Somerset coalfield as well as the Bristol Channel region. The downcutting of the ancestral rivers, like the proto-Usk and especially the proto-Wye may have received considerable impetus from the addition of melt-water. If the Stonehenge glaciation is the one recorded on the $\delta^{18}O$ isotope scale at 2.47 Ma then the Icelandic equivalent would be Flói-Holt I as defined by Einarsson & Anderson (1988, Fig. 2). This sets an older time limit for all the principal pre-Anglian glacial events of the region.

At this remote period the Cretaceous cover could have extended over most of Gwent, though the proto-Wye may have breached it locally. The two palaeodivides (Fig. 13.4) would still have been intact at the time of glaciation though the western divide may have been over-ridden by ice. The possibility that ice may have invaded the Bristol Channel simultaneously (Kellaway 1971) is not excluded.

Formation of the karstic depressions and red Clay-with-flints

Following the Stonehenge glaciation there is a period of 0.5 to 1.0 Ma, in which there is very little reliable information beyond that provided by the $\delta^{18}O$ curve. Within this long period of time during which the removal of parts of the Cretaceous cover was in progress and the stripping of the Triassic and Jurassic rocks proceeded in Gwent and the Bristol Channel region, the Lower Tertiary cover on the Chalk of Salisbury Plain began to break up as a result of the formation of large karstic depressions in which the older Eocene rocks (mainly Woolwich and Reading Beds, London Clay and Bagshot Beds) are preserved on Salisbury Plain and other Chalk downland areas. In adjacent areas of Normandy as at Valmont near Fécamp (Cavelier & Kuntz 1974) the deposits preserved in similar collapse structures include marine Pliocene sediments and the Sables de Lozère and there is a history of intermittent subsidence before a layer of red Argile à Silex (Clay-with-flints) covered the top of the collapsed material filling the sink.

On Salisbury Plain the evidence is less complete but in the Chalk downlands generally, the large solution structures are normally seen to pre-date the final spreading of the red Clay-with-flints over the Chalk terrain. The relationship of the Stonehenge erratics to the red Clay-with-flints is obscure as the Clay-with-flints is poorly developed and may be mainly residual in those areas where the bluestones were found and collected up by Neolithic man. Since the red Clay-with-flints normally consists of remoulded Eocene material and broken Chalk flints from which any residual Chalk has been removed by decalcification, it is possible that the feeble development of Clay-with-flints in the central area of Salisbury Plain reflects in some degree the erosion of the Eocene during and immediately following the conclusion of the Stonehenge glaciation. The contents of the Chitterne (Clay Pit Hill) pipe, as far as they have been determined, include white sands with flint pebbles which may be of Bagshot age. These include the 'Druid Sandstone' which Prestwich supposed might be the parent material of the large 'sarsens' of Stonehenge. This would imply that there might be a thin representative of the London Clay and the Woolwich and Reading Beds at greater depth. It is known that the pipe clay was worked in the pit but if any superficial pockets of bluestones or a thin cover of red Clay-with-flints existed in Neolithic times they have long since been destroyed by man.

Although the Clay-with-flints may have been formed by solution of the Chalk and in situ disintegration of the Eocene, its base contours do not coincide precisely with those of the undisturbed Eocene strata. This is clearly seen in some areas, notably at Seaford Head in E. Sussex, where the mantle of red Clay-with-flint is slightly transgressive and is accompanied by decalcification and collapse, due to the widening of joints in the underlying Chalk. Also significant is the inclusion of large masses of Greywether sandstone in the red Clay-with-flints as at Christmas Common near Aston Rowant, Oxfordshire (Plates 31 & 32) where the sandstones do not exhibit post-depositional hardening. This implies that they have not been exposed for a significant length of time to subaerial weathering. Some of the loamy or brickearth-like deposits associated locally with the red Clay-with-flints in Wiltshire contain hardened sarsens and boulders of Greywether sandstone. For the most part these fine-grained deposits post-date the formation of the red Clay-with-flints and their occasional interbedding with Clay-with-flints may be due to contemporary solifluction and resorting.

The presence of unhardened Greywether sandstone both in large karstic depressions containing disturbed Tertiary sand and in red Clay-with-flints in solution-widened fissures in the Chalk, indicate that the typical red Clay-with-flints is not a sedimentary deposit but has been produced by solution of the Chalk and subsidence of the Tertiary cover rocks. In view of the almost complete decalcification of the material including fresh broken Chalk flints the process is likely to have taken place at low temperatures in the presence of abundant melt-water. Any translatory movement is due primarily to subsidence or solifluction.

In most downland areas, including the Marlborough Downs and Salisbury Plain, there is little or no evidence to show that a continental-scale glaciation was responsible for the formation of the Clay-with-flints, though an ice cover would have provided a favourable low-temperature environment favouring dissolution of the Chalk. A possible alternative environment would be permafrost or a zone of very deep seasonal freezing. The development of permafrost with the subsequent establishment of a thermokarst regime combined with the solution of Chalk, or the development of a seasonal frost zone, would account for the presence of unhardened Greywether sandstone masses embedded in Clay-with-flints filling solution-widened fissures in the Chalk. There is, however, no evidence for the formation of contemporaneous frost wedges or large polygonal patterns in the red Clay-with-flints, and it is possible that the progressive disintegration of the Eocene cover under cold-temperate or sub-Arctic conditions with heavy winter snowfall, if maintained for a sufficiently long period, might have resulted in widespread collapse of the Eocene clays and silcreted sands into the underlying solution cavities. This would have been sufficient to protect the sandstone masses from atmospheric exposure. Many of these big sandstone masses have been found tilted at high angles and buried in Clay-with-flints, their subsidence having kept pace with the deepening of the sink into which the Tertiary strata have collapsed.

The high-level drift deposits of Bath

By the time that the deposition of the high-level Drift deposits of Bathampton Down and Farleigh Down took place, the destruction of the Mesozoic cover in Gwent must have been well advanced, though Cretaceous rocks still existed in the vicinity of Bathampton Down (Fig. 14.4). Varney (1921) supposed that a drainage route led eastwards from Bathampton Down to the Vale of Pewsey, but the writer inclines to the alternative view that it was probably directed towards Warminster (Fig. 14.4). At this time almost all the areas of Lower Carboniferous rocks west of Bath were still concealed by Mesozoic rocks though some of the more elevated areas of Old Red Sandstone in Gwent may have been bared. This accounts for the presence of Upper Old Red Sandstone and Coal Measures rocks in the high-level Drift deposit (Chap. 13).

All three rivers shown on Fig. 14.4 may have carried melt-water from Wales but the Monnow–Wye system may have been the dominant member. The dissection of the red Clay-with-flints on Salisbury Plain is likely to have been fairly well established by the time the high-level Drift deposits were laid down. The climate at this time is likely to have been cold or, at best, cold-temperate since there is evidence of considerable solution with the formation of caverns in the underlying Jurassic limestone on Bathampton Down at this time. Some of these may have been filled at least temporarily with ice.

The Northern Drift glaciation

At the time of the Northern Drift glaciation the Severn valley between Tites Point (near Purton) and Tewkesbury did not exist, or was a very shallow depression terminating some distance north of Tites Point, (perhaps near Gloucester) and draining northwards to join the Proto-Soar. This ancient river system is said to have originated on the northern slopes of Bredon Hill and continued northwards towards Leicester (Rice 1968). Work on the Wolstonian glacial deposits of the Midlands and the Severn–Warwickshire Avon terrace gravels shows that the southwest-flowing Avon which now joins the Severn at Tewkesbury did not exist in the early part of the Penultimate glaciation (Shotton 1977).

The Northern Drift gravels are not known to have crossed the eastern palaeodivide (Fig. 13.6) which was in existence at this time. The western promontory of the Cotswolds at Stinchcombe Hill near Dursley therefore formed a wider and continuous pre-Anglian barrier across the line of the existing Severn estuary at Tites Point, severance having taken place at a later date possibly due to rifting and fissuring by movement along the Malvern Fault Zone and the transverse Denny Island Fault which cuts it. The eastern side of the Malvern Hills and the Cotswolds was therefore drained, either northwards to the Proto-Soar or eastwards towards the Thames and London basin. Only the Palaeovalley was drained in a southeasterly direction towards Wiltshire.

Post-Beestonian–pre-Devensian glaciations

Of the remaining pre-Devensian glaciations there are, at present, no reliable indications of a Beestonian ice sheet in the Wessex–Severn Estuary region. The material carried by the Northern Drift glaciation was marked by the movement of much Permo-Triassic ('Bunter') material and in this sense the Northern Drift may mark an early stage in the phased movement of rocks from the exposed Permo-Triassic formations of the west Midlands and Welsh borders towards eastern England. Evidence for a direct relationship between the Northern Drift of the Cotswold and the Kesgrave gravels of Essex is lacking and the presence of other high-level deposits (and even glacial erratics) on the Berkshire downs, and the uplands adjacent to the Thames valley, indicates the possibility of other ancient ice masses having been present in the intervening areas. In addition, there is a distinct possibility that the Beestonian ice spread in a different direction from that taken by the Northern Drift, the evidence for this movement having been largely expunged by the later ice sheets which traversed the Midlands in Wolstonian times. The view taken here is that the Northern Drift is older than the Beestonian (Fig. 14.1).

For the sake of comparison the positions of some of the classified Icelandic glaciations (Einarsson & Albertson 1988) have been shown in Fig. 14.1. These appear to correlate well with the later British glaciations, but the period between the deposition of the Northern Drift and the Cromerian interglacial which is presumed to have succeeded it, is one for which there is little or no reliable local evidence. The immense changes which produced the Severn–Warwickshire Avon drainage system took place almost simultaneously with the breakthrough of the Bristol Avon by way of the Clifton gorge and this aspect is therefore briefly discussed below. There are, however, problems in relation to the status of the Wolstonian glaciation (Rose 1987), and this could affect the dating of the changes in the Severn drainage. It is also possible that unsuspected tectonic movements along the Denny Island Fault Zone may have affected stratigraphical interpretations.[1]

[1] There is some evidence for changes in the direction of ice flow in central England in Anglian times which may indicate that basement movements were in progress in Pleistocene times. Thus the Lower Boulder Clay (? Anglian) underlying the Chalky Boulder Clay of the S.E. Midlands (Hollingworth & Taylor 1948) contains rocks derived from areas to the North and N.W., while the Chalky Boulder Clay contains hard Chalk from Yorkshire and Red Chalk from Lincolnshire. Hard white Chalk and Red Chalk were conveyed as far west as the N.E. Cotswolds (Stretton-on-the-Fosse) by (? Wolstonian) Chalky Boulder Clay ice, so the change in direction of flow affected by a very large part of Central England. It is possible that these changes may be related to others recorded in the English Channel and North France, but much more evidence would be required in order to establish this relationship in sufficient detail.

The mid-Quaternary evolution of the River Severn

In pre-Anglian times the Proto-Soar drained northwards towards Leicester and the Trent valley (Rice 1968; in Shotton 1977, fig. 4) but was blocked by Anglian or Wolstonian ice and only partly re-excavated in post-Wolstonian times when the SW draining Warwickshire Avon came into being. By late Wolstonian times, however, water was flowing through a narrow gap between the Cotswolds and the Forest of Dean at Purton and Alvington with its floor at an altitude about 200 ft (60 m) A.O.D. At this level there are a number of flat-topped hills south of Berkeley on which small boulders of Upper Old Red Sandstone and other rocks have been found resting on Lower Lias, possibly indicating fluvioglacial or glacial action. In the succeeding period the erosion of the soft red Triassic marls and the even softer Lias clays and sands of the Severn valley proceeded with great rapidity, partly as a result of Devensian downcutting by water temporarily ponded above the Ironbridge Gorge. The great size of the valley above Tites Point near Purton is largely a reflection of the rapidity with which downcutting of soft rocks was achieved by the Severn and its tributaries. The older river gravels of the Severn below Gloucester (BGS Sheet 234) tend to be restricted to the western side of the main valley. Water first crossed the Palaeozoic rocks west of the Berkeley–Hagloe Fault at Purton and Sharpness where the Woolridge Terrace was laid down and any structural severance of this barrier must date from before this time. By Main Terrace times the greater part of the valley at Gloucester was covered by braided streams, but flow was restricted below Tites Point, limiting the subsequent development of the big meanders which formed upstream in late Devensian times. In view of the magnitude of the Wolstonian–Anglian glaciations, it is not difficult to envisage a barrier in the Severn valley above Tidenham in early Pleistocene times. Below Tidenham, however, we have to reckon with the presence of the ancient River Wye which was already well developed in Anglian times and able to receive water from the Severn and Bristol Avon.

The existence of the Woolridge terrace (Wills 1938, p 171) at Alvington and Purton on the west bank of the Severn north of Tidenham is sufficient indication of the existence of the Lower Severn in late Wolstonian times. But the terrace disappears completely a short distance south of Alvington and is not seen again. Some other patches of gravelly material classified by Welch & Trotter (1961) as river gravels occur on the west bank at lower altitude (Sheet 250). Terrace 4 of this group has been correlated with the Bushley Green terrace of the Severn and this may be correct, but the only continuous terrace linked to the Severn sequence between Magor and Purton (Tites Point) is Terrace 2 which is correlated by Welch & Trotter (1961, p 133) with the Kidderminster Terrace. This can be followed up stream to the Severn valley, extending from Tites Point to Gloucester (one inch Sheet 234) where the position of the Kidderminster Terrace is reasonably well established.

On the east bank of the Severn none of the Severn terraces has been found between Berkeley and Shirehampton. This may have resulted partly from later erosion. However, on the west bank between Woolaston and Sedbury Cliff where the western Palaeodivide is considered to have been intersected by the Severn (Fig. 13.6), there are no terraces older than the Kidderminster Terrace, which maintains an almost uniform but very low gradient (*c.* 1 ft per mile) and is reasonably well defined as far east as Magor (8 miles SW of Chepstow) where it has fallen to the level of the estuarine alluvium. West of the Magor it is concealed by younger estuarine deposits. On the east side of the estuary the correlative of Terrace 2 of the Chepstow area is the so-called '50 foot Terrace' of Shirehampton, of which very little is known. However, if it is to be correlated with the Kidderminster Terrace of the Severn then the evidence for the formation of the Avon Gorge can be fairly used to claim that the completion of a well graded channel in the gorge was finalized in Ipswichian or late Wolstonian times.

Of the terraces or terrace fragments on the west bank of the Severn above Tidenham, Terrace 4 may be the Bushley Green terrace of the Severn. Nevertheless, apart from Terrace 2 (Kidderminster Terrace) an altimetric correlation of the 'gravels' below Tidenham is of questionable value. The figure given by Welch & Trotter (1961) for the grading of the Kidderminster Terrace cannot be applied to the fragmentary deposits preserved around Beachley and the country extending westwards to Llanvache. Dr Brian Hawkins who has examined these deposits has observed very large sandstone boulders (about 1 ton in weight) preserved in the gravels near Chepstow and there are other indications that the so-called '100 foot Terrace' (Terrace 3) like its altimetric equivalent at Shirehampton may be of glacial or fluvioglacial not fluviatile origin. There are still many unsolved problems in this area, one being the origin of the deposit of pebbles containing abundant glacial erratics including igneous rocks, proved in the eastern entrance to the Severn Tunnel and by borings at Severnside (Leese & Vernon 1961). To Dr Hawkins and the writer it appears increasingly likely that the glacial features of the Severn estuary and the adjacent parts of the Bristol and Somerset Coalfield and the western Mendips are the product of at least two glaciations. In the younger one the glacial deposits of the low-lying area at Kenn were laid down by ice (? Wolstonian) invading the Nailsea Basin and carrying rocks from South Wales. To this may belong the morainic deposit at Stanton Drew which is composed of Old Red Sandstone, Triassic conglomerate, Lower Lias breccia and Dundry Freestone, all having moved in an easterly direction along the Chew valley. The glaciation which produced this moraine is clearly pre-Devensian and may be of Wolstonian age. The more elevated Court Hill Channel at Clevedon and the associated deposits of east Clevedon and the Flax Bourton gap may have been formed by ice loaded with sand from the floor of the estuary and may be of (? Elsterian) or early Anglian age.

Appendix 14.1

Etymology of Stonehenge: the nature and origin of sarsens and bluestones

Many of the names applied to geographical features in Wessex are of Celtic or Romano-British origin and have survived into modern times, but the ancient earthworks, megalithic monuments and older English settlements generally carry names of Saxon origin. Some of these have undergone drastic changes which are difficult to understand, thus the Saxon name for Avebury was *Weala–dic*, the dyke of the Britons (F & L Vatcher 1976). There is substantial evidence for the Saxon occupation of Avebury commencing about the middle of the 6th Century. It was at this time that the Anglo-Saxon invaders first saw the famous stone circle on Salisbury Plain which they called *Stan–henge* (Stonehenge), a name by which it has now become familiar throughout the English speaking world.

By the 13th Century *Weala–dic* had become *Waleditch*. Subsequently, perhaps during the Wars of the Roses, the name ceased to be used. In the 16th Century Leland failed to visit Avebury though he referred to it in his *Itinerary* as *'Aibyri'* and in later years it became known as Abury or, as John Aubrey spelt it in the 17th Century, Aubury. The curious fact is that though the burh continued to be occupied from Saxon times onwards, the monument was virtually unknown to the outside world when John Aubrey came upon it while out hunting in 1648. He encouraged King Charles II to inspect it and attended the King and his brother, later James II, when they visited Avebury and Silbury Hill in the autumn of 1663.

Stonehenge, though situated on what was formerly an open stretch of Salisbury Plain, has been a constant source of interest and speculation and has retained its Saxon name. As will be seen, it may never have been completely forgotten even though local activity was restricted to proceedings such as scraping the stones for the preparation of medicinal remedies,[1] or viewing the rising of the sun over the Hele (or Heel) Stone at the Summer Solstice.

[1] Celia Fiennes visited Stonehenge between 1685 and 1696 and saw the stones being scraped. Owing to atmospheric pollution the stones are now bare, but in former times they were lichenous.

1 Stonehenge. The derivation is from Anglo Saxon and given by O.E.D. as (OE) *stan* and *henge* as in *henge-clif*. The word *hengi* is used in Iceland where it is applied to sea cliffs which have been strongly undercut by marine erosion; the sense is of an overhang or something suspended in the air. This is a reference to the lintels which distinguish Stonehenge from other surviving stone circles. The name Stonehenge is therefore purely descriptive, having no reference to the original function of the monument. It has become customary to speak of stone circles as 'henges' but this is only justified where there is evidence that upright stones or wooden posts were supporting lintels.

2 Sarsens or Greywethers. The word *Sarsen* is usually applied in Wessex to residual masses or boulders of silcreted Tertiary Sandstone. In some areas boulders of Cretaceous sandstone have been called 'Sarsens' but the typical sarsens of the Chalk downlands are residual silcreted (i.e. silica-cemented) sandstones, usually of Woolwich and Reading, Bagshot or Barton age. The greatest accumulation of sarsens is seen on the Marlborough Downs at the present day (Brentnall 1946; Clark *et al*. 1967; Bowen & Smith 1977).

O.E.D. gives the derivation of sarsen as a variant of *Saracen*, but this is questionable. The name *Saracen* (or *Sarracene*) was applied to a Middle Eastern people and was in use as long ago as 893 A.D. It appeared in 1644 in the diary of Richard Symonds (ed. Long, 1859, p. 151) where it is applied to the Tertiary sandstone boulders of Fyfield on the Marlborough Downs. Symonds also recorded the name Greywethers (mis-spelt as 'Greyweathers'), a name formerly used in the Wessex downlands where the sarsens (or boulders) scattered on the hillsides commonly bear a strong resemblance to a flock of sheep.

The supposed connection between *Saracen* and '*Sarsen*' is said to be through the use of the sarsens in constructing 'heathen' megalithic monuments which were regarded as 'Saracens' or 'strangers of unfamiliar or ungodly nature'. The explanation has a pedantic flavour which ill-accords with the very down-to-earth usage of *sarsen*, as for example by ploughmen. The term is by no means restricted to the vicinity of the stone circles, and is applied to half-concealed and buried boulders in areas where large stone monuments do not exist. Strictly speaking the word sarsen has no precise lithological significance, though it is usually restricted to siliceous boulders.

In his 'Natural History of Wiltshire', written after the Civil War (ed. Britton 1847, p 44), John Aubrey refers to *'Sarsden'* or *'Sarsdon-Stones'* as being hardened by exposure to the weather and mentions the traditional method of breaking them up by the use of fire and water. By this means, or by excavating the sarsens and dragging them away to make up field boundaries, or for road metal, large areas were cleared for cultivation. Residual sarsens of Tertiary sandstone are found in Jurassic terrain at a considerable distance from the Chalk downlands, as for example at Nunney Combe near Frome and at Rode near Trowbridge. Some of these have been carried into their present position by solifluction. Others occur in coastal areas of Sussex in high-energy marine gravels of Ipswichian age, or elsewhere as erratic boulders in till, as in the Chalky Boulder Clay north of London.

The sarsens most dreaded by agriculturalists are not the plainly visible Greywethers which can be a useful source of stone for building and road-making in an otherwise stoneless area, but the half-concealed or buried sarsens that constitute such a serious obstacle to ploughing and are a source of danger when cutting undergrowth with a scythe. Hence the probable Saxon origin of the name, *sár* meaning sore, grievous or distressing, and *stan* a stone. Sarsens, or boulders of silcreted sandstone, once hardened by weathering, either on the surface or at shallow depth in the ground, are almost indestructible and some of them have become so hard that, to quote Aubrey, "no toole can touch them". Such intensely hard silica rock could not have been cut, shaped and perforated to make and fit the lintels of the Sarsen circle or the horseshoe trilithons of Stonehenge. Indeed some of the lintels provide evidence of erosion by rainwater runnels or meltwater channels which formed while the sandstone was still soft.

H. B. Woodward (1887, pp 448–450) gives a short but useful account of Greywether sandstone indicating its distribution and use, pointing out that in Surrey the unweathered sandstone 'is friable when first excavated, but it hardens by exposure'. As illustrating the extent to which it has been used for building purposes he observes that 'much of Windsor Castle is built of Greywether sandstone'. In Surrey most of the large sandstone masses found in the Aldershot area are of Barton sand age (Dines & Edmunds 1929, 121, 123); a useful account of the working of the sandstone in the Chilterns was given by Davies and Barnes (1953).

The sarsen fields of Wiltshire were described in the Geological Survey Memoir (Sheet 34) by Ramsay *et al* (1858). In this account attention is drawn to the presence of masses of 'Grey Wethers' far out on the lowlying surface of the Jurassic rocks near Swindon, the conclusion being drawn that the Eocene strata formerly extended northwest of their present limits on the Marlborough Downs.

Even at this early date it was recognized that 'the Grey Wethers belong to the more compact silicious patches of the Lower Sand Tertiary Strata' and that *'possibly they further hardened on exposure to external influences'*. In the light of repeated references by writers from John Aubrey in the 17th Century onwards to this characteristic weathering feature of the Greywether sandstone, the assumption that the trilithons and 'sarsen' circle of Stonehenge were built by pounding away at weather-hardened rocks ('harder than granite') brought by man from the sarsen fields of the Marlborough Downs to Salisbury Plain, is seen to have distorted subsequent conclusions relating to the source and treatment of the Tertiary sandstones used at Stonehenge.

It can be difficult, if not impossible, to distinguish between isolated weathered glacial boulders of Greywether sandstone and those of non-glacial origin. Glacially transported sarsens are found in Chalky Boulder Clay in the home counties north of London and some of these are striated. The Greywethers on Overton Down in Wiltshire have probably been directly derived from the distintegration of the Tertiary cover of the Chalk. Those seen on the lower slopes and in the lateral valleys at Fyfield and Lockeridge have been soliflucted into their present position in Devensian times. The streams of Chalk mud in which they moved downhill have long since been destroyed by weathering or solution, so the boulders remain as a post-Devensian lag deposit. As the Kennet valley is followed eastwards towards Marlborough the large hardened silcreted lumps become less and less common and smaller rounded sarsens, some no larger than a football appear in the stream bed.

There is, therefore, no simple recipe for determining the history of isolated sarsens or groups of sarsens stranded on Chalk downland without making reference to the material in which they were once entombed, or the evolution of the land surface on which they now rest. On the other hand, there is a very clear distinction to be made between hardened Greywether sandstones or sarsens which have been exposed to atmospheric weathering and the softer patchily silcreted Tertiary sandrocks (both in situ and in solution cavities in the Chalk), which have not undergone post-depositional hardening by atmospheric weathering. This distinction is fundamental to an understanding of the difference in the origin and use of the Tertiary sandstones at Avebury and Stonehenge respectively.

At Avebury all the Greywether sandstone masses used in the monument belong to the group of post-depositionally hardened silcretes. It was suggested by F & L Vatcher (1976, p 42) that the hardened rocks used at Avebury and found scattered over the Marlborough Downs are the product of the formation of a single Tertiary duricrust. It is not possible to pass over a reference of this kind which encapsulates a concept which could further confuse the archaeological approach to some of the rocks used at Stonehenge.

When Lamplugh defined the term silcrete in 1902 he applied it to the siliceous sandstones of which there are a number known in the Eocene succession in northern France and southern England. Although there are minor differences in the silcretion process which are reflected in the petrology of the silcreted sandstones, the general distribution and character of these masses shows that patchy silcretion was a very late-diagenetic or post-diagenetic one due to lateral migration of silica in solution, generally under hot, possibly tropical conditions. The silcretion of the Greywether sandstones took place in some cases before jointing was completed but locally the deposition of silica continued after the formation of the joints so that thin linings of chalcedonic silica were formed. Atkinson (1979, p 19) records the occasional presence of these linings in the Stonehenge sandstones which he describes as 'patches of a natural siliceous cement, like a thick layer of amber lacquer' which lines the joints.

As seen in sections in the Woolwich and Reading beds of Hampshire and Sussex and the Thanetian sands at the base of the Eocene at Dieppe, silcretion is clearly post-depositional and shows considerable lateral variation in intensity. In some respects it might be described as 'concretionary' as some of the silcreted masses of sand are highly irregular in form. The section in the basal Eocene at Dieppe illustrates the relationship of the sandstone masses in the Thanetian sands to the underlying Chalk and the overlying Eocene strata (Pomerol & Feuguer 1968, 121–124; figs 61–3). At no time however has the writer observed any alteration of the underlying Chalk or the basal Eocene strata, by lateritization or mass silicification or the formation of a pallid zone comparable with that of the duricrust of central Australia. It is possible that the application of the purely descriptive term silcrete to the South African duricrust (Lamplugh 1907) may have led to a tendency to treat the terms silcrete and duricurst as interchangeable. Both the Greywether sandstone of the Marlborough Downs and the silicified and hardened cap rock of the Australian duricrust may be silcretes, but this does not prove that the Greywether sandstone is a duricrust. The Australian duricrust is an unmistakable feature when seen from ground level or the air. It consists of a reddish brown hard siliceous cap rock which is the product of long continued weathering and chemical processes which are no longer operating in the region. It crowns a deep weathering profile including a very striking zone of pale kaolinized silica-depleted rock which forms the 'pallid zone' (Lanford-

Smith 1978); the superficial silicification affects a wide range of rocks from shale or clay to basalt and even granite. If anything comparable with the pallid zone ever existed beneath the Bagshot Sand or Woolwich and Reading Beds, it had disappeared before the sands were deposited and this process must therefore have long pre-dated the normally weak, patchy silicification of the Greywether sandstone. The same provision applies to other sandstones in southern England. Silicification of the sand was essentially due to the lateral migration of groundwater through the sediment and, as Senior & Senior (1972) have indicated, this process can take place at several levels within a sedimentary sequence, but the silcretes formed in this way are not in a strict sense duricrusts.

The fact is that the final atmospheric hardening of the Tertiary sandstones of southern England and northern France has taken place (and still takes place) under temperate conditions long after their initial silcretion.

At Avebury the monument was entirely constructed of naturally shaped weather-hardened sandstone masses of local origin. The sarsen boulders of the station stones and the Heel Stone, are also weather-hardened local sarsens. On the other hand those employed in the sandstone circle and trilithons of Stonehenge III were unweathered stone, recovered from collapsed masses of Tertiary sediments preserved in solution cavities in the Chalk. Only the Hele (or Heel) Stone and, possibly, the Station stones can be regarded as being naturally weathered boulders which have not been dressed by hand. The other sandstone masses used at Stonehenge may retain one or more of their natural joint or bedding faces but the majority have been worked or trimmed by hand. The working of the stones including the accurate shaping of the lintels of the outer circle and the jointing and fitting of the lintels of the trilithons in the inner horseshoe were carried out before the final hardening of the stone took place. Most of the worked sandstones were probably excavated from one or more karstic depressions in the Chalk, possibly including the one described by Prestwich (1898) at Clay Pit Hill near Chitterne. It is generally supposed that the Avebury sandstones are of Woolwich and Reading beds age but it is possible that like those used at Stonehenge they may be Bagshot or even Barton in age.

The siting of the Stonehenge monument may have been determined by the presence of the Hele Stone which was probably adapted for use before 2250 BC in its natural site (Stonehenge I). To convert it to a gnomon all that would be required would be to raise it into a vertical position. Its present inclination has resulted from subsequent movement due to solution, thermal expansion and contraction, and frost action on the Chalk. The top may have been roughly shaped but is too strongly weathered to provide any evidence. Analysis of the construction of Stonehenge I shows that the Heel Stone has been moved eastwards from its original position in response to the eastward movement of the rising sun (Fig. 14.A1). As the stone weighs about 35 tons this was a difficult operation and accuracy of siting and levelling may not have been fully achieved.

The presence of engravings of axe heads on some of the uprights and trilithons suggests that finely pointed bronze or flint tools were used to engrave the surface of the sandstones soon after their extraction for use in Stonehenge III and before hardening was completed. Subsequent weathering has subdued the outlines but has not been able to remove them completely. The movement of these big masses of partly consolidated sandrock would have required care but it is possible that both wooden rollers and rolling stones were used to achieve this. Rounded masses of sandstone found in stone holes have mostly been described as 'mauls' used for dressing the stones, but the larger ones may have been rolling stones. These are reminiscent of the stones used at Ggantija (Gozo).

At the present time there are relatively few Tertiary sandstone boulders visible on Salisbury Plain. Some may be seen built into the corners of old cottages and others are scattered over the Plain, though not on the scale seen on the Marlborough Downs. It is not certain whether this is, or is not an original feature. Salisbury Plain has been cultivated almost continuously since Neolithic times and this could only have been achieved by clearing away such boulders as may have made the land difficult to plough. Large numbers of sarsens were found in Bowls Barrow yet there are no sizeable sarsens to be seen locally and we must assume that they have since been used and cleared away. Somewhat similar evidence comes from other regions including the South Downs where sarsens are extremely rare on the seaward slopes. However, many large Tertiary sandstone boulders are preserved locally in the Brighton raised beach deposits and the overlying Combe Rock, showing that they must have been present in substantial numbers in Ipswichian and Devensian times.

Regarding the known distribution of disturbed Tertiary sediments preserved in major karstic solution cavities in the Chalk, the lack of surface evidence is even more striking. In the excavations on the M40 at Christmas Common near Aston Rowant, Oxfordshire, one complex cavity in the Chalk filled with red Clay-with-flints yielded three masses of Tertiary sandstone, one measuring $21 \times 11\frac{1}{2} \times 4$ ft ($6.4 \times 3.5 \times 1.2$ m), the same length as the Slaughter Stone of Stonehenge (see Plates 31 & 32). The others were of roughly similar volume and the sandstone was sufficiently soft to be

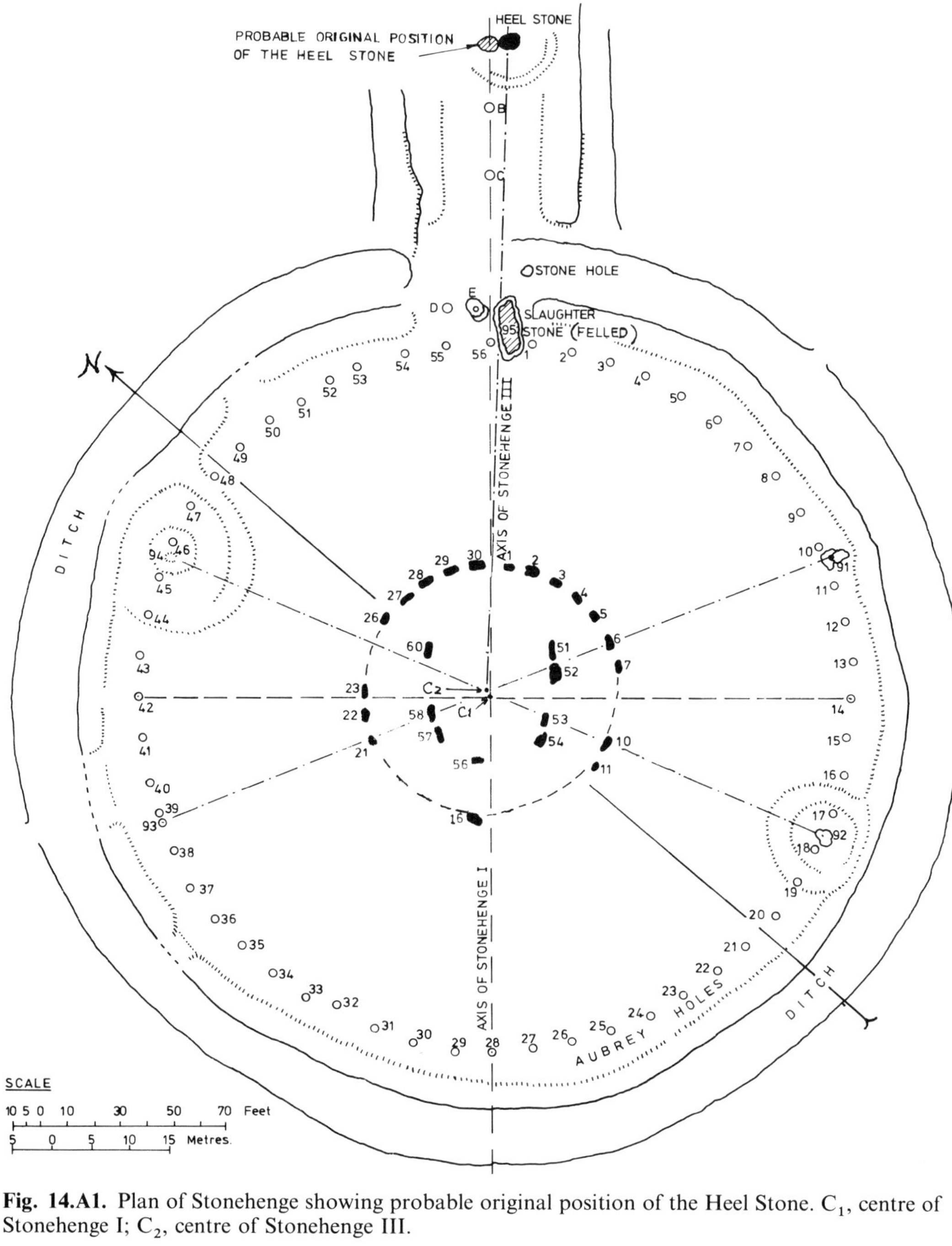

Fig. 14.A1. Plan of Stonehenge showing probable original position of the Heel Stone. C_1, centre of Stonehenge I; C_2, centre of Stonehenge III.

grooved with the aid of a bronze coin or cut with a piece of freshly fractured flint.[1] Many of these big sandstones preserved in solution cavities in the Chalk of the Chilterns are situated in areas where no sarsen is now visible on the surface. The big karstic depression at Clay Pit Hill, Chitterne (roughly 100 m in diameter and of unknown depth), would have been capable of supplying very large

[1] The appearance of stones 16 and 52 (Atkinson 1979, figs 5, 23B and 14) suggests that the soft friable sandstone was dressed before it hardened using a stone, flint or bronze adze. The presence of these markings on the side of Stone 52 shows that the dressing must have been done either during or before the construction of the Greywether sandstone circle.

quantities of silcreted sandrock for use at Stonehenge. Clay Pit Hill is about 15 km from Stonehenge, but it is possible that other concealed sources of sandstone may have been worked nearer the site of the monument. Once they have been back-filled with sand, flint rubble and chalk debris, ancient excavations can be very difficult to locate.

It is generally accepted that there is no field evidence to show that the curving avenue which links Stonehenge to the River Avon was used to convey heavy stones to the Stonehenge site. Avebury, Stonehenge and Stanton Drew all have avenues connecting the monuments to streams or rivers while circles at Marden and Durrington were constructed in direct contact with streams. It seems therefore that running water formed an important element in the activities taking place at all these sites and that avenues were constructed for cultural or ceremonial purposes.

3 The Heel and Slaughter Stones. Only two of the 'Sarsens' (or Tertiary sandstones) of Stonehenge are named. They are the Heel Stone, already mentioned as being a natural sarsen and possibly the key to the siting of the monument, and the Slaughter Stone, the survivor of a pair of stones situated between the Heel Stone and the outer circle. A third named stone, the Altar Stone, is a flat-lying slab-like mass of Palaeozoic micaceous sandstone which is one of the 'foreign stones' (or bluestones) described below.

There is considerable doubt as to the origin of the name Hele Stone (given as Heel Stone by some archaeologists who doubt the validity of the alternative form). The name 'Slaughter Stone' has also been questioned, as it is generally accepted that there is no archaeological evidence indicating that this stone was ever used for sacrificial purposes. The Slaughter Stone is a large stone about $21 \times 7 \times 3$ ft ($6.4 \times 2.1 \times 0.9$ m) which now lies prostrate in a shallow pit cut in the Chalk. It is one of a pair of stones which formed twin pillars at the approach to the monument (Atkinson 1979), the second stone being represented by stone hole E. It is considered that the twin pillars were not capped by a lintel. A line drawn from the present position of the Heel Stone to the centre of Stonehenge III does not correspond to the central axis of Stonehenge I. This suggests that the Heel Stone has been moved from its original position in response to an eastward shift in the position of the rising sun at the Summer Solstice (Fig. 14.A1)[1].

The spelling of 'Hele' was dismissed by Atkinson (1960, pp 29–30) as "an archaizing affectation of earlier antiquaries" and he gives its full name as the 'Friar's Heel' though he goes on to say that "the legend that the Devil threw the stone at a Friar, striking him on the heel, whose imprint can still be seen on the outer face of the stone may well be a legend invented to explain the name, for which there is no other known derivation". The implication that the name Hele (or Heel) Stone may be much older than this legend could well be true if, as the names suggest, the Heel Stone and the two Slaughter Stones both bear names of Anglo-Saxon (Old English or Old Norse) origin.

The greatest confusion regarding the name of the Heel Stone has arisen from Aubrey's obscure reference to the application of the name Heel Stone (or Friar's Heel Stone) to a fallen stone (14) on the west side of the monument (Newall 1959). That the name Heel Stone was 'transferred' in some way from this fallen stone to the most isolated and conspicuous stone of the monument, seems highly improbable. The superstitious tale told to John Aubrey, which he duly recorded, may well have been prompted by the appearance of an impression like the sole of a human foot (the so-called 'Friar's Heel') on Stone 14, and may have led him or his informant to confuse this with the original name of the Stone. There is however no obvious or logical reason why such an indentation on fallen Stone 14 should be described as a 'Friars Heel' and then 'transferred' to the Heel Stone[2].

Of all the stones in the monument the Heel Stone is the one most likely to have been identified individually by name. No other proper name has, as far as the writer is aware, ever been applied to it. Other isolated sarsens lying on Salisbury Plain also acquired traditional names. They include the ancient Cuckoo Stone (or Cuckoldstone) near Durrington.

Although the position and inclination of the Heel Stone is such that a highly accurate observation of sunrise may no longer be possible, it is clear nevertheless that it would have served to mark the rising of the Midsummer sun. Photographs showing sunrise over the Heel Stone on June 20, 1964 have been published by Gerald Hawkins (1970). The absence of one Slaughter Stone, and the deliberate downthrow of the surviving one, makes it impossible to assess the width of the original

[1] The axis of Stonehenge I (Fig. 14.A1) may have been determined by simple methods long before the construction of the Aubrey holes, using the Hele Stone and its shadow at the summer and winter solstice. This may account for the NE continuation of the straight part of the Avenue beyond the Heelstone.

[2] In some lights the Heel Stone shows a large natural indentation, vaguely resembling a giant footprint, but there is no obvious reason for calling this a Friar's Heel.

space between them, but since there was no lintel, they could, at an early stage (possibly before Stonehenge II), have constituted a kind of slot or backsight through which the sun was viewed rising over the Heel Stone.

The name 'Hele' may have been derived from the Anglo Saxon (or OE) *helan*—to shadow, or, *hele*—to cover, while the two Slaughter Stones may derive their name from OE *sleaght* which implies a sudden blow or *a sudden flash of fire or lightning*. In modern English the similar-sounding word slaughter (ON *slahtr*) is associated with the idea of violent or sudden death, but here it would have been the first flash of light from the rising sun which was the significant factor. Such a unique combination of names with Anglo-Saxon roots, linked in this fashion with the Heel Stone over which the midsummer sun is known to rise, is not mere coincidence. As will be seen below, an Anglo-Saxon origin for the name 'bluestones' is also indicated but this, like the Saxon name of Stonehenge (*Stanhenge*), carries with it no functional significance.

The writer prefers the derivation of the name Heel Stone by way of *hele* (OE) or *hela* (Old Frisian). The Wessex Saxons, still followers of Woden when they arrived in Wiltshire, would not have hesitated to substitute *Sunne* for *Helios* if the Romano-British people had been applying this word to the Hele Stone. The West Saxons had close ties with the Frisians and the use of the term *hele* in the older accounts of Stonehenge may not be as fanciful as has been suggested. Other closely related forms such as *helan* and *hael* (OE) are derived from the same root. OED gives their meaning as to hide or conceal.

The significance of the name lies in the fact that the centre of Stonehenge I was so placed in relation to the original situation of the Heel Stone that the latter controlled the direction of the axis of Stonehenge I (Fig. 14.A1). The two Slaughter stones, which were not linked by a lintel, could have formed a sort of backsight in which the Heel Stone (or gnomon) was visible with the midsummer sun rising above it. With the passage of time the easterly migration of the rising sun might have led to the felling of the eastern Slaughter Stone. The western one (Stonehole E) would have ceased to function. Once the sarsen circle of Stonehenge III had been built, the rising sun and the Heel Stone would have been visible in the gap between stones 30 and 1 of the sarsen circle. The western stone (E) may well have been left standing and could have survived to be seen by John Aubrey in the 17th Century.

The potential interest attaching to the origin of the names of the Heel Stone and the Slaughter Stone(s) is that they may imply that the *function* of these stones was at least partly understood by the Saxons. The suggestion would be, therefore, that this knowledge is likely to have been passed on to them by the Romano-British inhabitants with whom they came in contact after the Saxon invasion and conquest of Wessex. There is, however, nothing in the etymology of the Stonehenge names to suggest that there was any contemporary knowledge of other possible astronomical functions of the monument.

4 The Bluestones. The foreign stones or bluestones of Stonehenge (Judd 1902; Thomas 1923; Kellaway 1971) have attracted so much attention that only a brief description is necessary. As a result of recent studies (Thorpe *et al.* in press) the sources of the Bluestones have been reassessed. They are known to cover a much wider area than was formerly believed, possibly including North Wales (Appendix 2).

With the exception of the Altar Stone, the visible bluestones are igneous or metamorphic rocks of Lower Palaeozoic age with their ultimate source in Wales. They include hard ophitic dolerites which can be closely matched with those of Mynydd Preseli in Dyfed (Pembrokeshire). Many of the softer schistose or sheared bluestones which were originally visible above ground have broken down or disintegrated and are now known only as stumps or chippings which have been revealed by excavation. Some of the harder stones may be frost-severed pieces. Others are natural boulders, some having well marked zones of weathering, while a few have been striated and were therefore regarded by some archaeologists as being of glacial origin. The wide range of sizes suggests that the material is essentially morainic though it may have been outwash carried by fluvioglacial streams.

A boulder of dolerite of Preseli type weighing about 3/4 of a ton[1] was found in 1801 by William Cunnington in Bowls Barrow near Heytesbury (R. H. Cunnington 1975). When the long barrow was opened it was found that a great heap of stones had been piled on the floor 'in form like the ridge of a house'. Cunnington reported that 'the large stones came rolling down so fast upon us that we were obliged to desist from exploring further'. He carried away ten of the stones of which one, now in Salisbury Museum, proved to be a bluestone 'ye same to those of the upright stones in ye inner circle at Stonehenge'. The boulder is dolerite, part of the original outer surface is preserved and this is not weathered white but shows an abraded surface (Plate 33). The boulder has been damaged, either

[1] Owing to a typographical error this was given as '9' instead of 'a' ton (Kellaway 1971).

accidentally or in an attempt to break off pieces. It was taken from Bowls Barrow to Cunnington's house at Heytesbury where it stood for some time in his garden and was later moved to Salisbury Museum. Its unweathered condition suggests it may originally have been enclosed in a chalky deposit.

There is nothing in Cunnington's account or in the appearance of the boulder to suggest that this is a worked stone or that it had any special place in the burial mound. From the appearance of the rock it is likely that the damage is largely due to its treatment after its extraction from the long barrow (Plate 33). Had it been fractured before its removal, dark coloured dolerite would have been identified at once, but it would appear that it was not recognized until after it arrived at Heytesbury in company with nine other boulders, presumably all sarsens. To argue that, because bluestone boulders cannot now be found on the ground in the vicinity of Bowls Barrow, this dolerite boulder was therefore conveyed by human transport from the Preseli Mountains, is as absurd as to insist that the associated sarsens were carried to Heytesbury from the Marlborough Downs. The probability is that the boulders, sarsens and bluestones alike, were collected or dug up in the vicinity of the long barrow and were then piled up as Cunnington described them. There is nothing to show that the bluestone boulder occupied any special position in the heap or that it was necessarily the only one of its kind in the Neolithic long barrow.

If, like *Stan-henge*, the name *bluestone* is of Old Norse origin then its equivalent would be *blá-stan* and I am indebted to Dr Sigrun Gudmundsson who first drew my attention to the idiomatic use of the word *blá* (or *blau*) in Iceland, where it is employed not only as the name of a colour, but also to indicate a person or object of strange, or exotic character. Thus an African with a black skin would have formerly been very conspicuous in Icelandic society and might therefore have been described colloquially as a '*blau-man*' [1] or *blá-maðr* (plural *blámenn*). The equivalent English word is *bloman* (ME), now obsolete.

The Altar Stone is a pale greenish-grey micaceous rock of upper Silurian or Lower Devonian age not found in the Preseli Mountains but the slight sparkle of the mica would certainly qualify it to be considered as one of the bluestones in the sense indicated above. The general rather uniform colour of the sandstones which form the larger part of the monument serves to accentuate the singularity of the far-travelled bluestones. Was it the idiomatic use of the word *blau* which led Geoffrey of Monmouth (1136) to suggest that the stones of Stonehenge are African?

5 *The Altar Stone.* Although the Altar Stone appears to be a flat-lying 'bluestone' at the centre of the sarsen circle, it is generally considered to have been an upright stone which has collapsed (Atkinson 1979). The derivation of *Altar* is given by OED as L. *altere* (first recorded use 1000 AD) and it is unlikely that it would have been used by the 6th Century Saxon invaders. The date of the collapse of the Altar Stone is not known.

6 *The Station Stones.* Of the remaining stones only the four Station Stones are identified by name, the collective name being a term used by surveyors. The Station Stones are said to have considerable astronomical significance (Newall 1959; Hoyle 1966; Hawkins 1970; Newham 1972). Only two have survived, both likely to have been sarsens of local origin hardened by weathering before their emplacement, though one of them may have undergone some rough trimming by hand.

Appendix 14.2

An ignimbritic tuff-lava discovered at Stonehenge in 1921–1924

At the time of Colonel Hawley's excavations at Stonehenge in the 1920s a number of igneous and metamorphic rocks of Welsh origin were discovered. A few of these carried striations which appeared to be of glacial origin. Others had weathered exteriors which showed that they are natural boulders. One in particular attracted the attention of Mr R. S. Newall who considered it to have the form of a glacial erratic. This specimen was forwarded to the British Geological Survey (then the Institute of Geological Sciences) where it was sliced and examined by Mr R. K. Harrison, Chief Petrographer, whose report, dated 19 October 1971, is given below.

[1] The English saying 'Once in a blue moon' is an expression of the same concept and the source of this particular usage of the word blue. The last blue moon seen in England was in October, 1950.

RSN 18 ignimbritic tuff-lava

> 'This large, dark blue-grey, hard, flinty (? partly worked artifact) shows a white weathered crust up to 5 mm thick. The thin section shows a complex structure of very finely banded welded tuff (compressed foliated shards cemented by fine silica) with composite quartz grains and strings of dusty leucoxene, separated by patches of much finer grained, finely fluxioned glassy lava with patches of granular quartz. This specimen appears to represent a complex of originally viscous glassy lava and welded vitric tuff, all presumably of rhyolitic composition.'

The dimensions and general appearance of the rock are shown in Plates 35 & 36.

Comparison of this ignimbritic rock with other Welsh Ordovician volcanic rocks suggests that it is more closely related to the known ignimbrites of North Wales than to those of the Mynydd Preseli area of Dyfed. On the other hand none of the ignimbrites available for comparison provided a sufficiently good match to enable a link to be made with one specific locality. At the time, this conclusion appeared to be disappointingly vague and out of line with some of the more dogmatic statements made in respect of Stonehenge rocks. In the light of more recent work, however, Mr Harrison's cautious approach has been fully justified.

If the ignimbritic tuff-lava is indeed an erratic of North Wales origin and the ice which conveyed it is as old as it appears, then careful consideration must be given to the value of implying direct derivation from individual exposures or outcrops which exist at the present day. In the 2.47 Ma which may have elapsed since this Pliocene glaciation occurred, the mountainous parts of Wales (both north and south) have undergone substantial changes of relief. While, therefore, it may be justifiable to compare the Stonehenge rocks with suites of Welsh Palaeozoic igneous and metamorphic rocks *on a regional basis*, it may not be possible to give precise locations without having more information about the Pliocene relief and surface geology of Wales than is currently available.

The weathered zone seen on the surface of the ignimbritic boulder is similar to other weathered exteriors of specimens examined by Mr R. Sanderson (Appendix III). This type of weathering is to be seen at the present day in peaty or humic environments and could have occurred before the rock was moved in ice from Wales to Salisbury Plain. Reporting on some of the ophitic dolerites from Stonehenge, Mr Harrison commented that 'In order to ascertain the differences between the altered outer crust of pale dolerite and the fresher, blue-grey interior, X-ray diffraction photographs were taken of powders drilled from the respective zones in ENQ 2301, the outer crust here being about 5 mm thick. Mr B. R. Young reports that both samples contain chlorite, quartz, feldspar, pyroxene and amphibole. The crust contains more quartz and less feldspar than the interior. Kaolinite was not detected'. Superficial leaching by acid water has broken down the feldspar, leaving residual silica.

If the weathering does date from before the translation of the rock to Wiltshire then the stone must have been conveyed either by hand or in ice. If the bleaching and weathering took place after arrival at Salisbury Plain then the boulder might have been carried by fluviatile or fluvioglacial transportation, though the weakness of the rock (due to shearing) argues strongly against fluviatile transport from Wales, or rounding in any medium other than ice.

When found, the weathered boulder had been thrown away with chippings and other waste material. An attempt had been made to dress one end of the boulder but this, in Mr Newall's opinion, had failed because of the sheared condition of the rock. It would appear that this small boulder, already deeply weathered, would never have been of any practical value. To suggest it had been carried from North Wales to Wiltshire only to be tested and thrown away as worthless would imply an astounding lack of common sense and understanding of the properties of rocks on the part of the men who built Stonehenge. If, however, the bluestones were recovered locally from material scattered on the surface of the Chalk or were present in solution cavities, then the presence of inferior material is comprehensible. Having gathered up all the available bluestones, both from natural sources and from abandoned Neolithic structures, the Bronze Age builders of Stonehenge used the large ones for constructional purposes and tested the smaller boulders for the manufacture of implements. Those which were unsuitable were thrown away.

It is worth noting that Sir A. C. Ramsay (1863) observed that the foreign stones of Stonehenge include rocks which are petrologically similar to those of North Pembrokeshire and that others resemble rocks from Caernarvonshire and the Llandeilo Flag district of Montgomeryshire west of the Stiper Stones. This appears to be the first published suggestion that the bluestones may include rocks similar to those found both in southwest and North Wales.

Appendix 14.3

The petrography of rock fragments evacuated from sites at Stonehenge, including extracts from a report by R. W. Sanderson (formerly of the British Geological Survey), 1980

Among the material preserved in Devizes Museum are two collections of stone fragments and chippings made by H. Cunnington (1876–1886) and W. Gowland (1901). This material, totalling about 2500 and 500 samples respectively, has been subjected to a comprehensive analysis by Mr R. Sanderson; part of his detailed Report is summarized below.

Petrography

The fragments have been divided into seven broad groups of which five are known to exist, or have existed, as megaliths, and confirm the lithologies recognized by Judd (in Gowland 1902).

Group 1 Siliceous siltstones and quartzitic sandstones (Sarsen or Greywether sandstone). The fine-grained rocks commonly containing root-clasts. In view of the amount of Tertiary sandstone used in the monument these rocks are poorly represented.

Group 2 Dark green micaceous quartzitic sandstone or subgreywacke. These pieces are non-calcareous, or medium grain. They exhibit a somewhat variable degree of shearing in that some are rudely cleaved and others possess fine sericitic cleavage surfaces. This group is unknown as megaliths. Chippings are concentrated outside the bluestone circle, the quantity suggesting that 2 or 3 megalith-sized stones may have been present. This rock could not have come from any area nearer than Cornwall or S.W. Wales.

Group 3 Pale green, calcareous, micaceous sandstone. This rock possesses a variable fissility with mica-spangled bedding planes. The detrital mica is of coarser grain-size than the sericitic cleavage planes of Group 2. This is the rock of which the Altar Stone is formed and the chippings are concentrated in its vicinity.

Group 4 Dark fine-grained rhyolites and acid tuffs (ignimbrites), described as 'felsite' by the early authors. The chippings of this rock are concentrated in the vicinity of two of the four felsite stones (38 and 40).

Group 5 Soft, dark green cleaved basic tuffs. 'Schist'. This rock is known only from the stump of a single megalith in the bluestone circle between stones 32 and 33.

Group 6 Porphyritic albitized dolerite. 'Diabase'. Some of the fragments do not exhibit the feldspar phenocrysts, which commonly occur in the Preseli dolerites.

Group 7 Minor, miscellaneous rock fragments. In this group are included flint flakes, chalk, glauconitic sandstone and a number of foreign (to the Stonehenge area) limestones. None of this group is known to have been used as megaliths. Some of the fragments may have been introduced in post-Bronze age times.

In relation to the remaining standing stones, Group 1 is largely derived from the largest and most abundant Sarsen megaliths; Group 3 from the 'Altar Stone' and Groups 4, 5 and 6 from the 'Bluestones'. Patination due to weathering is seen on the rocks of groups 4 and 6 and is similar to that seen on the dolerites of the Preseli Mountains.

Table 14.A1. Relative proportions of rock fragments from W. Gowland's 1901 excavations (extract from Report by R. W. Sanderson, 1980).

	Rock Group	1	2	3	4	5	6	7
	Excavations							
Exterior	A & B	32.8	11.5	0.6	11.5	26.2	17.5	0
	I–III, VI–VII & Q	23.8	6.3	0.4	21.5	24.7	18.6	4.8
	F	33.0	2.1	0.5	33.0	17.8	12.8	0.8
	IV, V & VIII	31.0	12.2	5.8	13.7	20.6	14.4	2.2
Interior	D & E	41.3	2.8	6.7	12.3	11.2	26.3	0.6

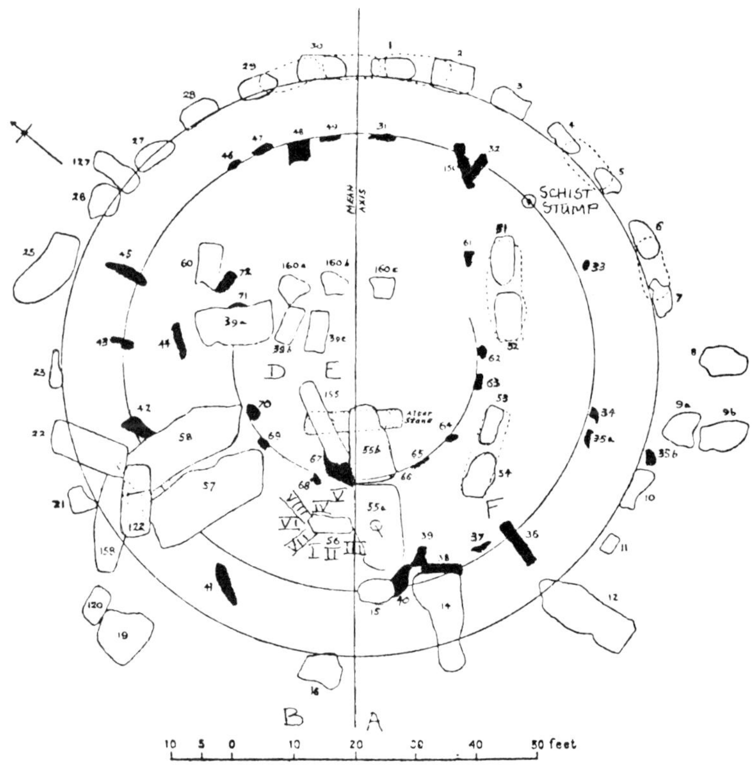

Fig. 14.A2. Plan of Stonehenge with the sites of Gowland's (1901) excavations marked in capitals and roman numerals.

Examination of Table 14.A1 indicates that the sarsen fragments, as might be expected, are the most abundant single group, although outnumbered by the 'bluestone' groups 4–6. Altar Stone chippings (Group 3) are concentrated in the vicinity of the Altar Stone, in excavations IV, V, VIII, D and E. Again, chippings of the felsite (acid tuffs & ignimbrites) of Group 4 occur most abundantly in the vicinity of two of the four diametrically opposed felsite stones (38 and 40, Fig. 14.A2) in excavations I–III, VI–VIII, F & Q. Perhaps of most interest is the apparent distribution of Group 2—the sheared micaceous quartzitic sandstone which has no known megalith counterpart. The highest proportions of this type occur outside the circle, excavations A & B, and between the trilithon and inner bluestone horseshoe, excavations IV, V & VIII. The Schist (basic tuff) group 5, is remarkably well represented, but is least abundant in the centre of the area (excavations D & E), the opposite of the distribution of the spotted dolerite (Group 6).

Additional note

There are several aspects of this work which are important to bear in mind. Firstly, the rock fragments and chippings are mostly small and their distribution is such that they cannot be regarded as having stratigraphical value. As Sanderson points out, the sites of Cunnington's excavations are not known and Gowland's sites are very restricted in their distribution. The extent to which these results can be considered as representative of Stonehenge as a whole is therefore uncertain.

Secondly, a point also mentioned in the report, the rock fragments almost certainly include some material which has been introduced fortuitously in historic times. Stonehenge was situated at a veritable crossroads of a number of old roads and trackways, used by travellers as well as farmers and persons resident in the area. The introduction of extraneous stones brought by gypsies who have camped and made fires on the site and of road making and building materials, is almost certain to have taken place. In addition there have been excavations made by those seeking buried treasure and by early investigators and these may well have led to contamination below ground. To add to these difficulties the activities of vandals throughout the ages are likely to have caused the formation of chippings which have become incorporated in the soil, having entered cracks and fissures in times of drought. For this reason the miscellaneous rock fragments forming Group 7 should, it is suggested, be regarded as being possible contaminants.

When due allowance has been made for these problems there remains a body of interesting evidence regarding some aspects of the petrography. Thus it will be observed that Group 1 which consists of Tertiary sandstone or silcreted sandrock (sarsen), is poorly represented as chippings in view of the vastly greater volume of the sandstone megaliths. If however the sandstones were trimmed elsewhere before erection and then finished on site, huge quantities of rough sandstone fragments would not be expected to occur at Stonehenge.

Group 2 is exceptionally interesting as it is numerically more important than Group 3 (Altar Stone). The rocks are quartzitic and micaceous subgreywackes probably of Lower Palaeozoic age and are most likely to be of Welsh origin. No megaliths composed of this rock have been found but the number of chippings suggest that the equivalent of two or three big stones might have been present. Material of this kind is not likely to have been a natural choice for a megalith and would not have been selected for the manufacture of implements. In this respect the rocks of Group 2 may have proved to be almost as useless as the small boulder of sheared ignimbrite described in Appendix 2. The re-use of old pre-shaped material made of good hard stone combined with soft sheared rock which would be unlikely to survive above ground for any appreciable length of time, shows how sorely pressed the builders of Stonehenge were to find sufficient stone. To the writer it appears obvious that the surface of Salisbury Plain and its abandoned Neolithic monuments had been ransacked for residual glacial erratics and no more could be found within easy reach of Stonehenge. Only its distance from Stonehenge may have saved the bluestone boulder of Bowls Barrow from expropriation.

There is one important aspect of these investigations which concerns the so-called 'Stonehenge Layer' regarded by Hawley (1921) as a stratigraphical horizon with chronological significance. This was strongly contested by Atkinson (1979) on the grounds that the chippings are not concentrated near the site of the worked stones. Sanderson's careful analysis shows, however, that in some areas of the monument the characteristic rocks are very strongly concentrated in the immediate vicinity of the parent stones as stated above. There is irrefutable evidence that some of the stones have been dressed on site.

Nevertheless, in view of the number of rebuildings which have taken place, there may also be truth in Atkinson's claim that some stone chips are widely distributed. It may well have been necessary from time to time to 'blind' the surface of the bare Chalk by scattering granular material, since periods of heavy rainfall or frost can quickly reduce Chalk to a sea of sticky mud. Both the Aubrey holes and some of the other excavations and stoneholes show evidence of infilling with Chalk rubble, in some instances followed by re-excavation. The reason for this may well have been the need to protect the Aubrey holes and the other features from excessive damage by the movement of heavy stones, while assisting the passage of massive sandstone monoliths using rolling stones which would only work on a firm even surface.

Acknowledgements. To the late Mr R. F. Newall and to Colin Bowen, Ron Harris and Robin Sanderson, I am indebted for much help. The format of Chapters 13 and 14 owes much to the advice of Alec Smith but many others both at home and abroad have contributed by way of information and discussion. Some are specifically mentioned in the text, but interest in the Stonehenge problem is so widespread that it is hardly possible to list all of those who have helped by constructive suggestions and in other ways since the publication of the 'Stonehenge' paper in *Nature* in 1971. For the opinions expressed here I accept full responsibility.

References

Atkinson, R. J. C. 1979. Stonehenge. Pelican Books, Harmondsworth.

Aubrey, J. 1656–1691. *The Natural History of Wiltshire*. Ed. J. Britton (1847).

Bishop, A. J. 1974. A Preliminary Report on the Middle Pleistocene Mammal-bearing deposits of Westbury-sub-Mendip, Somerset. *Proceedings of the Speleological Society*. University of Bristol. **13**(8), 301–318.

Bowen, H. C. & Smith, I. F. 1977. Sarsen Stones in Wessex: the Society's first investigation in the evolution of the landscape project. *The Antiquaries Journal,* **57**, 185–196.
Brentnall, H. C. 1946. Sarsens. *Wiltshire Archaeological and Natural History Magazine,* **51**, 419–439.
Cave, R. 1977. Geology of the Malmesbury District. *Memoirs of the Geological Survey of Great Britain.*
Cavelier, C. & Kuntz, G. 1974. Découverte du Pliocène Marin (Redonian) á Valmont (Seine-Maritime) dans le Pays de Caux". *Comptes Rendus de la Société Géologique de France,* **16**(6), 160–162.
Clark, M. J., Lewin, J. & Small, R. J. 1967. The Sarsen Stones of the Marlborough Downs and their geomorphological implications. Southampton Research Series in Geography, **4**, 3–40.
Cunnington, R. H. 1975. From Antiquary to Archaeologist: a biography of William Cunnington 1754–1810. Shire Publicatons, Princes Risborough, Buckinghamshire.
Davies, A. M. & Barnes, H. J. 1953. A Preliminary Survey of the Sarsen and Puddingstone Blocks of the Chilterns. *Proceedings of the Geologists' Association,* **64**, 1–9.
Dines, H. G. & Edmunds, F. H. 1929. *The Geology of the Country around Aldershot and Guildford.* Memoirs of the Geological Survey of England & Wales. HMSO, London.
Einarsson, T. & Albertsson, K. J. 1988. The glacial history of Iceland during the past three million years. *Philosophical Transactions of the Royal Society of London B* **318**, 637–644.
Gowland, W. 1902. "Recent excavations at Stonehenge". *Archaeologia,* **58**, 37.
Harland, W. B., Armstrong, R. L., Cox, A. V., Craig, L. E., Smith, A. G. & Smith, D. G. 1990. *A Geologic time scale, 1989.* Cambridge University Press.
Hawkins, G. 1970. *Stonehenge Decoded.* Fontana/Collins, London and Glasgow.
Hawley, W. 1921. Stonehenge: Interim Report on the Explorations. *The Antiquaries Journal I,* 39–41.
Hey, R. W. & Brenchley, P. J. 1977. Volcanic pebbles from Pleistocene gravels in Norfolk and Essex. *Geological Magazine,* **114**(3), 219–225.
Hollingworth, S. E. & Taylor, J. H. 1946. "An outline of the Geology of the Kettering District". *Proceedings of the Geological Association,* **57**, 204–233.
Hoyle, F. 1966. 'Speculations on Stonehenge'. *Antiquity,* **40**, 262–276.
Judd, J. W. 1902. "Note on the Nature and Origin of the Rock-fragments found in the Excavations made at Stonehenge by Mr Gowland in 1901". *In:* W. Gowland (ed.) *Recent Excavations at Stonehenge, Archaeologia.* **58**, 106–118.
Jukes-Brown, A. J. 1892. *The Building of the British Isles* (2nd Edition) George Bell, London.
Kellaway, G. A. 1971. Glaciation and the Stones of Stonehenge. *Nature,* **232**, 30–35.
—— 1974. Institute of Geological Sciences (investigations at Chitterne) Annual Report for 1973, pp 24–25.
——, Redding, J. H., Shephard-Thorn, E. R. & Destombes, J. P. 1975. The Quaternary history of the English Channel. *Philosophical Transactions of the Royal Society of London, A* **279**, 189–218.
Lamplugh, G. W. 1902. "Calcrete". Geological Magazine, **9**, 575.
—— 1907. "The geology of the Zambisi basin around the Batoka Gorge (Rhodesia)". *Quarterly Journal of the Geological Society of London,* **63**, 162–216.
Lanford-Smith, T. 1978. (ed.) *Silcrete in Australia.* Department of Geography, University of New England, NSW, Australia.
Leese, C. E. & Vernon, W. F. 1961. "Basal gravel in the Alluvium near Severn Beach". *Proceedings of the Bristol Naturalists Society,* **30**, 139–143.
Lucy, W. C. 1880. "On the extension of the Northern Drift and the Boulder Clay over the Cotswold Range". *Proceedings of the Cotteswold Naturalists Field Club,* **7**, 50–61.
Newall, R. S. 1959. *Stonehenge.* Her Majesty's Stationery Office, London.
Newham, C. A. 1972. *The Astronomical Significance of Stonehenge.* John Blackburn, Leeds.
Pomerol, C. & Feuguer, L. 1968. 'Bassin de Paris' *Guides Géologiques Regionaux.* Paris.
Prestwich, J. 1898. "Memoranda of the late Sir J. Prestwich". *Geological Magazine,* 1898, p 412.
Ramsay, A. C., Aveline, W. T. & Hill, E. 1858. "Geology of parts of Wiltshire and Gloucestershire, (Sheet 34)". *Memoirs of the Geological Survey of Great Britain.*
Rice, R. J. 1968. "The Quaternary deposits of central Leicestershire". *Transactions of the Royal Society of London.* A **262**, 459–509.
Richardson, L. 1933. "The Country around Cirencester". *Memoirs of the Geological Survey of Great Britain.*
Rose, J. 1987. 'Status of the Wolstonian Glaciation in the British Quaternary'. *Quaternary Newsletter,* **53**, 1–9.
—— & Allen, P. 1976. Middle Pleistocene stratigraphy in south east Suffolk. *Journal of the Geological Society,* **134**, 93–97.

——, —— & Hey, R. W. 1976. Middle Pleistocene stratigraphy in southern East Anglia. *Nature,* **263**, 492–934.

Sanderson, R. W. 1980. An examination of rock fragments excavated from sites within Stonehenge, Wiltshire. Petrology limit Report No. 4 Institution of Geological Sciences.

Shackleton, N. J. *et al.* 1984. Oxygen isotope calibration of the onset of glaciation in the North Atlantic region. *Nature,* **307**, 620–623.

Shotton, F. W. 1977. The English Midlands. Guide book for Excursion A2. International Union for Quaternary Research, X Congress, 1977 (ed. D. Q. Bowen).

Symonds, Richard, 1644. Diary of the marches of the Royal Army during the Great Civil War. Ed. C E. Long (1859), Camden Society.

Thomas, H. H. 1923. The Source of the Stones of Stonehenge. *Antiquaries Journal,* **3**, 239.

Thorpe, R. S & Williams-Thorpe, D. (in preparation). "The Scale of Megalith Transport".

Thorpe, R. S., Williams-Thorpe, D., Jenkins, D. G. & Watson, J. S. (in press). The Geological Sources and Transport of the Bluestones of Stonehenge, Wiltshire, UK. *Proceedings of the Prehistoric Society.*

Tresise, G. R. 1961. The nature and origin of chert in the Upper Greensand of Wessex. *Proceedings of the Geologists' Association,* **72**, 333–356.

Varney, W. D. 1921. The Geological History of the Pewsey Vale. *Proceedings of the Geologists' Association,* **32**, 189–205.

Vatcher, F. & M. & L., 1976. "The Avebury Monuments, Wiltshire". Her Majesty's Stationery Office, London.

Welch, F. B. A. & Trotter, F. M. 1961. Geology of the country around Monmouth and Chepstow. *Memoirs of the Geological Survey of Great Britain.*

West, R. G. 1977. East Anglia. Guide book for Excursions A1 and C1, International Union for Quaternary Research, X Congress 1977. (ed. D. Q. Bowen).

Wills, L. J. 1938. The Pleistocene Development of the Severn from Bridgworth to the sea. *Quarterly Journal of the Geological Society of London,* **94**, 161–242.

Woodland, A. W. 1971. The Llanbedwr (Mochras Farm) Borehole. Institute of Geological Sciences Report No. 71/18, pp 115.

Woodward, H. B. 1887. *The Geology of England and Wales.* George Philip, London.

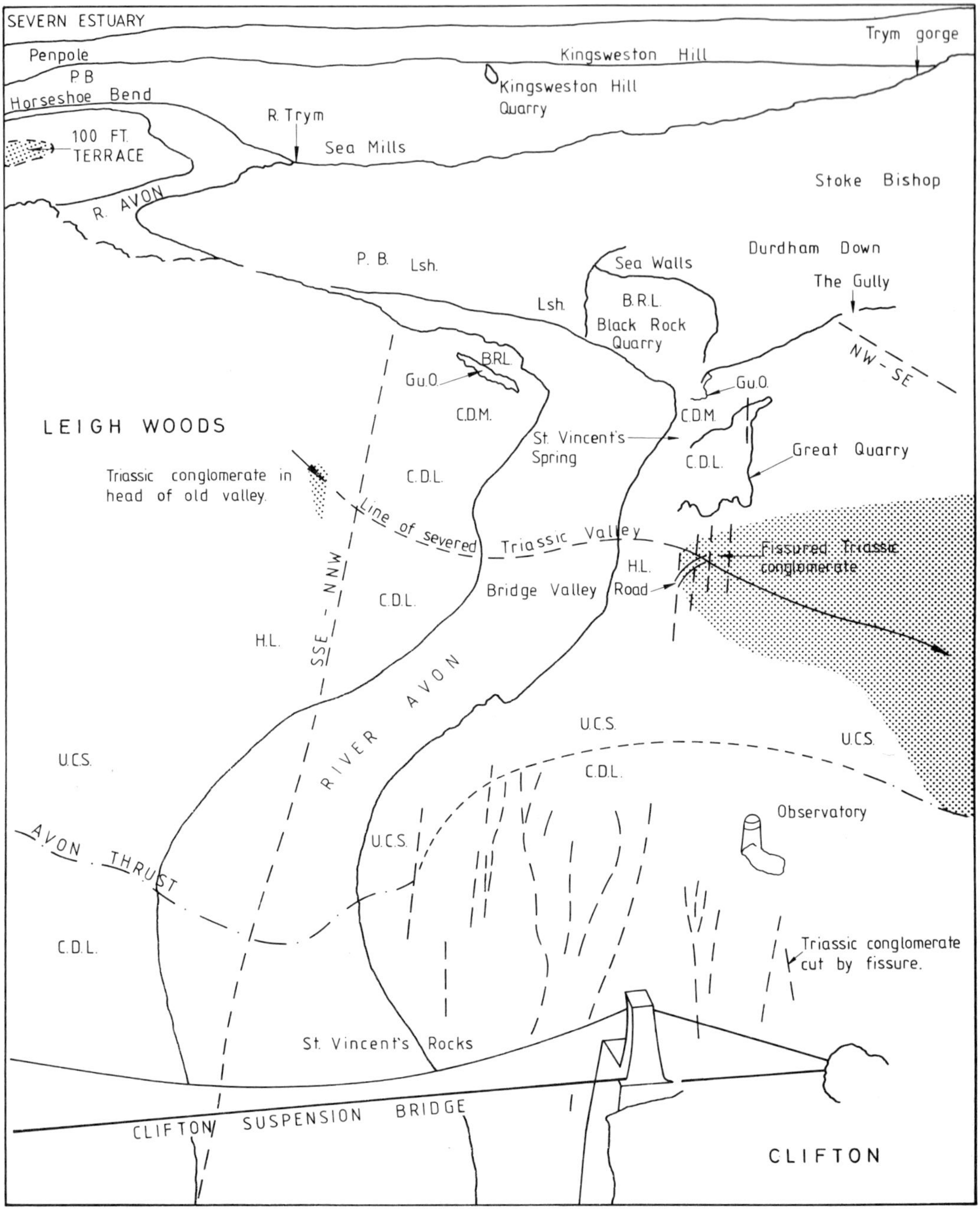

Plate A. The Avon Gorge at Clifton, Bristol (1953). View looking NW towards Kingsweston and the Severn Estuary. A Triassic valley filled with conglomerate was intersected by the fissure zone and partly destroyed at the time of the formation of the gorge. Triassic rocks in Leigh Woods–Clifton Down area are stippled. Fissures are shown by broken lines. A small thermal water spring issues near the foot of The Gully. The change in the direction of the gorge at Sea Walls may be due to the presence of Mesozoic

fissures with NW–SE strike, but the Gully and the valleys including the Trym may be part of the old Wye–Trym drainage system on the west side of the Palaeodivide. UCS, Upper Cromhall Sandstone; HL, Hotwells Limestone; CDL, Clifton Down Limestone; CDM, Clifton Down Mudstone; GuO, Gully Oolite; BRL, Black Rock Limestone; LSL, Lower Limestone Shale; PB, Portishead Beds.

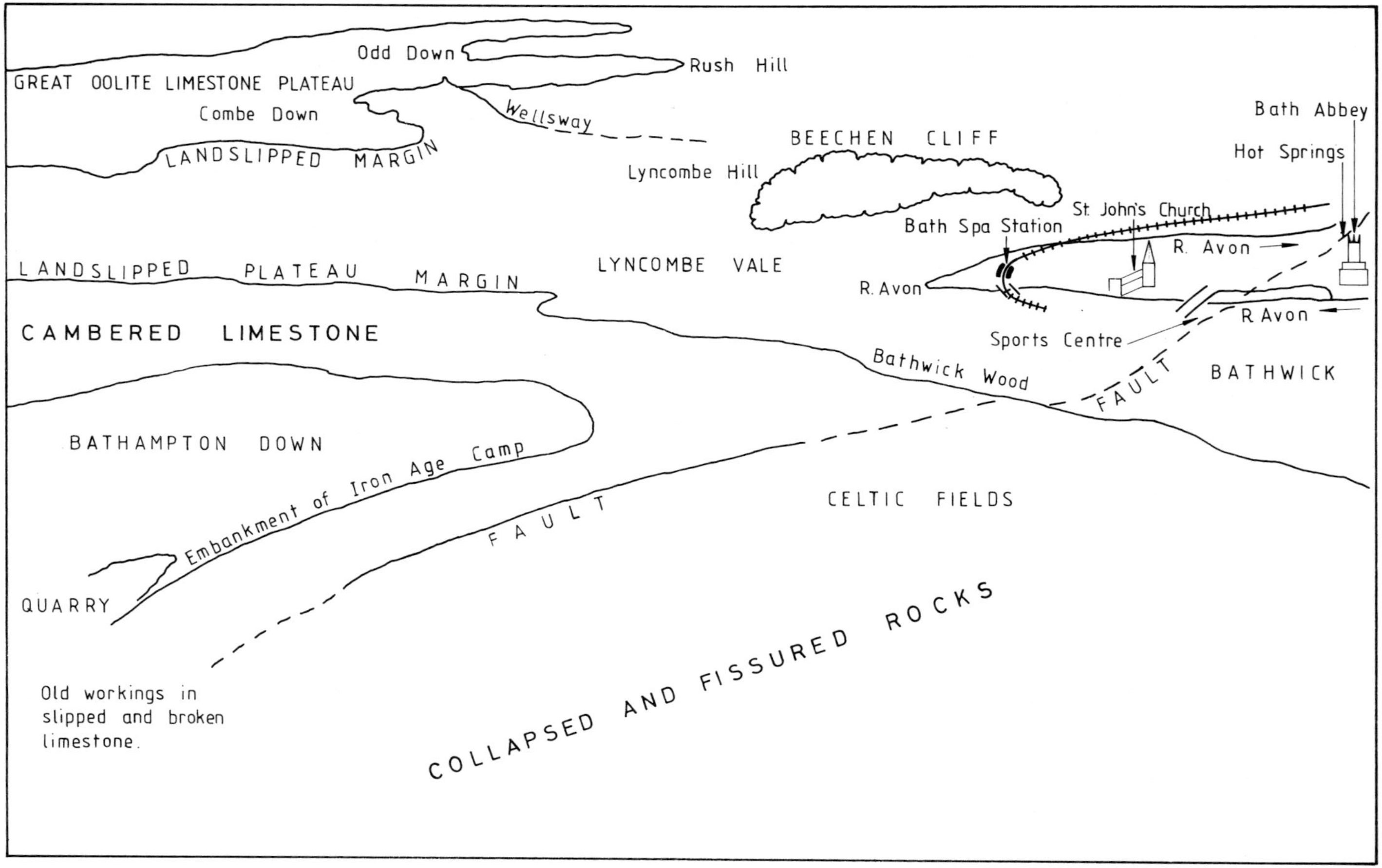

Plate B. Aerial view of the south side of the Avon valley at Bath (1948). In the foreground is the north end of Bathampton Down showing the inferred line of faulting and its possible connection with the Pennyquick Fault and the hot springs. South of the fault the Great Oolite is cambered towards Lyncombe Vale, the lower end of the camber terminating at the steep landslipped contact of the limestone and the underlying Upper Fuller's Earth Clay. The Great Oolite plateau at Odd Down and Coombe Down does not show such well developed cambers. It is difficult to estimate the throw of the fault at Bathampton Down owing to the disturbed nature of the rocks on the north side.

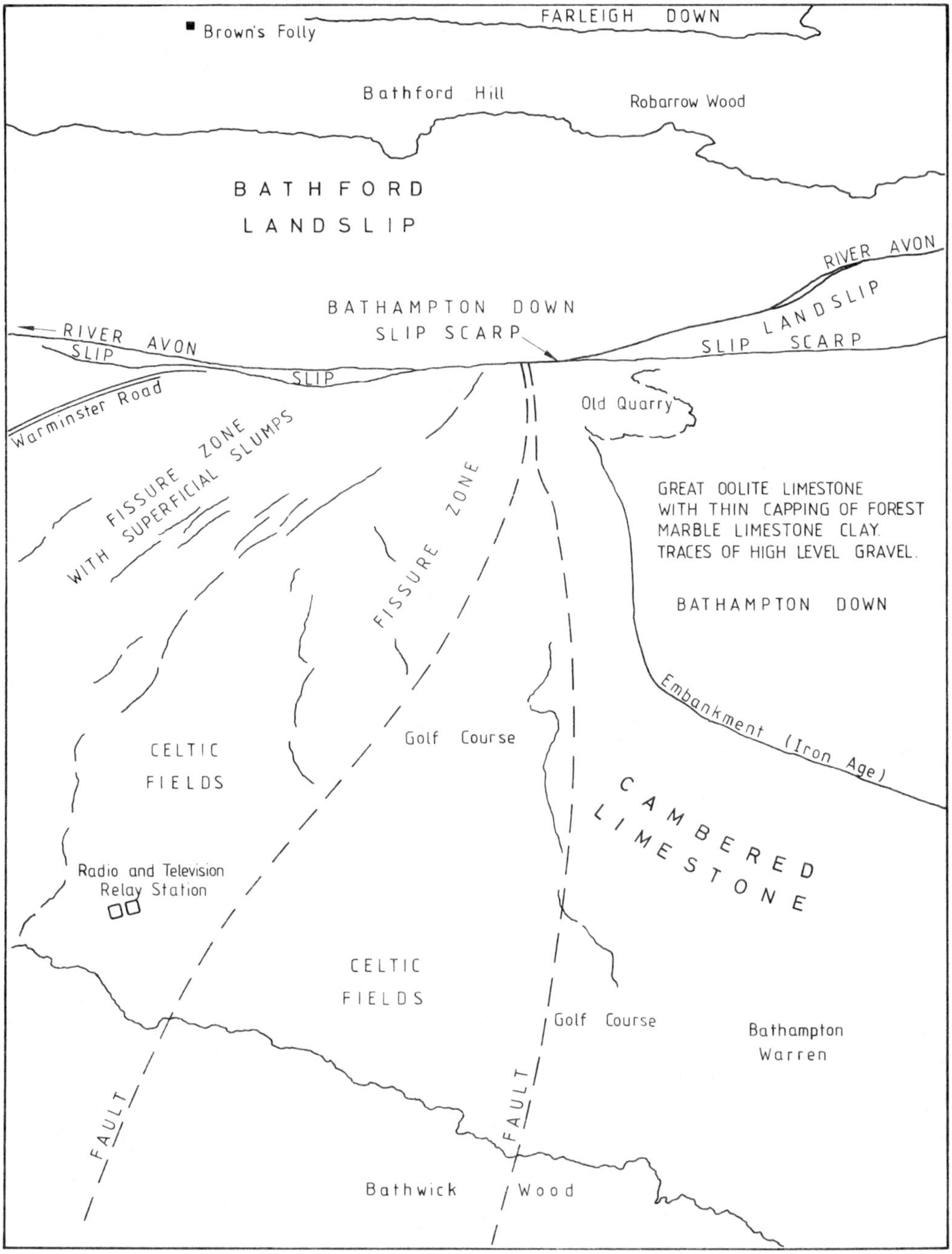

Plate C. Bathampton Down and the Avon valley above Bathford (1971). The Photograph is taken from Bathwick Wood looking ENE towards Farleigh Down and Brown's Folly. North of the fault the structure is obscured by rifting and collapse of the rocks in Quaternary times. Additional details of the structure at the NE end of the Down are given with Plate D. It is possible that a second fault may exist on the north side of the main structure.

The high ground east of the Avon valley includes Farleigh Down where there are patches of high level Drift deposits similar to those found on Bathampton Down. Downcutting of the Avon valley between Bathford and Limpley Stoke was severe in Devensian times and the big landslip on the eastern side of Bathampton Down extends beneath the alluvium filling the buried channel.

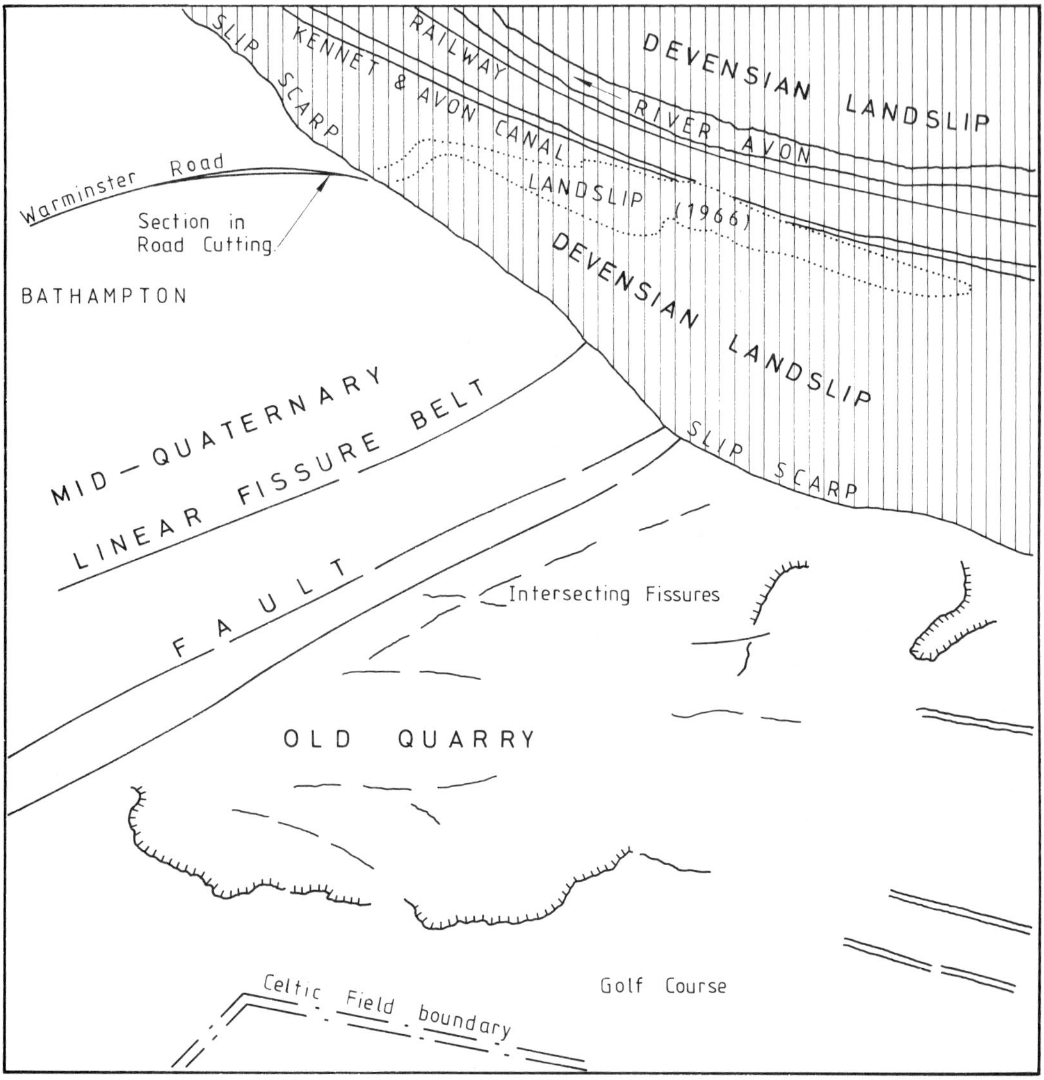

Plate D. The mid-Quaternary fissure belt on the NW part of Bathampton Down (1966). The photograph illustrates in more detail some of the features shown in Plate C. Material from road-widening operations was dumped on the metastable Bathampton landslips in 1966. This caused renewed movement which cut the Kennet and Avon Canal and threatened the railway between Bath and Bradford-on-Avon. The sections exposed in the Warminster Road on the outskirts of Bathampton proved steeply inclined masses of Inferior Oolite which had subsided into fissures in Midford Sand. Flint and chert from the high level Drift deposits mixed with Great Oolite Limestone and Fullers' Earth Clay were also present. The formation of the rifts therefore post-dates the formation of the high level Drift deposits and pre-dates the Devensian landslip which cuts the fissure belt.

Plate E. (above) Escarpment of the Lower Chalk at Oliver's Castle, near Devizes (1955). The western face of the chalk escarpment of the Marlborough Downs is indented by coombes with cirque-like heads formed by water draining out of the Chalk. Narrow gullies have been cut by spring melt-water draining off snow and ice fields on the plateau in late Devensian times when the bare chalk was frozen. Some of the gullies are superimposed on spurs separating the major indentations, implying that the coombes may have been filled with ice and snow in the early stages of gullying. The horizontal stripes are cultivation terraces (probably of Iron Age date). There has been virtually no scarp recession since their construction. © Crown copyright 1991/MOD reproduced with the permission of the Controller of HMSO.

Plate F. (opposite, top) Meanders of the River Wye above Chepstow. The picture is taken from above the Wynd Cliff looking in a SE direction towards the east bank of the Severn at Aust Cliff. The confluence of the Wye and the Severn at Beachley is marked by the Wye Bridge carrying the Motorway which continues by way of the Severn Bridge to Aust Cliff. For the most part the near cliffs are densely wooded. Those seen on the LHS are composed of Lower Drybrook Sandstone overlain by Drybrook Limestone. Although there is some evidence for indirect structural control of the course of the Wye between Chepstow and Beachley the incised meander pattern contrasts strongly with that of the Bristol Avon at Clifton. In relation to the Severn–Wye terrace sequence the Wye resembles the Bristol Avon below Sea Mills (Plate A).

Plate G. (opposite, bottom) The NW part of Bathampton Down. Shows the course of the fault (? Pennyquick Fault) where it crosses the plateau and passes into the area of landslipped and foundered ground at Bathwick Wood. South of the fault the major camber slope is inclined towards Lyncombe Vale. The plateau area in the foreground is probably composed of collapsed limestone resting on Fuller's Earth Clay. The embankment of the Iron Age hilltop enclosure is seen (top LHS) overlying the older Celtic field boundaries. None of the Iron Age structures shows evidence of appreciable post-Iron Age movement in this part of the hill. A lobate mud flow is seen on the lower slopes (RHS). © West Air Photography. Reproduced by permission.

Index